P9-DTJ-053

Changing Fortunes

CALIFORNIA STUDIES IN CRITICAL HUMAN GEOGRAPHY

1. *Changing Fortunes: Biodiversity and Peasant Livelihood in the Peruvian Andes,* by Karl S. Zimmerer

Changing Fortunes

Biodiversity and Peasant Livelihood in the Peruvian Andes

Karl S. Zimmerer

University of California Press
Berkeley / Los Angeles / London

University of California Press
Berkeley and Los Angeles, California

University of California Press
London, England

Library of Congress Cataloging-in-Publication Data

Zimmerer, Karl S.
Changing fortunes: biodiversity and peasant livelihood in the Peruvian Andes / Karl S. Zimmerer.
p. cm. — (California studies in critical human geography; 1)
Includes bibliographical references (p.) and index.
ISBN 0–520–20303–8 (alk. paper)
1. Human ecology—Peru—Paucartambo (Province) 2. Biological diversity—Peru—Paucartambo (Province) 3. Conservation of natural resources—Peru—Paucartambo (Province) 4. Quechua Indians—Agriculture. 5. Quechua Indians—Social conditions. 6. Agriculture and state—Peru—Paucartambo (Province) 7. Agriculture—Social aspects—Peru—Paucartambo (Province) 8. Paucartambo (Peru: Province)—Environmental conditions. 9. Paucartambo (Peru: Province)—Social conditions. I. Title. II. Series.
GF532.P4Z56 1996
306.3′49′098537—dc20 96–12931
CIP

Printed in the United States of America

1 2 3 4 5 6 7 8 9

The paper used in this publication meets the minimum requirements of American National Standard for Information Sciences—Permanence of Paper for Printed Library Materials, ANSI Z39.48–1984 ♾

To my teachers Robert Bieri and James Parsons

Contents

Acknowledgments ix

Chapter One **Fields of Plenty and Want** 1

Previewing the Fields 1
Overview 2
The Dilemma: *Seeds of Tomorrow* 5
Biodiversity and the Andes of Paucartambo 10
The Environmental Geography Approach 15

Chapter Two **The Great Historical Arch of Andean Biodiversity** 26

Ancient Domestication in the Eastern Andes 26
Sparse Biodiversity in Imperial Agriculture, 1400–1533 34
Rich Biodiversity in Commoner Subsistence 40
Colonialism: Coca and Crops in Paucartambo, 1533–1776 44
Haciendas and Communities, 1776–1969 55
Biodiversity and Long-Term History 65

Chapter Three **Transitions in Farm Nature and Society, 1969–1990** 68

Resource Paradoxes of the Land Reform of 1969 68
Diversification and the Postreform Political Economy 77
Synopsis: Biodiversity's Fate 84

Socioeconomic Differences and Dietary Change 90
Ethnicity, Power, and Biodiversity 97
Biodiversity and Recent History 105

Chapter Four **Innovation and the Spaces of Biodiversity** 107

Seeding Landraces 107
Farming the Landscape 117
Spaces of Biodiversity: Reinventing Flexibility 126
Absences of Biodiversity: Routes of Commerce 137
Farm Space as Key to Conservation 145

Chapter Five **Loss and Conservation of the Diverse Crops** 148

Fateful Places in the Paucartambo Andes 148
Ridding the Odd-Ripeners in the Northern Valley Cloud Forest 151
The Demise of Maize, Quinoa, and the Colquepata Wetlands 161
A Case of Conservation: Innovative Intercropping in the Southern Valley 172
Perceptions of Quechua Peasants 181

Chapter Six **Diversity's Sum: Geography, Ecology–Economy, and Culture** 186

The Place of Diverse Rationales 186
Voice of a Cultivator: Willful Words 188
Cultural and Moral Aesthetics 195
Ecological and Culinary Utilities 206
The Cusps of Recent Cultural Change 213

Chapter Seven **The Vicissitudes of Biodiversity's Fortune** 218

A Mixed Lesson: Less Certainty, Greater Flexibility 218
The Future of Sustainable Development and In Situ Conservation 226

Appendixes 233
Notes 251
Bibliography 273
Index 299

Acknowledgments

To conduct a research project in rural Peru between 1985 and 1990 and spend another five years at work in libraries, archives, and classrooms is to incur a mountain of debts. I am grateful to persons beyond enumeration in the Paucartambo Andes, the cities of Cuzco and Lima in Peru, Berkeley, Chapel Hill, and Madison. The research would not have been possible without financial support from the following: the National Science Foundation; the Social Science Research Council; the Fulbright Foundation, the University of California at Berkeley; the Department of Geography and the College of Arts and Sciences at the University of North Carolina, Chapel Hill; and the Department of Geography and the Graduate School of the University of Wisconsin–Madison.

My greatest debt for assistance is owed to scores of Quechua peasants and villagers in the Paucartambo Andes. Although they are too unprotected to name, the people of Paucartambo taught me about themselves, their livelihoods, and the nature of their farming with unstinting generosity, humor, and nearly as much patience. I am similarly grateful for the remarkable cooperation and welcome companionship of my field assistants, Leonidas Concha Tupayachi, Edgar Gudiel, Claudio Palomino, and Cornelio Cusi Huaman. Research in Paucartambo benefited from my association with the multi-institutional Changes in Andean Agriculture project of Stephen Brush and Enrique Mayer, and my tutelage there under the late César Fonseca Martel. I hope that this study meets the high standards set by my field mentors and that it aids the people of Paucartambo and other Andean regions whose hopes and dilemmas inspired it.

Geographer Mario Escobar, anthropologist Jorge Flores Ochoa, and sociologist Henrique Urbano graciously helped to guide my efforts in Cuzco. Ramiro Ortega Dueñas was invaluable in arranging my agroecological field experiments at K'ayra, the university's agronomy school, and its Center for the Investigation of Andean Crops. Other agronomy faculty, especially Hernán Cortés and Oscar Blanco Galdos, also contributed generously. I enjoyed a range of intellectual stimulation, logistical and moral support, and welcome diversions with other Cuzco and Paucartambo researchers, including Margot Beyersdorf, Brian Bauer, Aroma de la Cadena, Manuel Glave, Birgitta Härner, Bruno Kervyn, Eloy Neira, Sarah Radcliffe, Jorge Recharte, Ken Young, and Jorge Zamora. Deborah Poole and John Rowe introduced me to the historical archives in Cuzco and the interpretation of colonial paleography. Numerous librarians and archivists in various Cuzco and Lima collections provided professional assistance, even though they are among the most unrewarded of Peru's public workers.

In Lima I am indebted to geographers Hildegardo Córdova Aguilar and Nicole Bernex de Falen at the Pontificia Universidad Católica del Perú and the Universidad Nacional Mayor de San Marcos, and to Marcía Koth de Paredes at the Fulbright Foundation. From Raúl Hopkins Larrea at the Instituto de Estudios Peruanos (IEP) I learned much about the Andean beer barley business. Mario Tapia readily lent his advice about Andean biodiversity on a number of occasions. The International Potato Center (CIP) and genetic resources expert Zósimo Huamán facilitated my agroecological field studies. Hugo Fano, Douglas Horton, María Mayer, Carlos Ochoa, and Peter Schmiediche, also of CIP, offered ample time and counsel. Ricardo Sevilla Panizo of La Molina, the national agrarian university across the street from CIP, advised me about Andean maize. David and Laura Hess warmly welcomed me in their Lima home at frequent intervals.

Various intellectual communities in the United States helped to nurture and mature my research. I owe a colossal debt to graduate school professors at the University of California at Berkeley—including Herbert Baker, Brent Berlin, Roger Byrne, Allan Pred, John Rowe, and Michael Watts—and especially to my advisor, James Parsons. Subsequent discussions with faculty colleagues and students in my classes and seminars, first in the Department of Geography at the University of North Carolina, Chapel Hill, and then at the University of Wisconsin—Madison, helped and often challenged my thinking. I am fortunate to have found such supportive and stimulating communities in these institutions. I am especially thankful to Yi-Fu Tuan, Tom Vale, Bob Sack, Bill Cronon, William M. Denevan, and Fernando Gonzáles. The Cartography Laboratory at the University of Wisconsin—Madison crafted superb maps that surpassed this geographer's greatest expectations. Bruce Winterhalder, Frank Salomon, and Steve Stern gave pointed suggestions and sage advice. Others who have contributed freely to the research through comment, critique, and counsel include Robert Bird, David Douches, Jack Hawkes, Daniel Gade, Major Goodman, David Guillet, Miguel Holle, Gregory Knapp, Patricia Lyon, William Mitchell, Ben Orlove, Carlos Quiros, and Sinclair Thomson.

Research is not separate from one's personal life of predispositions and commitments. Although I cannot believe my study to be autobiographical, as one reviewer suggested, it is nonetheless indebted immensely to the teachers who have shaped and inspired me. Karl F. and Katherine B. Zimmerer taught me about change in diverse communities and instilled my interest in nature, albeit in urban settings. At Antioch College I learned not only biology and physics but that scholarship should be committed to making a difference beyond the academy. My most influential teacher there, biologist Robert Bieri, awoke a desire in me to transgress narrow disciplinary boundaries and work in the twin realms of biophysical science and the study of human beings as modifiers of nature.

Wes Jackson sharpened my focus on agriculture and strengthened my unslaked interest in transdisciplinary investigation during an undergraduate internship at the Land Institute in Salina, Kansas, in 1979. Gary Nabhan schooled me in ethnobotany and studies of the desert Southwest in 1981. In subsequent years James Parsons at the University of California at Berkeley showed me how geography specializes in the uniquely integrative tradition of study that I sought. He guided my efforts in Latin Americanist research and has continually inspired me with his example of scholarly excellence and affability. Artist Medora Ebersole, my wife, has taught me unceasingly about the creativity of the human spirit and its expressiveness in many everyday forms. Our children, Eliza and Stephen, have insisted that none of the above matters if it means missing playtime. Although I cannot repay fully the gifts of my teachers, I hope that research such as this study and my teaching and service to worthy institutions and causes that go with it can approach their high standards and prime examples.

1

Fields of Plenty and Want

Previewing the Fields

"Why are you so interested in these small potatoes?" asked Faustino, as we neared his neighbor Líbano's field known as the Big Hill. For the fifth time this season, we had climbed to the community grassland where Faustino worked in the plot of the well-to-do Líbano. Ascending the stony zigzag trail from behind Faustino's huts, I was accustomed to his question as much as to the footsore ascent. I gasped, "Well, Don Faustino, for one thing, I'm fascinated by the diversity of floury types in a potato field like Líbano's Big Hill," only to hear his ready query, "But why study them?" Faustino craned his neck to watch the flatbed truck trundling along the road below. "It sure looks like Líbano's truck, the Darwin, that's headed for Cuzco tomorrow," he said. "I'd better be hauling my sacks of beer barley to the village tonight."[1] Allowing his question to be left unanswered, Faustino swung the hoe into a nearby row, unearthing a gaggle of tubers and eyeing them for harvest.

From the lofty field, Faustino and I could scarcely glimpse the whitewashed walls and adobe tile roofs of Paucartambo Village. Wedged into a deep box canyon, the small country town was a way station on one of Peru's main trans-Andean highways. Unnerving narrowness and tortuous hairpin turns of the Paucartambo Road dictated a daily alternating of traffic flow between the major sierran city of Cuzco and the lowland Amazon frontier that stretches toward Bolivia and Brazil. On "entering days," minivans from Cuzco shuttled loads of curious tourists and researchers to Manú National Park and Biosphere Reserve. Manú spreads fanlike from mountainous Paucartambo, northeast into the Andean foothills and across more than two million acres of lush lowlands and mighty Amazon-feeding tributaries. Its exuberant biota lured the national and foreign visitors through the Paucartambo Andes. Many nature tourists and scientists en route to Manú Park appreciated their trip amid the rugged landscape and its picturesque patchwork of agricultural fields and rangeland.

Gazing through the van's dusty windows, some travelers no doubt garnered an impression of much variety in Paucartambo farming. They also took in vistas of farmers in the peasant communities. The Paucartambo people were said to live "timeless traditions and speak the language of the Inca," as eulogized in the Cuzco tourist brochures. Although a good number of the Quechua peasants in Paucartambo don Western-style clothing, travelers would often remark about their items of non-Western dress. Some women wear layers of embroidered skirts and the curious, flattopped hats, or *manteras,* that look like lamp shades, and every now and then men are seen wearing their pointed, tasseled, and sequin-plated wool caps. Glimpses could also be gained of unfamiliar farm tools, unusual plowing or harvest techniques, and the unique and recently publicized Andean crops (*los cultivos andinos*). To the visitor versed in environmental issues, those vistas conjured an image of farmers, and perhaps even whole farm communities, in edeniclike repose with the living heirloom crops of rich agricultural biodiversity.

First staying in Paucartambo in 1985, I was unaware of this pulse of environment-minded visitors. I did, however, carry a particular interest in how ideas similar to theirs were seeking to explain the nature of biodiversity in the farming systems of indigenous peasants and those of the Quechua people in particular. A renowned environmental resource for world agriculture, the diversity was poorly understood. My research goal was to examine its ecological character; its management by the local farmers; and the trends in its evolutionary fate shaped by a history of social, cultural, and economic changes. During more than two years of field research in the Paucartambo Andes between 1985 and 1991 and a similar commitment to archive and library research until 1994, I found that the role of diverse crops in peasant livelihoods at present and during the past was substantially different than asserted in current explanations. My findings on the environmental, social, and cultural roles of diverse crops offer a series of fresh insights for biological conservation and the urgent need to integrate it with economic development.[2]

Overview

This book examines the fortunes of diverse crops in the mountainous Peruvian Andes and the Quechua farmers who cultivate this rich and renowned environmental resource. By 1990, a lengthy history of agrarian transitions that began with the first Andean farming more than seven thousand years ago culminated in the partial but incomplete conservation of the diverse crop repertoire. Peru's Land Reform of 1969 and unbalanced economic development during the years from 1969 to 1990 have worsened the dilemma of biodiversity loss. The study tells a story of mixed fortunes belonging to four Andean crops—potatoes, maize, quinoa, and ulluco—and the farming systems and farmers that husband them. It offers a definitive assessment of a worsening extinction problem and the prospect for uniting biological conservation with agriculture that is economically sound and socially just.

A triad of central findings are at once ironic and instructive, inverting much conventional thinking about biodiversity, its conservation, and its role among the Quechua peasants of the Andes. First, contrary to the claims of full-blown genetic erosion, crop diversity in Paucartambo was not eliminated in the assorted social transitions beginning with Spanish colonialism. Ironically, a well-intended but misplaced belief in "genetic wipeout" has deflected attention from the real process and perils of genetic erosion as well as the prospects for conservation. In the second place the region's better-off Quechua in particular have still seeded the diverse crops, even while they also pursued the opportunities of modern commerce. This runs counter to the conventional wisdom in human environmental thought about resource use in "traditional" societies. Third, land use, places, and the role of diverse crops in the Paucartambo Andes have recently become more distinct; indeed, modernization has heightened the disparate relief of environmental quality among spaces and places, instead of leveling the differences among them.

Chapter one introduces the study's goals, its geographical compass, and its historical depth, by outlining the diverse crops of the Peruvian Andes and the Paucartambo region with its twenty-thousand-plus Quechua farmers. Diverse crops are seated at the foundation of growing and vital studies in multidisciplinary scholarship on environmental resources, economic development, and conservation dilemmas in Latin America and other regions of the less-developed world. A critique of the popular and influential film *Seeds of Tomorrow* is used to raise the main environmental and socioeconomic issues in this conservation topic. Critical reflection on the conventional wisdom portrayed in the film finds many of our current beliefs to be thinly supported assertions sorely in need of rigorous study.

Chapter two addresses the evolution of diverse crops amid the historical transitions in farming systems and peasant livelihoods of the Andes. Beginning seven thousand years ago, early Andean cultivators seeded an unmatched assortment of diverse crops in their tropical mountains. In the Paucartambo Andes the economic capacity of Quechua farmers to supply diverse crops for their ethical norms of a fit livelihood changed historically during ensuing eras of Inca (1400–1533), Spanish colonial (1533–1776), and late colonial and republican rule until 1969. Economic transitions after the Spanish conquest shifted the de facto conservation of diversity toward the better-off "haves" among the Quechua Indians, who benefited in concrete ways from the ecological and cultural advantages of their fit livelihoods.

Changing fortunes of the diverse crops fell even faster after the Peruvian government enacted far-reaching yet contradictory social and economic programs in its radical Land Reform of 1969. More than one hundred Peasant Communities were officially granted land in the Paucartambo Andes between 1969 and 1990. Still, as chapter three demonstrates, the dissimilar courses of de facto conservation and degradation via extinction remained allied to the better-off peasants and their poorer counterparts, respectively. The land reform in 1969 impacted the diverse crops in unexpected ways by adding pressures at the farm level that, on the one hand, have led many Quechua in Paucartambo to curtail their cultivation and to lose ground by their own standards. Well-to-do peasants and powerful villagers, on the other hand, have celebrated Quechua ethnicity and cemented social bonds with the wealth of diverse food plants.

The dynamics of farm space after the Land Reform of 1969, assessed in chapter four, sharpened the contrast between the landscapes of the Paucartambo Andes harboring diversity and ones without it. A quartet of local farm units—Hill (*loma*), Valley (*kheshwar*), Oxen Area (*yunlla*), Early Planting (*maway*)—were defined by the Quechua farmers and their communities. By creating the distinct units of land use, many farmers gained enough flexibility to seed the diverse crops while pursuing new commerce. Their creation of the four farm spaces not only enabled them to keep the diverse crops viable but also shaped the options for future development. The character of their farm spaces was of redoubled relevance to diversity's fortunes, since the units were used to guide the farmers' acts of dispersing and distributing their diverse crops across the landscape.

Chapter five examines how three places that are "hot spots" of diversity in the Paucartambo Andes—Challabamba (Plain of Maize Stubble), Colquepata (Step of Silver), and Mollomarca (Place of Snail Shells)—differed markedly in terms of whether their farmers could sustain the trove of crop plants after Peru's Land Reform of 1969. The array of place-based contrasts became sharper after the transformation of land tenure and the ensuing growth of commerce and semiproletarian status among the Quechua in Paucartambo. Forces such as crop-processing agribusinesses and urban food markets favored by national economic policies cast the fortunes of diverse crops into an ever more uneven geography, since the places of de facto conservation stood out from those of biological loss.

The Quechua in Paucartambo have factored a gamut of place-based criteria in determining the actual sums of diversity in their crop plants during the years between 1969 and 1990. In defining the composition of diverse fields the farmers pick their seed on the basis of factors dealing with ecology and economics, culture and cuisine, and a sense of moral aesthetics. Chapter six discusses how they invoke their concerns about diversity under the far-reaching and historically forged ethic of a fit livelihood. Diversity prescribed in their resource ethic has not been a fait accompli, however, because the Paucartambo farmers are chronically engaged in an energetic albeit contested reciprocity existing between themselves, their crop plants, and other groups in regional society. Their vital vocabulary of signs and symbols used to select the diversity of crops draws on the richness of Quechua terminology and local metaphors as well as religious practices.

In conclusion chapter seven examines our thinking on the past and present role of diverse crops in the livelihoods of Quechua peasants in the Andes. It evaluates how the ecological and social conditions behind the farming of diverse crops by the Quechua in Paucartambo are less fragile and fine-tuned than often assumed; instead they are more flexible and dependent on modified environments and changing management by farmers. The vitality of diverse crops in their farming raises a cautious hope for conservation integrated with sustainable development. By acknowledging farmers, such as the Quechua in Paucartambo, as architects of diversity's fortunes, although not solely of their own design, the conclusion recommends certain policies and programs that can contribute toward biological conservation and sustainable development.

The Dilemma: *Seeds of Tomorrow*

Since the nineteenth century when scientific crop breeding was born, the biological bits and pieces of diverse plants like the floury potatoes tended by Faustino in the Paucartambo high country have routinely underwritten the creation of improved crops. The diverse plants provide the genetic raw material to refine the yield, disease resistance, eating quality, growth traits, and other agronomic properties of the scientifically bred crops. Variation in the diverse species and the traditional folk varieties thus supplies a pool of resources that is the lifeblood of world agriculture and food supply. While most improved varieties are produced in industrialized, developed societies, the diverse crops are cultivated by indigenous and peasant farmers in Third World countries. Ties between the two sorts of crops are economically crucial, geographically complex, and sometimes hotly disputed.

One tie exposed to much recent publicity is the accelerating extinction of the diverse crops. Genetic erosion first gained front-page headlines in scientific literature in the 1970s due to the farsighted concerns of crop scientists Erna Bennett, Sir Otto Frankel, Jack R. Harlan, and Jack G. Hawkes. Scientific institutions such as the United States National Academy of Sciences also sounded the alarm of extinctions ravaging the diverse crops. By the 1980s and early 1990s the imminent threat of catastrophic genetic erosion was increasingly seen within the dire terms of an extinction crisis facing biodiversity at large (Frankel 1974; Frankel and Bennett 1970; Frankel and Hawkes 1975; Hawkes 1983; Harlan 1972, 1975a; NAS 1972; Soulé and Wilcox 1980; Williams 1988; E. Wilson 1988).

The dilemma of genetic erosion is worsened by our current difficulty in deciphering its causes and consequences. Without an adequate diagnosis of the problem, moreover, it cannot be confronted through an adequate means of conservation. Neither actual causes nor conservation prospects are well understood. Assertions of genetic wipeout due to the general forces of modernization and population explosion offer little guidance for correctives other than mounting vast stores of diversity or hoping for a similarly vague sustainable development. The dilemma of weak analysis then frustrates our major forums and stands in need of full-scale redress, notwithstanding a handful of focused case studies forged in recent years (Brush et al. 1988; Brush and Taylor 1992; Zimmerer 1991c).

We can reflect on a documentary film entitled *Seeds of Tomorrow* in order to illustrate the unfortunate dearth of knowledge about the causes, consequences, and conservation cures of genetic erosion. Produced and broadcasted as part of the prestigious Nova series on public television, the ninety-minute film was viewed first in 1985 by millions of people in the United States and Canada and subsequently on videotape by even greater numbers in classrooms and in other countries. Nova's *Seeds of Tomorrow* is a fast-moving examination of the extinction crisis facing agricultural biodiversity. It deploys a soundtrack mixed with indigenous music and arresting cinematography, cutting dramatically from rural redoubts in the Ethiopian highlands, the Andes of Peru, and the hills of southern Greece to a gleaming Los Angeles supermarket and a gigantic Monsanto chemical factory.

The film has single-handedly informed a large public at home and abroad for the first time about the genetic erosion crisis and its possible cures. Some Amazon-bound travelers traversing the Paucartambo Andes likely applied ideas gleaned from the gripping documentary while they gazed at the region's immense biodiversity. But the genuine alarm sounded in *Seeds of Tomorrow* belies its less-dramatic medley of vague assumptions and downright errors that epitomize the current dilemma. Numerous errors of both fact and interpretation crop up throughout the film. While it might be argued that such limitations merely show the imperfect state of current knowledge, the following errors in *Seeds* invite immediate redress because they could in fact imperil conservation efforts and unintentionally betray the praiseworthy purpose of concerned persons such as the film's producers and its participants.

One of the first substantive errors revealed in the unwinding of *Seeds of*

Tomorrow is the blanket assumption that cultural change in the rural backwaters of developing countries augurs the impending demise of diverse crops. This repeats an often-heard explanation that genetic erosion is driven inexorably by cultural erosion among people like Faustino whom I had accompanied to the Big Hill parcel. In the film a professional associate of the National Academy of Sciences intones that "people might cast off their traditional varieties, sort of like last year's automobile model." Yet counterevidence appears early in the film when anthropologist Ella Schmidt interviews a peasant potato farmer named Don Pedro in the central Peruvian Andes, about two hundred miles north of where Faustino cultivated: "Many people come to work for him [Don Pedro]. And they ask to be paid for this native variety because they looked [for] this very much all over and they can't find them. So he's one of the few who grow more native varieties and everybody knows him so they want to work with him."

The film's interview with Don Pedro suggests that diverse crops, at least in his case, have become cultivated like living heirlooms, more like a valued tradition than "last year's automobile model." Cultural change does not *necessarily* endanger the diversity of crop plants. Although it is untenable to believe that cultural changes never threaten the diverse crop plants and traditional folk varieties, the point is that the fate of diverse crops must be carefully examined rather than merely assumed.

Seeds of Tomorrow commits other similar errors of overstatement and unfounded generalization. One error surfaces in the claim that scientifically improved, or high-yielding varieties (HYVs), drive the inevitable doom of diverse crops. The film's narrator concludes ominously that "the very success of the new varieties is wiping out the genetic material they were fashioned from." His view of modern high-yielding varieties furiously vanquishing their diverse counterparts echoes many accounts in the crop scientific literature (appendix A). The casting of modern HYV crops in the Goliath-like role of fateful antagonist, however, reveals a limited grasp of how genetic erosion can occur. It is often overlooked that farmers make the decisions and take the actions that involve the high-yielding varieties and diverse crops. This unsurprising fact unlocks a few key questions that are left unasked by oversimplified analyses.

One question is whether farmers must curtail their diverse crops when they adopt the improved varieties. Simple one-for-one substitution has often been assumed; however, many Third World farmers have adopted improved varieties while still seeding diverse crops, innovating their earlier farming strategies even as they modernize (Brush 1986, 1987; Zimmerer 1991c, 1992a). Don Líbano, owner of the diverse floury potato field, exemplifies this dual form of livelihood. Together with the traditional floury potatoes he plants several plots of high-yielding varieties. Diversity-cultivating Don Pedro also makes this point in the film. Similarly, in another scene, geneticist Erna Bennett remarks about how wheat farmers in southern Greece partly switched back to their diverse types. Their message is plain: the farm-level processes of adopting and relinquishing crops need to be assessed carefully rather than merely assumed.

A different but no less weighty error in *Seeds of Tomorrow* deals with the ecological nature of crop diversity in general and famously diverse potatoes in particular. Voicing over panoramic vistas of mountain fields in the Peruvian Andes, the narrator intones, "The potato fields span elevations of thousands of feet. Hundreds of generations of farmers have saved from each field the choicest potatoes to plant again next year. *The result has been the precise tailoring, almost to individual plots of land, of so many varieties it is hard to believe they are all related*" (my emphasis). Yet the narrator's vow of "precise tailoring"—microenvironmental specialization—is unproven and until recently untested. In fact field and experimental studies on the potato crop's ecology and biogeography refute the alleged "precise tailoring" of diverse varieties found in the Andes (Zimmerer 1991a, 1994a; Zimmerer and Douches 1991). Equal cause for questioning the film's assertion on this point is that the specific characteristics of ecological adaptation in other diverse crop plants remain mostly unknown.

A final error worthy of note regards the historical interpretation of crop diversity that is expounded by *Seeds of Tomorrow.* The film recalls that "Nova went with Vietmeyer on a hunt for some of the foods that supported the Incas—*foods that have been left behind*" (my emphasis).[3] But what of the millions of farmers that currently grow and eat these plant foods on a regular basis and those who preceded them during the centuries since the Inca empire fell in the early 1500s? Imagine the factual blunder and blatant insensitivity of describing a quest for the "Blue Corn that the Anasazi culture left behind in Arizona." It is true, of course, that the existence of diverse crops like the Andean staples and the Hopi Blue Corn attest to farming pasts "without history" in the sense of being less documented and poorly known in comparison to their rulers (Wolf 1982). The difficulty of inquiry does not, however, excuse the misinterpretation of fact or the uninformed flight of fancy.

Brief critique of the ungainly errors in *Seeds of Tomorrow* is needed because successful conservation will hinge on analysis of the nature of the diverse crops and the character of the extinction dilemma undermining them. Pause on the impractical implications that flow from the flawed facts and faulty interpretation rendered in this well-meaning documentary. If cultural change is deemed to end inalterably in the extinction of crop diversity, then conservation of biologically rich fields would depend on the existence and probably the enforcement of cultural constancy—an inane proposition given peoples' right to elect change. It is also deluding to think of improved high-yielding varieties substituting quid pro quo for cultivation of diverse crops; if that were true, conservation of traditional varieties holds no hope in the presence of HYVs. If the full panoply of microenvironments is taken to be the absolute minimum for conservation, then anything else would be judged a biological compromise. Finally, if the Inca Empire was a sine qua non for the creation of this diversity, does conservation require its return?

The sizeable misconceptions outlined also infer that it is impractical to conserve biodiversity in the context of economic development and sociocultural change found in Third World farming. The same fateful verdict is echoed in the opinions of many policy makers who counsel conservation based solely on *ex situ*, or literally "out-of-place," storage in freezers and tissue culture (Plucknett et al. 1983, 1987; Williams 1984, 1988). Ironically, this claim is made by the two opposing sides of the hotly disputed issue of intellectual property rights. On the one hand, many believing that the diverse cultigens are a public good advocate storage in ex situ facilities as the sole means to conserve and protect the unimpeded free flow of diversity (Frankel 1974; Williams 1984, 1988). Similarly, several of the active proponents of proprietary rights for indigenous and peasant farmers also advocate ex situ storage as the fitting conservation solution, thus agreeing on the means of conservation but not the ends to be served (Fowler and Mooney 1990; Kloppenburg 1988; Kloppenburg and Kleinman 1987).

On the other hand, some claim that ex situ conservation alone is insufficient due to a variety of reasons related to the biology of crops and the social and technical features of storage (Altieri and Merrick 1987, 1988; Brush 1986, 1987; Cleveland et al. 1994; Nabhan 1985, 1989; Oldfield and Alcorn 1991; Prescott-Allen and Prescott-Allen 1982; Wilkes 1983, 1991). Detractors argue that ex situ conservation suffers because the evolution of diversity is dangerously frozen from many environmental influences including genetically compatible wild relatives. They point out that such stores could be ruined due to intentional targeting such as terrorism or coincidental mishaps such as power failures. They add that adequately sampling the full range of existing diversity is infeasible. The critics suggest that "in-place," or *in situ*, conservation based on continued farm production must therefore complement the centralized collections of stored resources.

Behind their enthusiastic advocacy of in situ conservation and its coupling with sustainable development, however, lies little analysis of the changing ecological, social, and cultural roles of biodiversity in Third World farming. While it is thus possible to criticize the weakness of ex situ programs, the policy recommendations that promote in situ conservation cannot offer much in the way of specific insights. The initial advocacy of in situ conservation may thus be thought of as a "first generation" of inspired thinking and general criticism (Brush 1989). Well-designed and detailed case studies must now introduce more rigorous analysis, mount revisionist critiques, and offer specific recommendations. The best studies can bring the insights and reasoning of science and scholarship to the exercise of policy-making on Third World environment and development issues. Such aims have guided various recent research efforts on other resources such as soils and vegetation that similarly are in compelling need of conservation (Blaikie 1988; Ives and Messerli 1989; Turner et al. 1990).

Biodiversity and the Andes of Paucartambo

More than ten million small-scale farmers in the central Andes of Peru, Ecuador, and Bolivia cultivate a prodigious diversity of crops. The peasants farming this mountainous spine that runs longitudinally across the tropics of western South America maintain one of the world's greatest shares of cultivated plants. Beginning in the late nineteenth century, plant geographer Alphonse de Candolle, botanist O. F. Cook, crop evolutionist Nikolai Vavilov, and later, geographer Carl Sauer and archaeologist Hans Horkheimer all marveled at the truly immense diversity of plant domesticates cultivated within small areas of the central Andes (Candolle 1908; Cook 1925b; Horkheimer [1960] 1990; C. Sauer 1950; Vavilov 1949–50, 1957; West 1982). The uplifted backbone of tropical South America, they realized, supports an unrivaled array of crop species and plant subtypes—known as landraces, or in common parlance as traditional folk varieties, or cultivars.

The small-scale peasant farmers of the Andean highlands sow at least forty crop species that evolved there prior to the onset of Spanish rule in the 1500s, and the introduction at that time of numerous European cultigens (Crosby 1986, 1991; Gade 1975; Harlan 1975b; Horkheimer 1990). If we add to this list the numerous plants domesticated in the easterly foothills, or *montaña,* the total of Andean crops rises to at least seventy species and perhaps to as many as two hundred (Cook 1925a, 1925b; MacNeish 1977, 1992; Pearsall 1978, 1993).[4] Even more vast is the number of diverse varieties belonging to the major Andean crops; cultivated potatoes contain as many as five thousand landraces (Huamán 1986), while maize in Peru may boast as many as six thousand cultivars. This bewildering diversity of potato and maize types in Peru surpasses any other country (El Comercio 1987; Grobman et al. 1961; Manglesdorf 1974). Plentiful minor crops that are uncommon or rare elsewhere enrich the biodiversity of Andean farming even further. They include the tuber-bearing ulluco, mashua, and oca; the grain-yielding quinoa and amaranth; and the leguminous tarwi, or Andean lupine.[5]

The rugged Paucartambo Andes, a Corsica-size region, straddles the eastern flank of Cuzco in the highlands of southern Peru (map 1).[6] More than twenty thousand Quechua farmers reside there in the region's one hundred-odd Peasant Communities. The Paucartambo farmers cultivate twenty-six species of main crops that have evolved for at least several thousand years in Andean mountains and valleys (table 1). From their varied crop roster, the diversity of

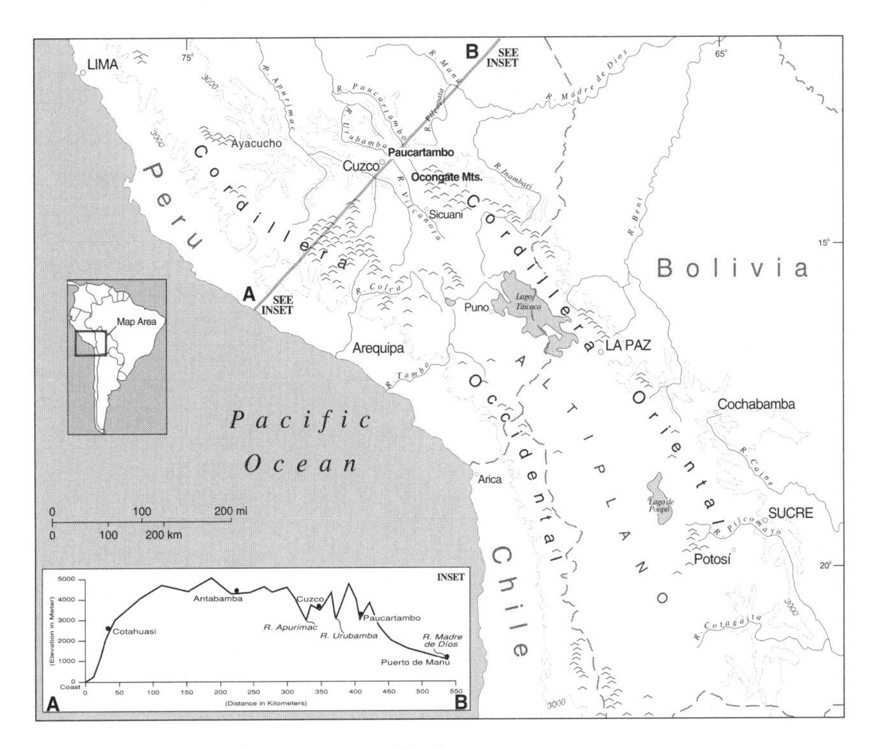

Map 1. The Central Andes of Peru and Bolivia.

four domesticates stands out: the staples of potatoes and maize and the minor and less known but nonetheless vital crops of ulluco and quinoa. Ulluco is a tuber-bearing relative of New Zealand spinach and quinoa is a cereal crop akin to buckwheat. The immense variety of the four crops flourishes due to the field environments and the farming activities of the region's Quechua cultivators.

"I know of no other place where agricultural occupation runs continuously through as large an altitude range," noted the touring Carl Sauer when he traveled to Cuzco in 1942 (West 1982). Although Sauer did not detour eastward to the Paucartambo Andes, its landscape would have affirmed his observation: a river of the same name cascades northward from the glaciers of the massive Ocongate Range, carving into the rolling high-elevation plateaus where towering glacier-covered peaks puncture the horizons east and west of the main Paucartambo Valley. The Río Paucartambo slices steep canyons of abrupt relief and varied montane habitats that box in the main river channels (map 1). The Quechua farmers who have tilled the tortuous terrain of Paucartambo fit the description offered by Sauer: they seed the diverse crops from the most deeply trenched channels below 8,850 feet (2,700 meters) to the lofty plateaus nearing 13,450 feet (4,100 meters).[7]

Table 1. Diverse Andean Crops of Paucartambo

English[1]	*Paucartambo Quechua*	*Scientific Name*
Amaranth	*achiwiti, kiwicha*	Amaranthus caudatus
Arracacha	*rakacha, viraka*	Arracacia xanthorrhiza
Achira	*achira*	Canna edulis
Chile pepper	*rokotu*	Capsicum pubescens
Quinoa	*kinwa*	Chenopodium quinoa
Winter squash	*sapayu*	Cucurbita maxima
Crookneck squash	*lakawiti*	Cucurbita moschata
Achocha	*achoqcha*	Cyclanthera pendata
Lucuma	*lukma*	Lucuma obovata
Andean, or bush lupine	*tarwi*	Lupinus mutabilis
Oca	*oka, apiña*	Oxalis tuberosa
Passionfruits	*granadilla*	Passiflora sp.
Avocado	*palta*	Persea americana
Common or kidney bean	*purutu*	Phaseolus vulgaris
Yacon	*yakon*	Polymnia sonchifolia
Potato or floury potato	*papa*	Solanum stenotomum
Potato or floury potato		S. goniocalyx
Potato or floury potato		S. x chaucha
Potato or floury potato		S. tuberosum subsp. andigena
Precocious potato	*papa chawcha*	S. phureja
Bitter potato	*papa ruk'i*	S. x juzepczukii
Bitter potato	*papa ruk'i*	S. x curtilobum
Mashua	*añu*	Tropaeolum tuberosum
Ulluco	*ulluku, papa lisas*	Ullucus tuberosus
Maize	*sara*	Zea mays

1. The table lists mainly field crops and omits numerous species of protected herbs and woody plants found in house gardens (Zimmerer 1989). English names for most Andean food plants are not standardized. The table and this book adopt the English names used in the recent survey of Andean crops by the National Research Council (1989).

Human experience with the diverse crops of the Paucartambo Andes began as early as seven thousand years ago, when early farmers nurtured the world's first domesticated forms of potatoes, common beans, and a handful of minor crop plants (Kaplan 1980; Lynch 1980; Lynch et al. 1985; MacNeish 1977, 1992; Pearsall 1978, 1992; Towle 1961). Ancient farmers guided a remarkable profusion of the crops and crop variants through their cultivation techniques and technologies. While biodiversity in the Paucartambo Andes still flourished during the twentieth century, the region's people were given cultural renown for their non-Western ways. A census of 1970 figured that the province stood second in Cuzco for its "Indianness" at 90.8 percent (Rowe 1947b).

Local scholarship claimed to find the "last Inca" in eastern Paucartambo (Flores Ochoa and Nuñez del Prado n.d.). Guidebooks advertised the region's religious fiestas and particularly that of its patron saint, the Virgin of Carmen, which in July drew thousands of adventurous onlookers to join the capital's two thousand or so villagers (Villasante Ortiz 1975).

The "Indianness" of the Paucartambo Andes, however, did not imply a geographic isolation of the region's people from outside economic forces and cultural influences. In fact, they had experienced periods of Inca domination (circa 1400–1533), Spanish colonialism (1533–1826), and republican rule by the Peruvian state (1826–present). The farming of diverse crops by the Quechua in Paucartambo thus rarely if ever existed beyond the pale of such pressures as taxes, trade, and tribute, and the impacts of imperial and national life. Several well-known sojourners were led to the Paucartambo Andes and penned impressions of its local landscapes and people. They included the polymath Italian geologist-explorer Antonio Raimondi, who prospected Paucartambo with little success in the 1860s, and the Swedish engineer Sven Ericsson, who later sought to make the region a gateway to the Amazon Rubber Boom (Raimondi 1874–1913; Villasante Ortiz 1952, 1975).

The repeated betrayal of economic bonanzas in Paucartambo belied a quiet exploration and undramatic extraction of its plant resources. As early as the 1920s the region was visited by enterprising collectors from the All-Soviet Institute for Applied Botany, which was at that time the world's foremost center of crop biological study and breeding. Under the leadership of potato specialist Sergei Bukasov and the direction of the Institute's founder Nikolai Vavilov, the 1927 Russian Potato Collecting Expedition ventured to Paucartambo in order to collect diverse Andean potatoes. Their collections in the region were reported to include the local potatoes *pitikiña, puka mama, pichucharo, sunch'u, chimaku,* and *alqay warmi* (Villasante Ortiz 1975, 47; see also Bukasov 1981; Vavilov 1949–50, 1957).

The visit of the Soviet team to Paucartambo followed closely on the peripatetic footsteps of O. F. Cook from the United States Department of Agriculture. An economic botanist, Cook had ventured to the Paucartambo Andes a few years before in 1915. He later authored a pair of landmark reports on the rich diversity of Andean crops that made them widely known to scientists in the United States and Western Europe (Cook 1925a, 1925b). Later, potato botanist César Vargas of Cuzco visited Paucartambo frequently and compiled a diverse collection of one thousand-plus potatoes described in his volumes on *The Potatoes of Southern Peru* (Vargas C. 1948, 1954; see also West 1982, 75–78). Vargas and others were aided in their efforts by Luís Angel Yabar, an eccentric Paucartambo villager dubbed locally as "Crazy Yabar" due to his obsession with the collecting and wildly unorthodox breeding of plants.[8]

National and international renown accrued to the biological riches of diverse crops in the Paucartambo Andes. During subsequent decades, leaders in the world of crop genetic resources such as Jack G. Hawkes of the University of Birmingham in England and Carlos Ochoa of the International Potato Center (CIP) in Lima likewise journeyed to the Paucartambo region in order to sample the riches of its renowned diversity (Hawkes 1941, 1944; Ochoa 1975). Social scientists also joined the flow of experts interested in the diverse plants farmed in the Paucartambo Andes. Robert Rhoades, formerly an anthropologist at the International Potato Center, recounted his travels in the pages of *National Geographic* magazine:

> In the Paucartambo Valley, high in the Peruvian Andes, only village elders remembered that a Soviet expedition had collected wild and cultivated potatoes there almost fifty years before. Most villagers had no idea how important their potatoes had become in other lands. They wondered why foreigners like me came here to learn about potatoes. I came as an American anthropologist, associated with the International Potato Center in Lima (Rhoades 1982, 672).

Later Hugo Fano and María Benavides, also of the International Potato Center, estimated that during the 1980s the Quechua in Paucartambo also ranked among the most stalwart producers of the minor crops like ulluco and quinoa (Fano and Benavides 1992). The studies of Fano and Benavides were possible, moreover, because Paucartambo was not torn by civil warfare between Shining Path guerillas and Peruvian antiterrorist police, unlike the southern Cuzco provinces bordering Puno and those western and northern ones closer to Ayacucho and Apurímac (Poole and Rénique 1992). These renowned crop collections and social science surveys undertaken in Paucartambo between the 1920s and 1990 did not, however, contribute much insight into how the diverse crop plants were being either managed and maintained or alternatively thinned or even extirpated. As the everyday domesticates of the Quechua commoners, the diverse crops of Paucartambo continued to furnish their personal needs without claiming much attention other than as a valuable source of breeding germ plasm.

A few traditions in Andeanist scholarship are so rooted in the study of diverse crops and Andean faming that they must be introduced with the Paucartambo region itself. The first tradition is epitomized in the classic essay, "Cultivated Plants of South and Central America," in the *Handbook of South American Indians*. Here Sauer treats these diverse crops as "living artifacts which give evidence of culture origins and diffusions" (C. Sauer 1950, 487). Sauer's chief sources included Cuzco scholars E. Yacovleff and Fortunato Herrera, who exhaustively searched historical documents, archaeological finds, and botanical-biogeographical monographs in order to compile taxon-by-taxon treatises (Yacovleff and Herrera 1934, 1935; see also Towle 1961). Daniel Gade, a self-described "historically oriented cultural plant geographer," combined similar sources with an added emphasis on the ethnobotany of contemporary agriculture. His studies center on the southern Peruvian sierra and especially the Urubamba-Vilcanota Valley, a large depression about twenty miles west of the Paucartambo Valley (Gade 1970, 1972a, 1975, 1981).

A second tradition in this sort of study sprang from ethnobotany, a subfield of anthropology, botany, and geography looking at the intricate relations between local people and diverse crops. As early as 1902 an ethnobotanical account of Andean potatoes appeared in the *Bulletin of the Lima Geographical Society* (Patrón 1902). Recently, ethnobotany cast in the archaeological mold has anchored major monographs and periodic reviews on early agriculture in the coastal foothills and in the high Andean sierra (Pearsall 1978, 1992; Towle 1961). Other studies based on "interdisciplinary ethnobotanical research" have focused on the management of diverse crops by Andean farmers at the scales of fields, communities, and regions (Bird 1970; Brunel 1969; Brush et al. 1981; Franquemont 1988; Jackson et al. 1980). Still a further ethnobotanical tradition is the legacy of agronomists in the region who bridged their career commitments to crop science with a seasoned and, in some cases, superbly gifted sense for the interrelations of food plants and Andean farm people (Blanco Galdos 1981; Cárdenas 1966, 1969; León 1964; Vargas 1948, 1954).

Well-founded concern about the threats posed to biodiversity in the central Andes and other Latin American regions establishes a third tradition. In an essay in the *Journal of Farm Economics* published in 1938 Sauer presaged this concern by decrying the extinction of irreplaceable crops as a result of the economic transformations in peasant farming (C. Sauer 1938). Later studies have documented the peril of particular crops: Gade examined the decline of the Andean lupine, or tarwi (Gade 1969, 1972b); taxonomist Ochoa assessed the disappearance of diverse potatoes (Ochoa 1975); and in recent analyses of the Peruvian Andes including Paucartambo Stephen Brush has compared the fate of diverse potato farming to rice growing in Southeast Asia (Brush 1986, 1989). Agricultural economics were employed to show how some Andean farmers adopt the improved high-yielding varieties that may, or may not, lead to the loss of diverse potatoes (Brush and Taylor 1992).

The Environmental Geography Approach

Traditions in the Andeanist study of diverse crops mostly eschew theory and its application to their special topic. Yet in judging the fortunes of the diverse Andean plants one confronts issues that escape the rigor of broader thinking only at the peril of unsound analysis. Indeed, the current dilemma of weak apprehension suffers from this oversight. Error-prone analysis and assumptions that plague statements like those in *Seeds of Tomorrow* can begin to be redressed via a selective use of related knowledge, both general and theoretical in scope. One such area consists of theory related to the ecology of agriculture (agroecology) and its concepts about biodiversity. Another area is that of broadly defined social theory that can be applied to biodiversity in settings such as the Paucartambo Andes.

A general framework for this task belongs to the thinking on human environmental relations that has long inspired geography, anthropology, sociology, and biological ecology. Geography and its core of environmental geography offers a relational perspective of interactive mutuality on the fluid ties between nature and society. The environmental geography perspective promises both the accenting of anthropogenic impacts on the diverse crops as well as the roles of those plants in human life. The model of regional political ecology, widely utilized in environmental geography, is especially apt for framing the interactions of diverse crops and cultivators. The regional political ecology approach seeks to assess resource change and degradation by integrating theoretical concepts from biological ecology, political economy and regional development, and the cultural ecology of resource management (Bassett 1988; Black 1990; Blaikie 1985, 1988; Blaikie and Brookfield 1987; Bryant 1992; Hecht and Cockburn 1990; Peluso 1992; Sheridan 1988; Zimmerer 1991d, 1994a, 1995).

Regional political ecology is strengthened selectively in the following section to guide the study of diverse crops. The concatenation of concepts from agroecology and social theory discussed below is not intended to put forth a Grand Theory but rather to craft a useful fusion.[9] Definitions of a few key terms about agricultural biodiversity must preface the discussion (table 2).[10] *Landrace,* a basic unit of this diversity, is preferred to *cultivar* and *variety,* terms whose strict meaning connotes scientific breeding. The term *landrace loss* is more accurate in many cases than the widely used *genetic erosion*, because the latter—although evocative—misstates the nature of biological consequence.[11] Finally, the people referred to as *peasants,* also called *farmers* and *cultivators,* by definition furnish a significant share of their own subsistence while also working in market commerce. Peasants like the Quechua in Paucartambo thus differ from purely capitalist or commercial farmers as well as from the truly miniscule number of pure subsistence growers, or so-called primitives (Wolf 1966). Also, *peasant* is translated as *campesino* in the Paucartambo Andes, a term that is widely voiced and not considered derogatory.

Agroecology and Biodiversity

Regional political ecology emphasizes the use of ecological theory to assess the deviation and typical degradation of environmental resources from a "natural" condition (for example, see Blaikie 1985; Blaikie and Brookfield 1987). The diverse crops differ, however, because human effort is essential for their survival and well-being. For example, a profound contrast distinguishes the diverse common beans that are in need of replanting by humans and the seeds of

Table 2. Common Terms for Crop Diversity

Term	*Definition*
Agricultural biodiversity	The biological diversity of crop plants, usually estimated at the taxonomic levels of species and landraces.
Landrace (or land race)	"Variable, integrated, adapted populations . . . recognizable morphologically" (Harlan 1975a, 618).
Landrace loss	The local disappearance of a landrace from a geographical place or region.
Genetic erosion	"The loss of germplasm from a population" (Brush 1989, 23).
Peasant	A farmer who raises food for his or her household's consumption as well as for commerce (Wolf 1966; Scott 1976).

wild beans that are naturally long-lived (Nabhan 1989). A fertile source of ecological concepts describing crop-cultivator ties is the field known as agricultural ecology or agroecology (Altieri 1983; Anderson 1952; Bayliss-Smith 1982; Clawson 1985; Cox and Atkins 1979; Loucks 1977; Netting 1974; Tivy 1990). In order to decipher and mentally unknot the crop-cultivator ties in the diverse agriculture of Paucartambo the agroecology approach is also taken to include a host of affiliated studies that have focused on the farm-related behavior of Andean cultivators (Brush 1977; Brush et al. 1981; Brush and Guillet 1985; Gade 1975; Guillet 1992; Mayer 1979; Mitchell 1991; Webster 1973).

The ample literature in agroecology must be scrutinized carefully, however, due to its uncritical and often unstated assumption of environmentally fine-tuned adaptation in the diverse crops. Farm critic Wendell Berry recapped the adaptationist proposition in his review of crop-cultivator ties in the famously diverse Andean potatoes, which he was contrasting with mechanized farming in the United States: "[Andean potato] varieties . . . are delicately fitted into their appropriate ecological niches" (Berry 1977, 177). Berry's synopsis of fine-tuned adaptation in the diverse potatoes summed up numerous case studies and overviews of Andean crop-cultivator relations; they include Vargas (1948, 9), Brush and Guillet (1985, 24), Guillet (1992, 61), Mitchell (1991, 39), and Webster (1973, 119). The Andean researchers' claims of the adaptation of diverse crops to microenvironmental niches were consonant, in turn, with the broad statements made on such crops. Jack Harlan, a doyen of crop diversity studies, wrote, "Different cultivars will be grown for different purposes or *to fit different ecological niches of the agricultural system.* . . . Each [landrace] has a reputation for adaptation to particular soil types according to the traditional peasant soil classification" (Harlan 1975b, 138, 163; my emphasis).

To be sure, diverse crops are adapted to the farm environment, otherwise they would not exist; however, we must be careful not to assume fine-tuned environmental adaptation. More than one decade's worth of recent ecological and evolutionary findings cautions that a high degree of niche specialization is not so common in many organisms and ecosystems as was earlier thought and, therefore, it cannot be justifiably presumed (Brown 1981; Gould 1982; Gould and Lewontin 1979; Zimmerer 1994a, 1995). Extending the cautionary insights of current ecological thinking to diverse crops, the character of crop-cultivator ties must also be apprehended without making a priori assumption of niche specialization. Surely an unquestioning belief in the adaptationist credo beset Nova's *Seeds of Tomorrow,* which alleges "precise tailoring, almost to individual plots of land." In the Paucartambo Andes several diverse crops and landraces, most notably potatoes, are found to exhibit a broad-based style of adaptation to varied farm environments (Zimmerer 1991a, 1994a; Zimmerer and Douches 1991).

Rejecting the adaptationist assumptions in much agroecology also means grasping that crop-cultivator ties are not only environmental outcomes but also the consequences of cultural practice. This does not deny that ecological and agronomic rationales motivate much work devoted to diverse crops by peasant farmers worldwide, including the Quechua in Paucartambo (Zimmerer 1991b).[12] The farmers are also motivated, however, by rationales aimed at nonecological objectives in seeding their diverse plants. Consumption criteria play a key role. Farmers in the Paucartambo Andes grow crops to yield a variety of customary foodstuffs in their version of a "decent life," an idea that taken broadly also pillars our own lifeways (such as portrayed in George Orwell's *Wigen Pier* and in current notions of a living wage; Orwell 1958).[13] Their expectations about diet and cuisine are incommensurate with purely physiological measures such as the "starvation minimum fixed by nature" that inform strictly economic approaches to agriculture (Todaro 1989, 315).

A nonadaptationist perspective on crop-cultivator ties also encourages us to consider the attitudes of farmers and their symbolic expression, topics of great pertinence to the role of environmental resources in general (Cronon 1983; Tuan 1993; Worster, 1988, 1993). Especially relevant to farming among the Quechua people in Paucartambo is the idea of a cultural or moral aesthetic used regularly to assign meaning and value to everyday material objects.[14] The Quechua farmers are particularly expressive in applying cultural or moral meanings to the stunningly diverse landraces found in potato farming (Zimmerer 1991b). Use of language through naming expresses a parade of cultural meanings that help give voice to their identities as Quechua farmers. Although their farming is symbolically rich, the Quechua in Paucartambo do not treat it as a purely indigenous activity. Indeed, one function of their naming customs is to express a micropolitics of "everyday forms of resistance" at odds with the region's dominant groups, such as village "patrons."[15]

Our critique of the adaptationist assumptions in agroecology must extend to a still fuller grasp of crop-cultivator ties. Too often agroecological studies have portrayed diverse crops solely as an end product of peasant farming; diversity in peasant crops is *also* a means to farming. By adopting the reflexive standpoint, one can begin to ask how diverse crops fit into farmers' allocation of resources and their systems of land use. Focusing reflexive questions on diverse crops in the context of land use also alerts us to the shaping effects of farm life, a fact of existence whether in the Paucartambo Andes or a Midwestern dairy district. In the case of Paucartambo the means that have been used to farm diverse crops are rooted in a lengthy history of plant domestication and agricultural evolution; varied habitats in this region as well as other eastern Andean enclaves gave rise to an exceptional share of the world's current crop diversity (Anderson 1952; Harlan 1975b; Hawkes 1983; C. Sauer 1952).[16]

The context of diverse crops in the land use of Paucartambo raises the concept of "verticality," or "ecological complementarity," which has often been used to explain the allocation of resources by farmers and their management of cultivars in the Andes of Peru, Bolivia, and Ecuador (Brush 1977; Dollfus 1981, 1991; Knapp 1991; Mayer 1985; Murra 1964, 1972, 1985a, 1985b; Salomon 1985, 1986b). Verticality refers to "a lasting Andean preference . . . for direct access to multiple [elevation-related environmental] zones" (Salomon 1985, 511). A geographical concept of land use, verticality holds that Andean farmers and communities have chosen to combine land use at an array of elevations, both in past periods and at present. Much historical and ethnographic evidence supports the prevalence of the verticality style of land use; however, the agroecological nature of verticality, its rationales, and its relation to the diverse crops are less agreed upon.

One version of verticality asserts that Andean farmers—most being agropastoralists, since they herd livestock as well—match land use to the optimal sites for growing their suites of diverse crops and farm animals. In this view verticality is taken to result from the agroecological matching of elevation-related climate with crop and animal ecology. The agroecological view was grounded in an influential eco-climatological sketch of the central Andes that shows an upper Zone of Temperate Tuber Plants and a lower Zone of Corn Cultivation (Troll 1958, 1968; see also Dollfus 1981, 1991; Murra 1985b; Pulgar Vidal 1946). Subsequent studies showed stacks or tiers or horizontal bands of elevation-related land use at work in various villages and Peasant Communities in the Andes of Peru and Bolivia (Brush 1976, 1977; Godoy 1984; O. Harris 1985; Mitchell 1991; Murra 1972; Orlove 1977b; Platt 1982, 1986). Subtropical fruits were seen at the base, topped by successive elevation-defined layers of maize, wheat (and sometimes barley), potatoes and tubers, and, finally, grazing land at the highest elevation of the mountains. This agroecological view has typically held that farmers match their diverse crops to vertical climate-generated microzones (Mitchell 1991, 39).

Convincing criticisms have pointed out, however, that the vertical design of tightly stacked and tiered bands of land use often does not distinguish landscapes of the central Andes. In most cases the idealized view appears to hold only at spatial scales covering expanses far larger than local areas (Godoy 1984; Mayer 1979, 1985; Mayer and Fonseca 1979; Mayer and de la Cadena 1989).[17] The critical empirical findings have chimed in with a small chorus of general critiques of the agroecological version of vertical land use (Mayer 1985; Salomon 1985). Both lead us to recognize that assuming land use is matched to crop and animal ecology is tantamount to yet another adaptationist fallacy. The ill-founded assumption of adaptationist land use has produced a frequent philosophical failing: most accounts of vertical agroecology show land use as static and merely as a sort of spatial container for farming. Land use in the adaptationist view is solely an artifact of agroecological specialization; it is not also thought of as a means to a variety of ends.

By critiquing the adaptationist flaw in vertical agroecology, we do not suggest that environments fail to influence land use and diverse crops. Rather we are urging a fuller grasp of land use in a sense similar to our revisionist critique of adaptationist flaws in the common concept of crop-cultivator ties. It recommends a closer consideration of the function of land use in the politics of peasant life. In the Paucartambo Andes a primary use of land is to assert control over territory; the function of land use in controlling territory is thus close to the original concept of verticality in Andean land use that was coined by John Murra (Murra 1964, 1972, 1985a, 1985b). Recently, Enrique Mayer has proposed renewing the earlier emphasis on the role of land use in the political control of territory. Mayer's proposal calls for doing away with the adaptationist-inspired model of vertical agroecology and replacing it with a broader consideration of land use based on what he calls production zones:

> I propose that we superimpose the concept of production zone as a man-made thing on top of the natural variations of the environment. When we think of production zones as man-made things, rather than as "adaptations" to the natural environment, our attention is directed to how they are created, managed, and maintained. Then the importance of the political aspects of *control by human beings over each other* in relation to how they are to use a portion of their natural environment will again come to the fore. (Mayer 1985, 46–47; my emphasis)

Not surprisingly, production and political concerns alone do not prefigure the use of farm resources; consumption criteria again play a key role. The

Quechua in Paucartambo voice expectations about a fit livelihood, which includes a suitable cuisine derived in substantial part from their diverse crops.[18] Since they produce much of their own food, their culinary norms expressed in a concept known as *kawsay* serve to encourage the growing of diverse crops. Furthermore, since the concern for cuisine insinuates crops and therefore farm resources, the kawsay concept may be seen as a sort of resource ethic (Zimmerer n.d.). Although routine and widely shared, the concept is also changeable and has been historically redefined as well as contested between the Quechua farmers and their rulers. Recent attention to the role of culinary norms and expectations about consumption suggest that this force may be as important as any in the verticality of land use, both past and present, in the central Andes (Larson 1988; Salomon 1985).

The broad view of land use must be careful to look at biodiversity in the full array of farming and other economic spaces worked by peasant farmers. Some spaces in rural landscapes contribute disproportionately to the concentration of this resource. Gardens, and especially the house or dooryard varieties of peasant people, are a favored candidate for genetic resource conservation since they often are biologically diverse (Alcorn 1984; Altieri and Merrick 1987; Anderson 1952; Wilkes 1991). Yet diversity-rich spaces like these are typically part of substantially larger suites of other cropping and economic strategies being conducted by farmers. Accounts of the diversity-rich farm habitats that overlook the other spaces of rural livelihoods are at risk of falling prey to flawed "just-so" stories of adaptation.

Finally, the nonadaptationist perspective recommends putting our concept of land use into the context of places or locales rather than isolating it at the scale of single farms or solitary communities. Place-based differences have long been recorded in geographical studies of environments and resource management. Indeed, the updated promise of a place-based analysis of soil erosion and the salience of contrasts at the scale of place helped lead to Piers Blaikie's early formulations of regional political ecology (Blaikie 1985, especially pp. 5–6; Blaikie and Brookfield 1987). Places, like land use, moreover, are made up reflexively through the activities of people in space (Entrikin 1991; Pred 1984; Sack 1980, 1986). In the Paucartambo Andes the nature of land use and the character of vertical production areas a la Mayer have formed in conjunction with the region's main places. Such place-based distinctness of land use in the region has proven critical to the mixed fortunes of diverse crops.

Adopting these nonadaptationist perspectives guides my contribution on agroecological matters in the study of diverse crops in Paucartambo. In rejecting the adaptationist assumption of microenvironmental specialization I find that several diverse crops in the region exhibit a broad base of environmental adaptation. They are not confined to narrow niches, in other words, but rather occupy broad areas defined by the farming systems of Quechua cultivators. The production areas do not create a perfectly stacked tier of fine-tuned vertical zones. Instead, the units of land use are found to cover broad ranges of elevation, overlap with one another, and show much proof of farmers' creative prowess as modifiers of nature. Biodiversity's agroecological role in the farming of the Quechua peasants in Paucartambo is thus more flexible but less certain than past studies have shown. Flexibility and uncertainty in farming have greatly influenced their culture history and the conservation of their diverse crops.

Social Theory and Biodiversity

Greater flexibility and less certainty in the farming of diverse crops also stems from their multifaceted social roles, taken to include social, cultural, and economic activities. In the Paucartambo Andes the diverse crops play a truly complex repertoire of parts in farm life. They are a socioenvironmental stratagem for subsistence, cultural icons that affirm identity, and economic stocks that support the Quechua farmers in their commercial ventures. Nowhere is the Nova film *Seeds of Tomorrow* more mistaken than in claiming the crops to be "foods that have been left behind." While some Quechua in Paucartambo have curtailed their diverse crops, their choice was not preferred. Landrace loss does indeed occur, but its process is as complex as the roles played by diverse crops. Assuming them "left behind" or "replaced" is not uncommon, however, which we see attested in some widely known statements of the leading authorities on "genetic erosion" (appendix A).

Pitfalls in the social analysis of environmental change are rooted in unfounded ideas of modernization among peasant societies of the Third World. The influential yet flawed Modernization Theory in particular advances the idea of socioenvironmental change as socially even and historically linear (Watts 1987, 1993). One failure of this theory is its belief that the growth of markets and commerce will uniformly convert the environments of peasant and indigenous production from traditional into modern. This tenet underlies the scenario of landrace loss intoned in *Seeds of Tomorrow*: "People might cast off their traditional varieties, sort of like last year's automobile model." To overcome the shortcomings of Modernization Theory and errors of interpretation such as those in the Nova film requires a broad social theory focused on ideas relevant to the human environmental nexus of crop-cultivator relations. Such studies suggest that landraces and subsistence-growing, whether in the central Andes or other semiproletarian societies of peasant and indigenous people, do not typically change in the unchecked way asserted by Modernization Theory.[19]

Social theory aids in creating more complex models and narratives of environmental change by drawing on insights from a variety of historical and ethnographic studies. The key social units among peasant and indigenous people are usually extended families, or households, and members within them—

especially women—who mostly work the diverse crops and perform related cultural labors, such as cooking and livestock-raising as well (Carney and Watts 1991; Collins 1986, 1988; Deere 1982, 1990; Reinhardt 1988; Sheridan 1988). Groups of households and communities may also be protagonists in actively changing the nature of subsistence environments. The modernizing households and communities of Third World peasants typically keep some customs in their subsistence-provisioning, although they often modify them. Even reports of the radical transformation and rampant degradation of the subsistence of peasants hint at socially uneven processes of environmental change (Grossman 1984; Nietschmann 1973; Watts 1983).[20]

The contribution of social theory to the study of uneven change in resources begins by asking about its immediate causes (Blaikie 1985, 1988; Blaikie and Brookfield 1987). On the one hand, a disproportionate share of diversity-deserting Quechua farmers in the Paucartambo Andes have come from peasants in the middle of the socioeconomic spectrum. The ranks of diversity-sowers, on the other hand, tend to diverge toward both the more well-to-do and the most destitute (Brush and Taylor 1992; Zimmerer 1991c, 1992a). Social inequality in diversity's fortunes conforms, therefore, to disparities in economic status among peasant households. (This conformity is not perfect, however, due to mitigating effects that include spatial pressures and cultural customs.) Since numerous historical and ethnographic studies demonstrate the commonness of wealth-based groups within the peasantries of the central Andes and other developing regions, the disparities may shape a variety of environmental change (Collins 1988; Deere 1982; Deere and de Janvry 1981; Larson 1988; Lehman 1982a, 1982b; Mallon 1983; Orlove 1977a; Orlove and Custred 1980; Sheridan 1988; Stern 1982).

The cultural customs of farming among indigenous and peasant people are not unchanging (Wolf 1982). Like economic changes, moreover, the course of cultural change is likely to be socially halting. In the Paucartambo Andes the wealthiest Quechua are recasting the diverse crops during recent decades as a sort of traditional luxury, at once a status-earning custom as well as an item beyond the reach of many (Zimmerer 1991c, 1992a). In so doing they are remaking a cultural custom and its meaning, albeit quite differently than what occurred with the imposition of Spanish colonialism in the 1500s when diverse Andean crops were suddenly recast as *comida del indio,* or Indian food. This is hardly the first instance of such cultural "reinvention" in the Andes (Allen 1988; Weismantel 1988; see also Hobsbawm and Ranger 1992), although it is among the first analyses with an environmental emphasis (Gade 1993; Zimmerer 1993b).[21] A sensitive social theory reports that the making of cultural identity, which includes the cultivation of diverse crops, must be studied in careful and historical detail rather than assumed to result from the ironclad mechanics of modernization.

A useful theory must offer an analysis of how farmers themselves experience social inequality and the salience of new cultural meanings. In fact, the more well-to-do Quechua in the Paucartambo Andes have adeptly taken advantage of their culturally high-status foodstuffs, forging a compelling link between ethnicity, power and inequity, and biodiversity (Zimmerer 1991c, 1992a). While farming many fields for commerce, the better-off Quechua have gained a variety of benefits from their diversity-rich, self-provisioning plots. An environmental dichotomy in the form of biodiversity conservation versus loss has thus introduced a powerful new set of terms for onging economic development and social change (Netting 1993). Environmental parameters seen at work in historical transitions suggest that insight can be drawn from the vital traditions of Sauerian geography and environmental history. The former focuses on the historical modification of nature and resources in Latin America in particular (S. Cook 1949; Johannessen 1963; Parsons 1971; C. Sauer 1938, 1958; Zimmerer 1995), and the latter highlights the processes of social change and economic development (Cronon 1983; Worster 1988, 1993).[22]

Finally, the perspective of a broad social theory can be used to analyze the relation of markets to biodiversity. This theory has sought to make sense of recent economic trends in developing countries, especially those of Latin America and Africa, which show scant or no improvement in peasant incomes and prospects for development (Deere 1990; de Janvry 1981; de Janvry et al. 1989; Reinhardt 1988; Roseberry 1993; Wilson and Wise 1986).[23] The Quechua of Paucartambo have gained little from Peru's economic development. Acclaimed nationwide land reform in 1969 resulted in an environmental paradox, since the legal autonomy of the landholding of peasants clashed with stiffening economic pressure on their resources. In expanding and diversifying their commerce in response to unstable markets and unfavorable government policy, the peasants of the Paucartambo Andes pursued a livelihood strategy shared by millions of semiproletarian farmers in Latin America (Deere 1990; de Janvry et al. 1989; Reinhardt 1988; Roseberry 1993). At the farm level, such changes in commerce were never far removed from adjustments in subsistence, which have included the depletion of significant biodiversity in Paucartambo (Zimmerer 1991c, 1992a).[24]

The relation of markets to biodiversity also raises the issue of the circulation of diverse crops through the exchange activities of peasant farmers. Whereas Modernization Theory posits a self-correcting capacity of markets to circulate environmental goods in response to rising demand and scarcity, a broad social theory leads us to consider also the cultural and historical making of market commodities (Orlove 1977a; Polanyi 1957).[25] It suggests that markets and other exchange mechanisms should not be thought of as either inherently adverse or intrinsically amenable to biodiversity. Since the Inca period, nonsubsistence activities within the regional economies of the southern Peruvian sierra and the Paucartambo Andes have offered only a small fraction of the diverse crops (Zimmerer 1993b). Faced with this changing series of nonsubsistence demands, the Quechua farmers during six hundred years or so have often innovated adjustments in their diversity-rich subsistence, which never rested like a vestigial relict.

2

The Great Historical Arch of Andean Biodiversity

Ancient Domestication in the Eastern Andes

The mother tongues of Quechua and Aymara spoken by eight to ten million people in the central Andes—from southern Colombia to northwestern Argentina—have never uttered a general word meaning *crop,* or *food plant* (Bertonio 1956; Beyersdorf 1984; Brush et al. 1981; Cusihuaman 1976; Gade 1975; González Holguín [1608] 1952; Hawkes 1947; La Barre 1947; Lira 1945; C. Sauer 1950; Zimmerer 1991b). Omission of the generic term *crop* in the ample agricultural vocabularies of Andean languages seems unlikely. The sophisticated techniques of farmers, their humanized planting landscapes, and, not least, the prodigious variety of crops appear to the contrary. This puzzle is resolved, however, by paying heed to the Quechua and Aymara lexicons: each crop and every landrace within the plenitude of thousands is given a specific name; the generic term *crop* can be omitted. By naming their food plants so exhaustively, the farmers are able to voice a litany of multi-faceted specificity—agroecological, culinary and nutritional, and cultural-symbolic.

Among the specifics elicited by this naming are ecological traits that evolved over millennia in Andean mountain fields. Many production traits in the suite of potato and maize mainstays and the minor quinoa and ulluco crops took shape especially in the eastern ranges of the central Andes. Bordering the tropical forests of the Amazonian lowland, the mountains of the eastern Andes set a key scene in the early history of agriculture. The oriental uplands, peak-studded plateaus, and deeply sunk valleys mainly belong to the easterly Cordillera Oriental chain—the Eastern Cordillera (map 1). The Río Paucartambo;

the upper Madre de Díos, including the Río Sandia; the upper Beni; and the Río Caine, as well as the middle reaches of the huge Río Apurímac and Urubamba-Vilcanota Valleys, carve major recesses of the eastern Andes. Uniting edges of the Cordillera Central—the Central Cordillera—and the Andean foothills, or montaña, the eastern Andes form a cohesive ecological unit known as the Eastern Slopes, or Los Vertientes Orientales (Weberbauer 1945).[1]

Early agriculture on the Eastern Slopes and the unique evolution of several crops there left many wild crop relatives bearing a family resemblance to the domesticates (Harlan 1975b; Hawkes 1983, 1990).[2] Uncultivated plants of the Eastern Slopes, such as wild potatoes, ulluco, and quinoa, derived from the same ancestral precursors or progenitors that gave rise to the domesticates. Domestication, or "ennoblement" as authors once wrote, converted some of the wild ancestors into full-fledged crops, while other descendents spun off lineages as uncultivated relatives. Many progenitors were cloistered in the uplands and valleys of the Eastern Slopes, including the Paucartambo Andes (Zimmerer 1989). In fact, at least half the food plants farmed in the region evolved from wild ancestors whose present-day members are found across the Eastern Slopes (appendix B.1). Rugged terrain and varied habitats thus shaped the ecological specifics of the earliest staffs of life.

A bias has plagued our crediting the eastern Andes, such as Paucartambo, in crop domestication and agricultural evolution. Nearly all investigations have dotted the artifact-rich and more easily reached valleys of the Pacific Coast and the Western and Central Andean Cordillera, while rarely entering into the eastern Andes (Towle 1961).[3] Indicative of the bias, the oldest crop remains known to archaeologists in the Andes, and indeed in South America, were unearthed in mountains allied to the Pacific Rim rather than the Amazon Basin. Sites at Guitarrero Cave (Guitar Player) in the Santa Valley about one hundred and fifty miles north of Lima, Tres Ventanas Cave (Three Windows) south of Lima, and Ayacucho, did not reach as far as the Eastern Slopes (Engel 1970; Harlan 1975b; Hawkes 1983; Kaplan 1980; Lynch 1980; MacNeish 1977, 1992; Pearsall 1978, 1993). However, the antiquity at seven thousand or more years of the earliest crop remains—potatoes, beans, chile peppers—excavated in the Western and Central Cordilleras belies prior domestication. The still earlier domestication of the crops likely unfolded in the eastern Andes, with the domesticates subsequently dispersed to sites such as Guitarrero Cave (Harlan 1975b; Smith 1980, 115).

In being a home for ancient agriculture the eastern Andes housed an immense array of field environments under its peaked ridges. While exploring the southern Peruvian sierra and making it known in the world of United States' foreign affairs, the geographer Isaiah Bowman saw a monumental spectrum of environments, both physical and human, that spanned "the greatest contrasts in . . . the Andean cordillera" (Bowman 1916, 68). This environmental diversity beneath the serrate eaves of the eastern Andes was of special significance to the ecology of early food plants in two ways. Heterogeneity favored diversity in the wild progenitors of crops and thus posed a plentiful choice of potential domesticates to the first farmers (Anderson 1952; Harlan 1975b; Hawkes 1983; C. Sauer 1952; Vavilov 1949–50, 1957). Habitat variation subsequently enriched the diversity of ecological attributes after crop domestication. A contemporary of Bowman, the eminent historian and breeder of potatoes Redcliffe Salaman, seized the diversifying role of environmental complexity in evolution of the domesticates. "The great [Andean] variety in the elevation and soil environment . . . led to the establishment of specialized types" (Salaman 1985, 10).

The major font of habitat diversity in the eastern Andes was climatic. Although climate there at the dawn of domestication differed from the present day, it nonetheless was generally similar by varying chiefly along three gradients: elevation, a transverse axis cut perpendicular to the main mountain chains, and north-south or latitudinally (Gómez Molina and Little 1981; A. M. Johnson 1976; Troll 1968; Winterhalder 1994). Climbing the elevation gradient, temperature dropped, perhaps similar to the present-day rate of 3.2°–4.5°F/1,000 feet (5°–7°C/1,000 meters). Precipitation levels rose upslope at an irregular rate. Across the transverse gradient, the windward slopes facing the Amazon were rain-drenched and cloud-shrouded, while semiarid "rain shadows" covered the western versants. Along the latitudinal gradient, the least dramatic, variation accrued via disparate regimes of rainfall that tracked the overhead sun. Complexity caused by the major axes of climate-driven variation created a "three-dimensional arrangement of climatic types, vegetational formations and landscape regions" (Troll 1968, 34).

The three-dimensional climates of the eastern Andes framed early farming with stunning variation over short distances. The resource-scouting Isaiah Bowman was impressed: "Between [the valleys and higher ranges] are the climates of half the world compressed, it may be, between 6,000 and 15,000 feet of elevation and with extremes only a day's journey apart" (Bowman 1916, 122). Here travelers and residents alike walked from a glacial ice cap to a subtropical grove of orange-bearing citrus trees in a single day. More recent studies calculate that as many as one-third of the world's one hundred or so biologically significant climates continue to be packed into the eastern Andes (ONERN 1976, 2, 31; Tosi 1960).[4] Within the Paucartambo Andes alone are squeezed no less than thirteen of the world's chief climates. They range from semiarid desert to the perennially humid tropics to tundralike regimes of snow and year-round cold. Great swings of temperature each day, from below freezing at night to shirt-sleeve weather at midday, likely prevailed in the past as well.

Contrasting climates in escarpments of the eastern Andes no doubt gave play to a kaleidoscopic mix of plant habitats. A striking variety of vegetation is found today, although its details likely differ from that of the past. At present, four principal groups of vegetation carpet the core area of the Paucartambo

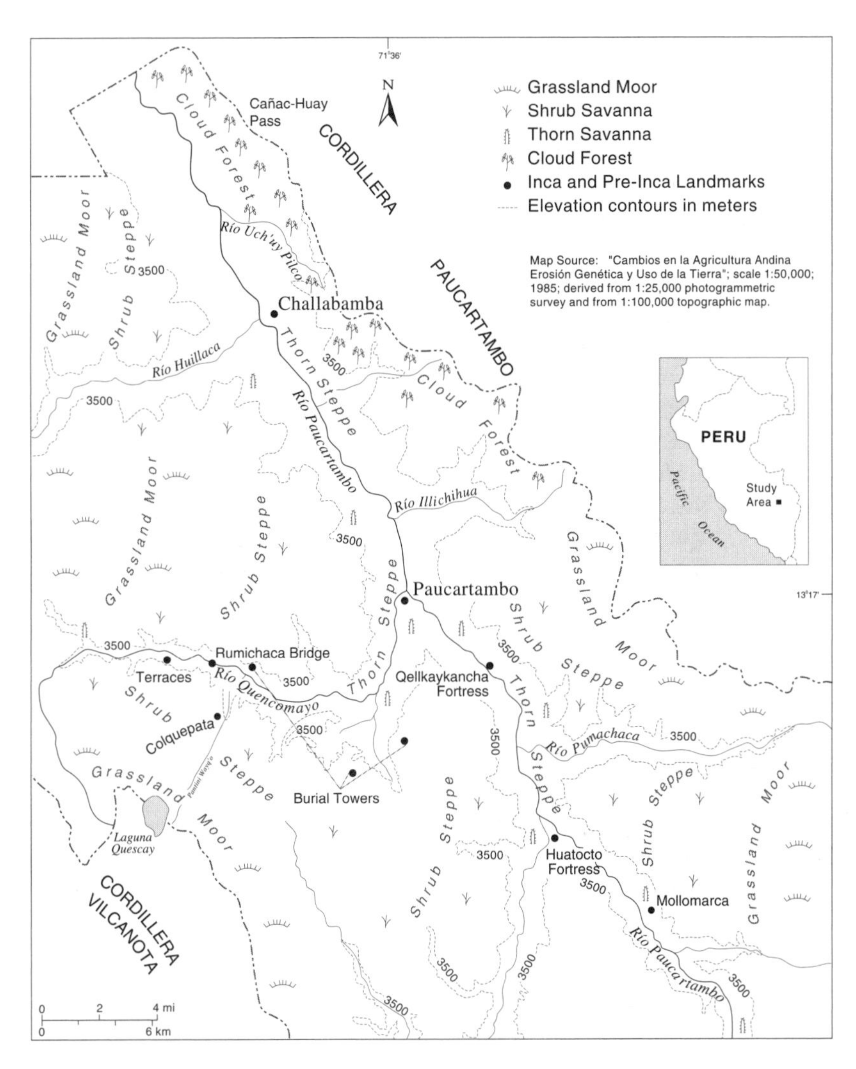

Map 2. Vegetation and Pre-European Settlement in Paucartambo.

Andes from below 8,850 feet (2,700 meters) to above 13,450 feet (4,100 meters) (map 2). The widely varied quartet—a grassland moor of prairie and bogs, the shrubland or shrub savanna of shrubs mixed with grasses and scattered small trees, the thorn scrub or thorn savanna of spinescent shrubs and trees mixed with grasses, and the montane cloud forest known as the Eyebrow of the Jungle, or *ceja de la selva*—likely made up vegetation for several thousand years. Based on present-day studies, we know that the four vegtation types are a mosaic, grading into one another and overlapping and contoured only loosely by elevation (Zimmerer 1989; Zimmerer and Langstroth 1994). The celebrated *q'euña,* or *Polylepis racemosa* tree, for instance, occurs not only in the grassland moor but also in the shrub savanna and even in out-of-the-way ravines within the thorn savanna (appendix B.2). Elevation limits of each vegetation type at present vary by as much as 500 feet (150 meters) due to local factors. Similar patterns probably adhered in the earlier floral landscapes of the region.[5]

Edaphic, topographic, and geomorphologic contrasts also enriched the immense wealth of natural variation in the eastern Andes. Soil types differed irregularly in response to climate, vegetation, and geologic bedrock. Slope failure including landslides—which the Quechua call *wayq'o*—as well as massive debris flows and slumps triggered still more complexity. Various observers, including Isaiah Bowman, Carl Troll, and Augusto Weberbauer, commented how the cropped slopes were undermined by a chronic instability. Slopes of the eastern Andes may have been particularly unstable; due to a weak parent material of sandstones and shale, slope movements there probably abetted the course of plant domestication by selecting for wild plants with large propagules—be they seed, fruit, or tuber. After domestication, the Sturm und Drang of geomorphologic disturbances churned the array of arable settings and pressured early farmers to find a mix of fit crops (Anderson 1952; C. Sauer 1952; Vavilov 1949–50, 1957).

Abrupt topography of the eastern Andes sculpted a host of discrete habitats for vegetation subtypes. Aquatic plants steeped in bogs, marshes, and glacial lakes. Ravines out of the reach of fire and later livestock harbored the tallest trees. Herbs and grasses were scattered across rocky outcrops. Bunch grass carpeted many deeper soils. Human inhabitants began modifying the already varied Andean landscape at least ten thousand years ago. At first, they mainly gathered wild plants; burned shrub, grass, and forest land; and hunted game; later, they farmed and herded (Dollfus, 1981, 1991; Lynch 1980; MacNeish 1977, 1992; Pearsall 1978, 1993; Troll 1968). Andean peoples impressed distinct marks on vegetation and soils in their heavily inhabited alps. Modifications due to land use, in some cases indelible, escaped few settings: aquatic ecosystems supplied water and food, ravines sheltered trees for fuel and building, the rocky outcrops could be grazed, and deep soil invited agriculture. The varied environments were habitats for humans as well as artifacts of their endeavors.

Ancient farming in the eastern Andes deeded a dual legacy to the fortunes of diverse crops. In one inheritance it bequeathed the basic materia prima of biological diversity in terms of copius variation. A second inheritance passed on the specific traits within diversity's raw material since early farmers were already sifting and winnowing the crop variants, ultimately sowing some but not others. Once selected, the main crop traits imposed a series of requirements on the farming systems that followed, for the diverse crops could be nurtured only within certain agroecological bounds.[6] Two main needs of the nurtured crops—the need for particular field habitats and the need for farm tasks, such as planting and harvest, to be exercised at particular times—faced each succeeding generation of Andean farmers. Farmland and labor-time needs accented the pivotal part of the mountain farmers then, as now, in bestowing the resources for survival of their diverse crops.

Farmland and labor-time wants of several Andean crops shared a pair of conspicuous traits that were formed during the early millennia of agriculture. The crops evolved via the extension of their range of growing environments. In each species farmers adapted new landraces to fields at successively higher elevations in the Andean landscape.[7] The desiderata of crops in terms of farmland thus encompassed a range of elevation-related environments. With respect to the labors of farming, the crops exhibited a noticeable diversification of maturation period—the time lapsed from planting until harvest. Newly evolved landraces could be cropped in the shorter growing seasons at upper elevations, often by abbreviating the time-to-ripening. The twin adjustments to alpine fields and shortened growing seasons would prove to contour the ensuing history of agrarian change. We can consider such changes manifest in the early evolution of Paucartambo's four premier food plants—potatoes, maize, ulluco, and quinoa.

Following domestication, potatoes evolved into the principal crop of the Andes and eventually into the foremost vegetable worldwide. The evolution of potatoes witnessed both the upslope protraction of range and the divergence of ripening time. The complex of seven domesticated species grown in the Paucartambo Andes—one less than the crop's full set—presumably evolved from one or perhaps two wild potatoes, *S. canasense* and *S. leptophyes* (table 1; P. M. Harris 1978; Hawkes 1978, 1990; Hawkes and Hjerting 1989; Horton 1987; Ochoa 1990; Ugent 1970). The middle slopes of Paucartambo and other eastern Andean regions likely sited potato domestication since such wild potatoes abound there: wild *S. canasense* is found in "south Peru, common in depts. Cuzco and Puno, possibly in dept. Ayacucho, at 2,900–4,100 m [9,500–13,450 feet], in rock gravelly slopes, field borders and roadsides" (Hawkes and Hjerting 1989, 138), while the wild *S. leptophyes* covers a similar range of habitats. Based on the present-day occurrences, both wild potato species sprouted in comparable habitats in Paucartambo (Zimmerer 1989).

Early domesticated potatoes evolved rapidly by absorbing much diversity, as though they were genetic sponges sweeping upslope to fields each at a higher elevation (Hawkes 1978, 1990; Hawkes and Hjerting 1989; Ochoa 1990; Ugent 1970; see also C. Sauer 1950, 516–17). Genetic introgression with other species including wild relatives propelled much of this progress. By seven thousand years ago, when early mountain farmers at Tres Ventanas near the crest of the Western Andean Cordillera discarded the remains of a domesticated potato, many traits were already molded. Maturation periods had diverged in conjunction with the upslope protraction of growing range. Potato species of the middle elevations (9,500 feet/2,900 meters–13,300 feet/4,050 meters), such as *S. stenotomum, S.* x *chaucha,* and *S. tuberosum* subsp. *andigena,* matured in medium-length cycles (seven to nine months), while the frost-resistant *S.* x *juzepczukii* and *S.* x *curtilobum* reached term one to two months more slowly in the highest fields (12,950 feet/3,950 meters–13,450 feet/4,100 meters). The peculiar *S. phureja* could be cropped in three months and thus could cope with the rampant diseases of the most humid Eastern Slopes.

Maize also cropped up in the eastern Andes as early as several thousand years ago. Although the original cradle of maize domestication was likely set in Mesoamerica, the unique cereal diffused rapidly and with great success in South America (Iltis 1983). By five thousand years ago the maize crop was transported and traded southward to the Santa Elena peninsula on the Pacific coast near the equator and subsequently to the Andean valley of Ayacucho in the Central Cordillera (MacNeish 1977; Zevallos et al. 1977). By three thousand years ago maize was being cultivated in the vicinity of Guitarrero Cave in the Western Cordillera (Smith 1980). The maize excavated from the high-elevation archaeological sites at Ayacucho and Guitarrero Cave presumably grew nearby in lower fields at or below 11,500 feet (3,500 meters). Cultivation at the middle elevations of the central Andes attested that the maize crop had protracted noticeably upslope from its earlier lowland confines.

In order to crop maize at higher elevations the ancient farmers scouted for early landraces (Towle 1961). The long-maturing maize typical at sites near sea level could not finish in the truncated growing seasons of Andean elevations. By dint of necessity the maize types that were dispersed to mountainous regions, such as Ayacucho and Guitarrero Cave, ripened several months earlier than the sea-level counterparts (Grobman et al. 1961). A similar scenario of adaptive changes in maturation earmarked the maize variants known as Cusco Amarillo, which evolved a distinct short cycle in the Andes near Paucartambo when it was "brought 700 meters [2,300 feet] up from the Urubamba Valley to the nearby heights of Marras" (Grobman et al. 1961, 52). The diversification of ripening time in general, and increased earliness in particular, recorded evolutionary successes that were to reset repeatedly the ecological coordinates of the Andean maize crop.

The tuber-bearing ulluco—modest in reputation outside the Andes but a prime tuber crop in its eastern Andean home—similarly formed agronomic traits in its protraction up the Eastern Slopes. Although the crop's earliest fragments were unearthed in the western Tres Ventanas cave where they lay buried for at least seven thousand years (Engel 1970), the wild progenitor of ulluco likely ranged across middle elevations of the Eastern Slopes in southern Peru and northern Bolivia (Cárdenas 1969). The uncultivated ancestor presumably evolved into the present-day plants of wild ulluco in Paucartambo, known as

Fox Ulluco (*atoq lisas*), which pocket rocky slopes near the main river valley at 9,500 feet (2,900 meters) (Zimmerer 1989). According to the renowned Bolivian botanist Martín Cárdenas, the wild ulluco generally grips the middle slopes between 8,850 feet (2,700 meters) and 11,810 feet (3,600 meters), elevations that likely bracketed its domestication (Cárdenas 1969, 58). Subsequent evolution of ulluco extended its range upward of 13,100 feet (4,000 meters), and thus supplemented its ecological specifications.

Quinoa, or *Chenopodium quinoa,* the top Andean grain-bearer behind the introduced maize, also tethered new traits in its spread upslope. Domesticated quinoa also was first noted west of the Eastern Slopes, in this case by seven thousand years ago at Ayacucho in the Central Cordillera (MacNeish 1977); however, the crop's likely progenitor was a wild quinoa, *Chenopodium hircinum,* which presumably abounded in the eastern valleys and at the middle elevations below 9,850 feet (3,000 meters) (MacBride 1936, 470; H. Wilson 1978, 1988). This wild progenitor occupied a similar niche in the Paucartambo Andes based on the present-day occurrence of its descendents (Zimmerer 1989). Domestication of the wild quinoa gave rise to a group of cultivated quinoas referred to as valley types, which occupy fields between 9,850 feet (3,000 meters) and 11,800 feet (3,600 meters). From these, farmers bred a number of high-elevation landraces that vaulted upward of 13,100 feet (4,000 meters) (Aguirre and Tapia 1982; Tapia and Mateo 1992). Diversity in the quinoa crop therefore aped potato, maize, and ulluco by evolving landraces that were moved upslope in the eastern Andes.

Paucartambo's flank of the eastern Andes possessed notably fit field habitats for the four diverse crops. An evolutionary scenario known as differentiation-hybridization added further to the great diversity instilled by dispersal upslope and changes in maturation. Crops undergoing differentiation-hybridization became more genetically distinctive through regional adaptations during periods of geographical isolation. Cross-regional seed exchange then renewed hybridization that combined to create still other new variants (Harlan 1975b; Manglesdorf 1974). Through trade and migration between Paucartambo and other Andean regions, early farmers brought crops in and out of contact for millennia, thereby driving the cycles of differentiation-hybridization. Maize diversity offered the classic and perhaps best-known example of differentiation-hybridization. Traits of maize and the other diverse crops thus evolved in conjunction with the geographical mingling of people and products.

Prior to Inca rule beginning about A.D. 1400, the exchange of crops between early farmers in the Paucartambo Andes and their counterparts elsewhere waxed and waned repeatedly. Unfortunately the exact evolutionary role of the interregional crop exchange has remained mostly unknown except in glimpses gained through the archaeological record. We may surmise that the extraregional exchanges of trade goods and tribute were strengthened when the Paucartambo

Andes fell to larger polities such as the Huari Empire, which ruled from its capital near Ayacucho between roughly A.D. 400 and A.D. 800 and from its fortress in the Lucre Basin at the western edge of Paucartambo. Later the region likely felt the political impress of the southerly Aymara Kingdoms and the Tiwanaku Empire or at least the cultural influence of the Altiplano peoples. They built dozens of still intact one- to two-meter funerary towers, or *chullpas,* that were typical of the austral influence.[8]

Conquest of Paucartambo and much of the eastern Andes by the Inca Empire hailed an unprecedented geographical integration of Andean territories and eased the purposeful introduction of food plants into new regions. The tuber-bearing oca, for example, was probably dispersed northward to the Andes of present-day Ecuador under the aegis of Inca rulers (Kelly 1965). At the same time, the Inca governors imposed a social order on agriculture that defined a distinct role for crop diversity. Fathomable to a surprising depth in archaeological and historical materials, the farming systems of Inca rulers and those of peasant commoners offered a riveting contrast of great consequence to biodiversity.

Sparse Biodiversity in Imperial Agriculture, 1400–1533

One legacy of Inca rule for diverse crops was the chasm opened between diversity-rich farming for subsistence and diversity-poor cropping that was destined for nonsubsistence, or so-called "surplus," purposes. The remarkable divide in the social role of biodiversity that was wedged between subsistence and surplus during the Inca period became a long-lived precedent. Details of the divide were thus important not only during Inca rule over Paucartambo (circa 1400–1533) but also during succeeding periods. Its dimensions were set in the contrast between agriculture run by the Inca rulers, imperial agriculture, and that of the commoners, peasant agriculture. Agriculture in the Paucartambo Andes, located close to the empire's center in Cuzco, would inescapably evidence the imperial-commoner contrast in biodiversity.

The Paucartambo Andes fell to the Inca early in the chronology of imperial conquest by the Cuzco-based power. By 1400 or so the sixth Inca ruler, Inca Roca, had subdued the highland territory as far to the east as the "Paucar-Tambo river" (Cobo [1653] 1956, vol. 2; Rowe 1945). An invading army under the emperor's son Yawar Waqaq was reported to encounter feeble opposition from the Poques people living in the mountainous redoubt of Paucartambo (Garcilaso [1609] 1987).[9] New military successes of the Inca solidified their already firm control over nearby strongholds, such as the daunting fortresses at Ollantaytambo and Pisac in the Urubamba Valley. The mighty Inca soon strung a chain of thirteen settlements into Paucartambo. By annexing the region, the imperial rulers of the Four Quarters, or *Tawantinsuyo,* as they called themselves, gained control over a main eastern Andean rampart—the heart of the Eastern Quarter, their Antisuyo. By 1450, their sprawling territory of Andean mountains and coastal lowlands united the center of present-day Chile with southern Colombia.

The stone-paved Antisuyo Road ran eastward from the Inca capital in Cuzco through Paucartambo. Bridges spanned the Quencomayo River at Rumichaca (Stone Bridge) and the main Paucartambo Valley in a narrow box canyon where an Inca storehouse, or *tambo,* gave name to the site and to the whole region (map 2; Cieza de León [1553] 1853, 437). Two fortresses built in the Paucartambo Valley at Huatocto and Qellkaykancha, together with the laying of the Antisuyo Road between Cuzco and Paucartambo, firmly fastened the eastern frontier to the empire's center. Military installations and supply points along the route secured the cultural and economic integration of the Paucartambo territory with the Cuzco heartland. Because the region straddled the most exposed flank of the Inca center, impetus for its cultural and economic integration was sensible.[10] Royal fields of coca nestled in the montaña foothills raised even further the stakes of Inca geopolitical interest in the Paucartambo Andes.

One indicator of Inca cultural control was the cartography of Quechua-named places imprinted on the Paucartambo Andes (map 2). This toponymic legacy merits mention, although it was inherited partly from the Quechua-speaking Poque and may have been amplified after the Spaniards' Conquest. Caveats aside, the Inca rulers undoubtedly reinforced Quechua toponymy and coined new names such as the capital *Paucar*-storehouse. The principal river also was denoted Paucartambo in a probable effort to erase its earlier designation as the Mapacho.[11] Downstream a prime maize-growing enclave located where the shrub savanna mingled with the cloud forest was named Challabamba, or Plain of Maize Stubble (*challa-bamba*). Mollomarca, or the Place of Snail Shells (*molle-marca*), was settled in the southern Paucartambo Valley nine miles upstream from the paucar storehouse. In the westerly interior of Paucartambo Colquepata, or Step of Silver (*qolqe-pata*), perched on a topographic shelf above the zigzag gorge of the Quencomayo, or Zigzag River (*kenko-mayu*).

Tawantinsuyo ruled the economies of Paucartambo and other newly conquered regions for the most part by drafting the labor of commoner peasants in a corvée tax known as the *mit'a* (Murra 1980; Rowe 1947a; see also Hastorf 1990). Officials of the empire and its religion—the Inca cult—invested their legions of taxed labor in agricultural pursuits, to which the drafted peasants brought a skilled knowledge honed in their own experience. The skillful commoners toiled under Inca overseers in constructing staircase sets of dressed-stone field terraces, transporting and creating fertile field soils, and devising impressive irrigation works. They converted the barely arable waste of steep mountain slopes into highly productive parcels for cropping and into powerful signs of imperial mastery over both nature and the commoner subjects (Denevan 1986; Donkin 1979; Murra 1980; Rowe 1947a; Treacy 1994). When modern observers such as O. F. Cook marveled at the previous pinnacle of agriculture in the Andes, they were admiring above all the grandiose artifacts of the Inca state and its cult (Cook 1925a, 1925b).

The renowned diversity of crop plants in the Andes was a centaur, however, in the grandiose embellishment of farm resources showcased by the Inca rulers. Unlike the other prizes of Inca agriculture, the diverse crops clustered in the farm plots of peasant commoners rather than in the monumental fields of the Tawantinsuyo rulers (Zimmerer 1993a). This irregular feature of peasant farming under Inca rule was suggested by the ethnohistorian Murra:

> One [agricultural system] is old and autochthonous: the Andean highland growing plants domesticated in that area laboriously adapted to high mountain conditions, grown on fallowed land and dependent on rainfall. . . . The other is newer, imported, and based on maize (essentially a warm-weather crop, clinging to the lower and protected reaches of the highlands and in need of irrigation, terraces, and fertilizers to survive in Andean circumstances). (Murra 1980, 12)[12]

The biogeographical fact that diverse crops were holed in the farming of peasant commoners originated in the system of economic organization imposed on Inca territories. A system of "dual production" divided the Inca economy and its agriculture into two sorts in the conquered regions such as Paucartambo (Larson 1988, 314; Mayer 1985; Morris 1981; Murra 1980; Rose 1947a; Salomon 1985, 1986b). In the first type of production peasant subjects met their primary food needs through farming with resources readied by households and local ethnic aggregations later termed *señoríos.* These commoners obtained some subsistence goods through trade exchanges, most frequently small-volume items such as salt and hot capsicum, or aji chile peppers (*ají*). In the second arm of the Tawantinsuyo economy Inca political and religious rulers kept permanent workers and drafted mit'a or corvée labor from the subject populace of rural regions. This division of the Inca economy into subsistence activities and state-run efforts, with meager marketing if any, arbitrated the fortunes of crop diversity.

The state-run economy and its legions of mit'a labor draftees undertook a broad pair of endeavors. In one pursuit the Inca rulers ordained various outposts of nonfarm production in select regions, such as gold and silver mining in ore-rich districts, textile-weaving near high pastures stocked with royal herds of alpaca and llama, and imperial cultivation of coca in the eastern Andean foothills. The state-run economy of Paucartambo specialized in the growing of coca shrubs and the handpicking, drying, and processing of its leaves for use as a mild stimulant. In the second stream of its surplus economy the Inca lords designed the farming systems of imperial agriculture. They spread royal fields throughout all the realm's regions, a broader and more comprehensive area than those covered by the nonfarm activities. While resplendent with dressed-stone terraces, irrigation systems, and fertile soils, the farming systems of the Inca rulers served a series of special objectives that clashed irreconcilably with the diverse crops.

Foremost in the focused efforts of Inca farming loomed the large-scale growing of one crop, maize, which the Inca elite paraded for political purposes, exalted for religious ceremony and ritual, and acclaimed for its nutritive value (Murra 1960, 1980). In embracing maize the farming systems of the Inca rulers mostly omitted dozens of other diverse crop species. The sought-after maize required warm climates and prospered most profusely at elevations below roughly 10,150 feet (3,100 meters), where it was comfortably free of frost. Many sites sunk low in the main Andean valleys did, however, measure marginally for maize-cropping in terms of soil and water resources. Inca rulers dedicated large cadres of labor conscripts to transforming the steep and semi-arid slopes of thin soil into staircases or stepped benches of fertile and generously watered fields (Denevan 1986; Donkin 1979; Murra 1980; Rowe 1947a; Treacy 1994). In the densely settled Colquepata area of interior Paucartambo majestic maize-growing terraces of trimmed stone silhouetted the arterial Antisuyo Road from Cuzco that wound eastward through the Quencomayo Valley (map 2).

Maize farming operated by the Inca rulers fulfilled the political aim of making highly visible the transformative power of the Inca state and its religious cult. They marshalled massive throngs of conscripted labor and invested unendingly in highly visible infrastructure in order to erect this symbolic message. Their unequalled dedication of infrastructure to maize cultivation was only secondarily the consequence of an ecological imperative. Maize did, in fact, yield moderately without technologies such as elaborate terraces and with only the most rustic forms of irrigation and fertilizers (Bird 1970; Brush 1977; Grobman et al. 1961; Treacy 1994; Zimmerer 1991b). On this point, the influential analysis of Murra failed inasmuch as he thought the maize crop was "in need of irrigation, terraces, and fertilizers to survive in Andean circumstances," such that "its large-scale cultivation became possible only when the state made it its own" (Murra 1980, 12). Contrary to Murra's ecological reasoning, however, countless generations of Andean farmers predating the Inca period and many since cultivated maize without Inca-style infrastructure (Bird 1970; Brush 1977; Denevan 1986; Grobman et al. 1961; Zimmerer 1991b). Nonetheless, Inca rulers devoted unmatched attention to the maize crop.

In keeping with its special aims the state-run farming of maize favored a small subset of the overall variety in the highly diverse crop. The Inca focused on a few landrace varieties with special traits. Their interest in a handful of key agroecological types was distilled in the most definitive study of Peruvian maize by agronomist Alexander Grobman and his coauthors: "[Inca] state-organized varietal allocation to specific corn-growing areas is likely to have been practiced" (Grobman et al. 1961, 38). According to the Grobman team, one principal property in this varietal allocation by Inca rulers was ripening period. By pairing varieties of the right maturation period to the growing seasons of the empire's fields, they could plant maize widely. In the main areas of state-run maize agriculture between 8,850 feet (2,700 meters) and 10,150 feet (3,100 meters), as few as two or three landraces would have granted the sought-after spectrum of ripening periods (Bird 1970; Gade 1975; Grobman et al. 1961; Zimmerer 1991c).

Special storage and consumption traits demanded by the Inca reinforced the restriction of surplus farming to a small number of maize landraces. Storage quality mattered a great deal since maize made up one of the major contents of walled Inca granaries, each called a *qollqa.* Authorities designed a special circular qollqa for the maize crop. Stored maize served not only as a staple for foot soldiers in the state's military but also as a base of its redistributive economy. By proportioning food to their subjects in times of need, the Cuzco-based conquerors banked a great deal of political legitimacy. Concerned for this reason about perishability and possible losses in storage, the rulers dictated cropping of the flinty, hard-shell types belonging to the scientific race Morocho, which, due to a stony coat, repels storage pests. Morocho's heavy weight per unit volume meant, moreover, that these landraces could be stored and transported more efficiently than the scores of other maize types. The Morocho maize was slated as a principal foodstuff of the royal army (de Acosta [1590] 1940, 300; Cobo [1653] 1956, 1:160; Morris 1981; Yacovleff and Herrera 1934, 236).

The aims of Inca rulers with respect to consumption also narrowed selection to a preferred and quite particular handful of maize types. The Inca cult brewed maize into a frothy beer—well-known today as *chicha,* an Arawak term carried by the Spaniards, or as *aqha* in Quechua. Their brew was widely sanctified in rituals and ceremonies. The religious rulers could brew the tastiest and most potent maize beer using landraces with a high sugar content, such as those of the taxonomic races Cusco Cristalino Amarillo and Chullpi, as well as colorants derived from Kulli (de Acosta [1590] 1940, 299; Cobo [1653] 1979, 219; Grobman et al. 1961).[13] The Inca state also enshrined maize beer in its ceremonies. Authorities hoped that their concoctions would attain a cache above the everyday quality of brews that the peasant commoners fermented from less specialized ingredients. A select suite of well-suited landraces would assure the rulers a prestigious beverage. Beyond beer-making, the Inca elite also exalted other landraces for specific purposes. For example, their religious cult lorded the exceptionally large kernels of Cusco Gigante, or Giant Cuzco maize, in public ceremony (Grobman et al. 1961, 298).[14]

A paltry sum of diversity prescribed by narrow criteria also distinguished the potato crop of Inca-run farming. The tuber commanded a sizeable scope of state production since it supplied the staple foodstuff for redistribution to commoners in the case of critical food shortages. Conscripted laborers piled them high in unique rectangular storehouses with stone floors (Morris 1981). Potatoes processed into freeze-dried forms known as *chuño* and *moraya* predominated in the special style of storehouse, since they preserved for as long as one year. On this count, both archaeological investigations and the chroniclers of the early colonial period who observed and inquired into the last remnants of the Inca's extensive apparatus for food storage and redistribution have agreed (Cieza de León [1553] 1853; Garcilaso de la Vega [1609] 1987, 250; Morris 1981; Murra 1980; Guamán Poma de Ayala [1613] 1980, vol. 1:338).[15] The tons of freeze-dried potatoes stockpiled in such warehouses added crucially to the military advantage of the empire's armies (Murra 1980; Troll 1958). Fresh potatoes also filled the warehouses, although they offered less advantage.

Quantity, not diversity, was given priority in the large-scale farming of potatoes commanded by the Inca state. Massive quantities preferred for freeze-drying consisted of a relatively small number of landraces belonging to the so-called bitter species of *S.* x *juzepczukii* and *S.* x *curtilobum.* Termed bitter potatoes due to an acrid taste when fresh, the two bitter species are sexually sterile and as a consequence have given rise to no more than two dozen landrace types of which only a handful are found in a single region (Hawkes 1978; Hawkes and Hjerting 1989). In fact, the bitter species *S.* x *curtilobum* includes no more than a few landrace types. Even the quantities of nonbitter potatoes stored in a fresh state in the Inca warehouses likely amounted to a modest number of landraces. The conspicuous contrasts in preservation distinguishing the nonbitter potatoes presumably inspired the Inca authorities to select one or a few of the less perishable kinds for state-run production (Morris 1981, 341).

While meager diversity distinguished most Inca agriculture, at least a few state-run storage depots were mentioned to house a robust array of the crop products (Betanzos [1551] 1968; Cieza de León [1553] 1959; Cobo [1653] 1979). This exceptional storage differed with respect to its function and to the likely source of the stored crops. The chronicler Juan de Betanzos based in Cuzco, for instance, noted that some former Inca warehouses near the city contained maize and chuño as well as the hot capsicum, or aji chile pepper, common or kidney beans (*frijol* or *poroto*), Andean lupine or tarwi (*tarwi*), and quinoa (*quinua*). The diverse-crop caches, Betanzos surmised, were of a special character. They fed workers during the rebuilding of Cuzco and were stocked by the indigenous authorities (*señores, caciques, kurakas*) of rural regions around the city rather than through production run directly by the Inca state. Such stores did not reflect the characteristic style of agriculture conducted by the Inca state and its cult.[16]

In sum the state-run sphere of Inca agriculture mostly raised a few sorts of crops and a handful of landraces in order to meet the specific agroecological, storage-related, and consumption needs of Tawantinsuyo authorities. Its narrow biological base appeared at odds with the imperial works of terracing, irrigation, and soil management; yet the stark contrast was more apparent than real. In each aspect of agriculture the Inca authorities strove to construct arresting contrasts between their farming systems and those of the peasant commoners. Agriculture did indeed soar to an apogee under Inca rulers, but it was in large part a symbolic performance of powerful dominion over the Andean landscape and its inhabitants. Diverse crops, like the everyday and decidedly nonmonumental forms of much irrigation, soil management, and terracing, resided in those farming systems *not* supplying surplus for the state, that is, the cultivation of peasant commoners.

Rich Biodiversity in Commoner Subsistence

"The Indians of the Peruvian Viceroyalty," as the Spanish priest Bernabé Cobo wrote in the 1600s, had formerly been the peasant subjects of the Inca empire. Peasant commoners, they had mastered an "art of agriculture" that moved Cobo: "The art of agriculture consists of tilling the soil, planting, and nurturing, for all plant types while paying careful attention to climate, location, and other considerations. This, more than any of life's other necessities, has been perfected by the Indians of the Peruvian Viceroyalty" (Cobo [1653] 1956, 250–51).

Commoners of the Inca empire had seeded a far greater suite of diverse crops and landraces than their rulers. The chief chroniclers of the Inca past in the central Andes plainly beheld that bias in the biological basis of farming under Tawantinsuyo. Pedro de Cieza de León, El Inca Garcilaso de la Vega, Felipe Guamán Poma de Ayala, and Bernabé Cobo agreed on this fundamental fact, although they dissented in their general historical interpretations. The magisterial text and illustrations of life under the Inca and Spaniards that was authored by the indigenous nobleman and chronicler Guamán Poma de Ayala entitles one section "Indians," in which mention is made of over eleven crop species and more than one dozen distinct landraces of maize and potatoes (Guamán Poma de Ayala [1613] 1980, vol. 2:980–84). Although not attempting to total the full diversity tilled by commoners, Guamán Poma de Ayala depicted its concentration in their fields and cuisine.

Pedro de Cieza de León, Bernabé Cobo, and Garcilaso de la Vega likewise portrayed the cornucopia of diverse plants as the "crops of the Indians" and the "foods of the Indians" in their accounts of Inca-era agriculture. The three, who all knew Cuzco and its nearby highlands, were cognizant of how the agriculture of commoners contrasted with state-run farming. They described state farming as "of the Inca state" or "of the Inca cult," literally *del Inca* or *del culto* (Cieza de León [1553] 1853; Cobo [1653] 1979; Garcilaso de la Vega [1609] 1987). The three seasoned observers shed more light on the situation of commoners when they interchanged the expressions "crops of the Indians" and "foods of the Indians." What the peasant commoners grew of course comprised the bulk of their diets; yet the farm-level production of diverse crops by commoners under Inca rule was neither a fait accompli nor a faithfully unchanging essence of their peasant livelihoods.

Instead, the wealth of crop diversity was generated through a pair of creative conditions. The first condition dealt with consumption: "The Andean peoples must achieve the *levels of consumption defined by their cultures as adequate* through the articulation of complementary productive zones at varied altitudes and distances" (Salomon 1985, 511; my emphasis). Adequacy implied not only a certain quantity or "level" of dietary intake but also its quality predicated in norms about the expected variety of customary foodstuffs. Later, in the early period of Spanish colonialism during the 1500s and 1600s, such livelihood norms—albeit altered since the Inca epoch—were defined for outsiders by the first Quechua dictionary as *cauacachiqqueyoc* (having one). As inferred in the above statement by Frank Salomon, however, the role of livelihood expectations among commoners already existed prior to Spanish records.

Equally important as the cultural meanings were the farm resources sought by peasant commoners to cultivate the materia prima of their "adequate consumption." While subjects aspired to a varied diet derived from the diverse plants, such production hinged on the capacity of households and the señorío ethnic groups to muster the appropriate complements of farmland, labor-time, and other resource inputs, such as tools, animal-manure fertilizers, and storage devices. The livelihood norm inscribed in the ethical concept of a fit livelihood, in other words, did not guarantee diversity ipso facto. On the contrary, the inputs used in the farming of commoners lessened the raft of resources that was likely to be dispensed to the Inca state and the cult for their own creations. Diverse crops claimed in the cultural customs of peasant commoners under Inca rule were thus caught palpably between two strong forces, vital local habits and the unprecedented demands for state production.

The farm resources needed to furnish the local customs of expected cuisine were not reducible to mere quantities of land and labor. In fact, the ecology and lengthy evolution of diverse Andean crops stipulated certain qualities of the key resources. Farmland necessary to till the diverse landraces of subsistence-bearing maize, for example, did not shore up the ability of the farmers to seed diverse potatoes, or vice versa. In the same way distinct desideratum in terms of labor-time was demanded from the existing regimes of farming. Early maize ripening in five months fielded a crop calendar quite unlike slow-growing bitter potatoes with a ten-month cycle. Many ties of the diverse crops to habitats and farming systems were rooted in the rugged terrain of the eastern Andes. Whether the peasant commoners subject to Inca rule in Paucartambo and other regions could outfit themselves with the long-evolving diversity thus hung on their harboring the necessary armada of farm resources.

Land granted by the Inca state to kinship groups or communities and to individual families sufficed to produce the foodstuffs of customary subsistence (de Acosta [1590] 1940, 300; Cobo [1653] 1979, 211, 213; Polo [1561] 1940, 133).[17] While the changing size of peasant families spurred occasional adjustments in the state's land allotments, commoner subjects and local authorities, as well as Inca rulers, shared or at least became resigned to a well-defined apprehension of the resources needed for subsistence. The shared apprehension specified not only an ample quantity of land but also a suitable spectrum of field environments. In the Paucartambo Andes a document dating from the late 1500s detailed how subsistence-growing farmland under the Inca traversed the warmest canyon bottoms below 10,150 feet (3,100 meters), where even the slowest-ripening maize could bear fruit, to sites as high as 13,300 feet (4,050 meters), which were planted with only the frost-tolerant bitter potatoes (AAC 1595). At least ideally, commoners could thus access the full range of arable land called for in the diverse crops.

Labor-time counted no less than land in deciding whether the farm households and ethnic groups under Inca rule could provision the accustomed diets of diverse crops for subsistence. The commoners expended a good-size amount of labor in order to nurture, cajole, and sweat their accustomed repast from mountain sides and valley walls (Cobo [1653] 1956). At great expense, they routinely plowed, mounded, and otherwise worked the difficult soils using only wooden hand tools such as hoes and the *chakitaklla* foot-plows, or foot-levered hoes. They toiled unendingly to make their fields more arable by building rustic terraces and rudimentary irrigation canals, humble designs compared to Inca ones but nonetheless effective as well as labor-demanding. Regarding the care of cultivation, their farming was more akin to gardening or horticulture than field cropping. Chronicler Cobo had sensed that commoner farming was both an art and a necessity; yet the everyday aesthetic that was expressed in the farming and cultural tastes of commoners bore fruit only at a great expense of their labors.

The scheduling of labor-time mattered as much as its overall amount to assure the success of the diverse crops. Commoners aimed to allocate labor-time in keeping with the seasonally driven march of farm tasks. Potatoes, maize, ulluco, quinoa, and the other taskmasters had evolved distinct landraces that diverged with respect to ripening. Viability of these diverse crops was imperiled if the peasant tillers could not find the labor-time to care for staggered farm tasks and cultivation calendars. Felipe Guamán Poma de Ayala keenly noted this existential tie between labor scheduling and the diversity of landraces and crop species (Guamán Poma de Ayala [1613] 1980). His discourse on maize cropping, for instance, featured the fast-maturing *michika zara* (or *maíz temprano,* literally Early Maize), sown during July and August in order to ripen as early as January. Potatoes similarly provided early forms, widely known as *maway papa* and *papa chawcha*, as well as slow-cycle ones with long dormancy that were referred to as *siri* (Sleep).

Whether labor availed to farmers in the right quantity and quality for the diverse plants depended on demands of the corvée labor draft, or mit'a. Inca officials levied the corvée on local authorities, known as kurakas, who were responsible for regularly recruiting a quota of workers calculated in accordance with population size (Murra 1980; Rowe 1947a). Forced labor recruitment naturally drew from the ranks of commoners, and being drafted by the corvée presumably took priority over their allocation of labor-time to subsistence purposes (Cobo [1653] 1979, 211, 212; Mitchell 1980; Spalding 1984, 310). Many persons were sent to the Paucartambo foothills, or montaña, to tend the empire's coca fields and secure the supply of leaf.[18] While a small reserve of captive lowlanders toiled there yearround, the Inca rulers supplemented them with corvée workers most likely drafted from highland Paucartambo.

The Inca exaction of labor-time through the corvée, however, did not substantially curtail the careful tilling of diverse crops. The empire's claim on labor apparently spared its subjects enough so that they could work when required in their subsistence farming. Individual families shouldered the bulk of decision-making and labor-time allocation in this effort, although relations with other households and the local ethnic group also impacted the overall availability of labor for their plantings. Neighbors and kin, for instance, were regularly recruited by peasant subjects using local forms of labor exchange, typically swapping one day's labor in so-called *ayni*, which could lessen the constraint of labor shortfalls during brief spells of bottlenecking in the cultivation calendar (Cobo [1653] 1979, 213; Rowe 1947a). Most peasant commoners under Inca rule thus welded together the art and necessity of their craft, using a battery of resources and diverse crops.

Still their recourse to farmland and labor rarely availed beyond the amounts needed to nurture accustomed livelihoods. "No one receives more [land] than precisely the amount necessary to subsist" noted Cobo (Cobo [1653] 1956, vol. 2:121). The chronicler Juan Polo de Ondegardo echoed this judgment with respect to labor-time in deducing that, beyond fulfilling the Inca demands and those of customary subsistence, the inhabitants lacked time to "undertake their own things" (Polo [1561] 1940, 142). Notwithstanding their demonstrable defense of livelihood rights, the peasant commoners hardly wielded a surfeit of resources given the ironclad limits on local allotments. Indeed, the allocation of land and labor resources to agriculture and other economic activities in the Paucartambo Andes and other regions under Inca rule was not optimal in its efficiency (Knapp 1991, 187). A fairer estimate was that political resolution over the control of land and labor left the Inca subjects with limited but nonetheless sizeable and sufficient sums of resources for the cropping of diverse foods.

The imperial flourish of the Inca during the 1400s and the early 1500s reformed the great arch of agricultural evolution that for several millennia had taken shape in the Andes. Although the Inca period endured but briefly compared to the arch's earlier span, Inca transformations set a new trajectory for the fortunes of diverse crops. The formidable rulers fully restructured farming, in a sense reinventing it. Contrary to certain views currently in vogue, it was their subjects but not their state that sustained the legacy of biodiversity. The Incan transformation of farming wholly rearranged both farm space and the biogeography of crops. Diverse plants flourished in the well-tended but ordinary fields and storage depots of commoners, while Tawantinsuyo and its cult celebrated a monumental agriculture that was purposively different and politically prepotent. As the material and symbolic weight of their monuments crumbled after the Spaniards' Conquest, the legacy of a livelihood rich in diverse crops was to persist through once again being altered.

Colonialism: Coca and Crops in Paucartambo, 1533–1776

A triad of shifts in the role of diverse crops followed the onset of colonial life in the Paucartambo Andes and in its economic bonanza as the new Peruvian viceroyalty's center of coca growing. First, the onset of Spanish colonialism transformed the basic terms of culture and ethnicity that enveloped the diverse crops. The encounter between colonial rulers and Andean people, including the Quechua in Paucartambo, gave rise to the ethnic category of "Indians" and the culinary category of "Indian food." Second, the policies of the viceroyalty of Peru designed to address Indian farming were self-contradicting, some promoting subsistence based on diverse crops and others upending it. Ultimately colonial policies and economic institutions helped to re-create the fundamental contrast between the diversity-rich subsistence of commoners and other diversity-poor farming. Concurrently, they made poignant the biological gulf between the more well-to-do Indian peasants that grew diverse crops and their poorest counterparts.

Spanish conquest of the Andes began three hundred miles north of Paucartambo in 1532; Iberian soldiers and priests, with their toting retinues, climbed the Cajamarca mountains in search of gold and silver lodes and the human souls of an Inca Empire torn hopelessly by civil strife and a raging epidemic of fatal smallpox. With the fall of the paralyzed Inca rulers, the Spaniards could establish a new viceroyalty of Peru, which by 1535 encased the territories of present-day Chile, Argentina, Bolivia, Peru, Ecuador, and Colombia. While consolidating territory, Spaniards in the Andean "New World" were at the same time bent on more prosaic forays into the world of foodstuffs. In one search for descriptive analogies the soldier-scribe Pedro de Cieza de León likened the potato to the truffle, chestnut, and poppy, while in another he imagined quinoa similar to Moorish Chard: "Potato, which is like the truffle, and when cooked is as soft inside as boiled chestnuts . . . foliage like that of a poppy . . . quinoa, another very good grain . . . like Moorish chard (Cieza de León [1553] 1959, 44)."

The strange plants were not mere artifacts of vanquished lifeways, however; even while Cieza de León was discovering these new victuals—and trying to describe them—the fate of the unfamiliar food plants was being recast in the new context of Spanish colonialism. Leading chroniclers in the sprawling colony agreed that the bulk of crop diversity belonged to Indian agriculture (Cieza de León [1553] 1853; Cobo [1653] 1979; Garcilaso de la Vega [1609] 1987; see also Guamán Poma de Ayala [1613] 1980). Both Bernabé Cobo and Garcilaso de la Vega dwelled in Cuzco so that their observations could be safely applied to Paucartambo. They lived, moreover, during the seventeenth century, well after the onset of colonial rule. In the aftermath of the Spaniards' Conquest this farming of foodstuffs to be consumed by Indians, described as "of the Indians," or *de los Indios,* was clearly demarcated from that "of the Spaniards" (*de los españoles*). Diverse crops flourished mostly—although not fully—on the Indians' side of the ethnic and social divide. The biological and cultural schism that opened in the early colonial period was partly due to the fact that the "food of the natives did not find favor with the masters," but the forces at work were more than taste preferences alone (C. Sauer 1952, 152).

Influence of colonial policy and economic institutions on the diverse crops was more complex. Los Andes de Paucartambo, the colonial-era sobriquet for Paucartambo (Andes from the Inca term *Anti* for lowland tribes and the Inca's eastern territory *Antisuyu*), boasted a booming coca economy in its montaña foothills east of the Cordillera Paucartambo and the Cañac-Huay Pass (map 3). Coca-growing, like silver mining, which formed the other jewel of the viceroyalty's economy, was carried out with workers drafted via a colonial labor corvée. Francisco de Toledo—viceroy of Peru between 1569 and 1585 and the foremost author of its colonial policy—patterned the Peruvian mita closely on the Inca mit'a.[19] Unlike the Inca state, however, the Peruvian viceroyalty proportioned most laborers to the pursuits of mining and coca-growing, and few at all to agriculture in the highlands. As a consequence, the question of whether the diverse crops existed in mita-run agriculture was mooted; however, that absence only hinted at the issues acting on diversity. The sheer size of the colonial coca economy and its proximity to highland Paucartambo was of immediate consequence to the diverse crops.

The province, or *corregimiento,* of Paucartambo held the rare status of being administered directly by the Royal Council of the Indies in Spain rather

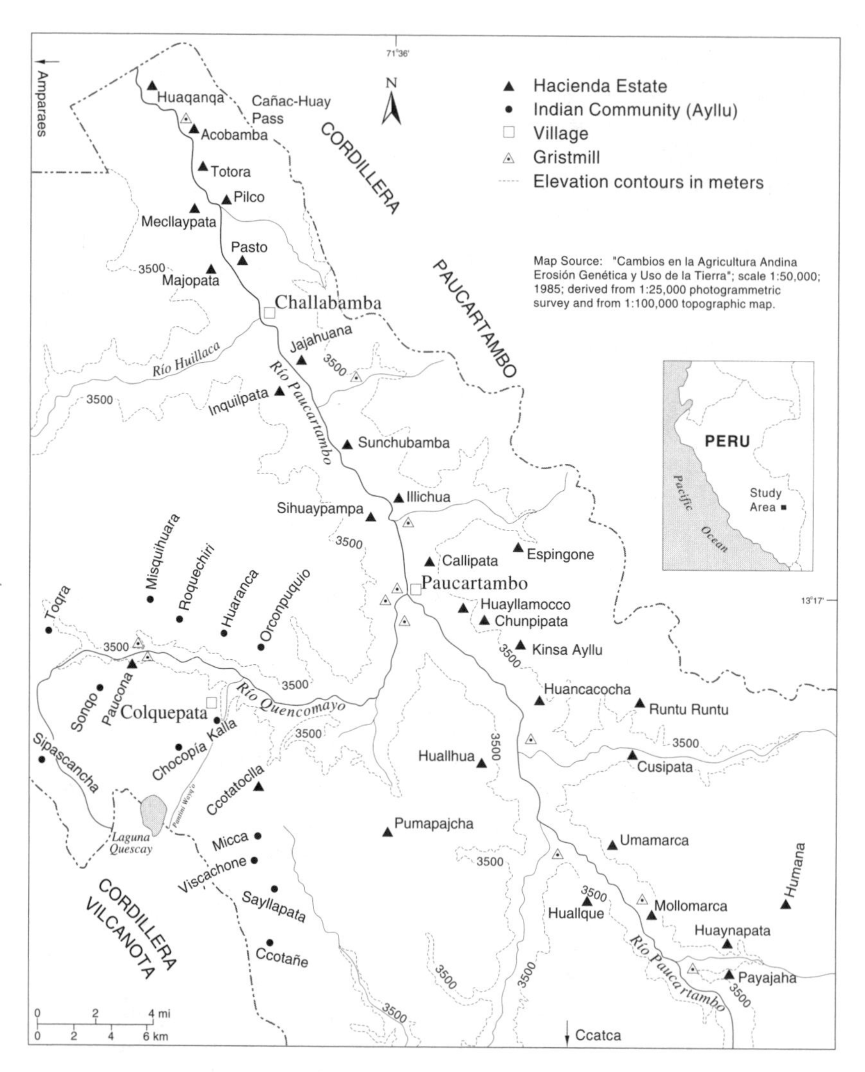

Map 3. Paucartambo in the Colonial and Republican Periods (1540–1969).

than the Peruvian viceroyalty. In 1565 crown-ruled Paucartambo was inhabited by the populations or ethnic groups known as *ycayguagua, catca, colquepata, challbamba,* and *guacanqa* (BN 1587, A370). Elevated status of the region and the fate of its people derived from the political and economic importance of coca-growing. Only the city of Cuzco, but none of the eighteen corregimientos in neighboring areas, shared this royal status with Paucartambo.[20] Los Andes de Paucartambo wielded a near-monopoly over production for the newly booming coca trade. The coca trade, which linked Paucartambo and surrounding Cuzco more to present-day Bolivia and Argentina than to Lima and coast of Peru, expanded severalfold in the decades following conquest. As late as 1700 the region still accounted for eighty-five percent of coca traded in South America (Brisseau 1978; Glave 1983; Mörner 1978, 81, 1985).

After a military expedition of 1539 to Paucartambo's montaña foothills, the Spanish commander Hernando Pizarro had claimed the famed coca-growing estate that formerly belonged to the last Inca emperor, Huayna Capac (Cieza de León [1553] 1959). Within two decades, the demand for coca leaf from the Paucartambo montaña boomed due to the demand of Indian miners at the rich mountain of Potosí in present-day Bolivia (Blakewell 1984). Tens of thousands of miners, many mita workers, chewed the leaf daily for its mild physical stimulus and alleviation of hunger and fatigue. Mining of the rich silver deposits turned Potosí into the most populous city in the Americas and one of the largest in the world by 1620. In the premier trade route of all Peru caravaning merchants bought coca in the Paucartambo montaña and the trading villages of its highlands, in order to haul it south over five hundred miles to Potosí (Cieza de León [1553] 1853, 440; Glave 1983).

At their source, goods and labor in the prosperous coca business flowed between the montaña foothills and highland Paucartambo along the vertiginous Cañac-Huay Trail that wound—snakelike or snail-like, depending on the chronicler's choice of metaphor—across the eastern Andean escarpment. The precipitous pack trail left an indelible, fear-stained impression on travelers that was penned by Garcilaso de la Vega, the Cuzco-raised chronicler who sought to interpret the Andes for a European audience: "To enter the valleys where coca is grown, a pass called Cañac-Huay is used. It has a drop five leagues long on an almost perpendicular slope, the mere sight of which inspires fear and horror, not to mention the climbing of it, for it goes up winding first to one side then to the other like a snake" (Garcilaso de la Vega [1609] 1987, 222).

Under a petition to the Crown a small group of Spaniards founded the settlement of Paucartambo where the coca trail from the montaña bridged the Paucartambo River at a distance of six miles or so beyond the awe-inspiring Cañac-Huay Pass (plate 1). They petitioned on grounds that the new settlement be "a trading center for buyers of coca that come from Collao [the Titicaca plain] and Charcas [south of Titicaca in present-day Bolivia including Potosí]" (Maurtúa [1558, 1572] 1906). Located near the former Inca storehouse, the village of Paucartambo seated the provincial government and a *repartimiento,* or tributary unit, with a nearby agricultural hinterland that in 1583 counted 1,200 inhabitants and 242 tribute payers (Escobedo Mansilla 1979, 236).[21] The village itself consisted mostly of creoles and Spaniards, containing a total of eighty persons in 1690, probably not much changed since the late 1500s (Villanueva Urteaga [1693] 1982, 254). In that year the population of greater Paucartambo—including the repartimiento of the same name plus

Plate 1. The capital of Paucartambo (1981 population: circa 1,620), wedged along the main river, is today an entrepôt for many of the region's twenty-thousand-odd Quechua peasants. In the 1550s it was made capital of the colonial coca-growing center of Los Andes de Paucartambo. The village served as a nucleus for manorial haciendas until the Land Reform of 1969. Since then its villagers have dominated commerce through their control of marketing, government institutions, and transportation, along with their alliances with better-off peasants.

two or three others—numbered 6,250, nine out of every ten persons being Indians (Mörner 1978, 19).

Indian peasants in Paucartambo found their lives threatened, and in many cases tragically terminated, by brutal injustices of the coca-growing bonanza in the nearby montaña. Labor-intensive cultivation of coca shrubs on the estates of Paucartambo's foothills consumed a steady stream of fresh workers. Benefiting from the state-run mita, the owners of coca estates drafted heavily from the highland Quechua-speakers of Paucartambo. The coca mita drew primarily from this province, since the other Cuzco provinces sent workers to the mine mitas for Potosí and the mercury operations at Huancavelica near present-day Ayacucho (Mörner 1978, 1985; Wightman 1990). Quechua from Paucartambo who worked in the coca mita died in massive numbers due to overwork and rampant disease. In fact, so many Indians perished on the coca estates that Viceroy Toledo outlawed the use of the labor draft for new plantations beginning in the 1580s, although owners successfully forced labor from the highland Quechua throughout the next century (Levillier 1925; Wightman 1990).

Tribute payments to the viceroyalty—another foundation of the colonial economic policy—were a means by which Indian labor continued to flow to the coca estates even after weakening of the coca mita. Colonial officials ordered the tribute-paying Indians of the Paucartambo repartimiento to submit their annual tax in the form of coca leaf (Escobedo Mansilla 1979, 236; Gade 1979). The Indians of Paucartambo could procure the coca for colonial taxes most readily by laboring for in-kind payment on the dreaded montaña estates or in the transport, drying, and packaging of the leaf. Tribute demands for coca thus worsened still further the catastrophic mortality of Paucartambo Indians. By 1689, Paucartambo's demographic tragedy was apparent in a regionwide male-female ratio of 53:100. The lowest of any Cuzco province, the ratio was plain proof of the deadly outcome of laboring on the coca estates (Mörner 1978, 23). The tragic widespread loss of Indian lives destroyed livelihoods and customs of resource use in its wake.

Tribute demands that were placed on the Quechua in Paucartambo pressured for the production of other crops in addition to coca. In general, colonial administrators exploited the territories of the *encomienda* (a territorial lease of tribute-payers) and the repartimiento to mandate that the Indian farmers of the region render them the main crops recently brought from Spain, especially wheat and barley (Bueno 1951; Crosby 1986, 1991).[22] Reviewing the role of the two European cereal crops during the early colonial era, Gade found them diffused mostly by way of rulers and markets. In the case of wheat it was clear that "Indians grew wheat only because it was among the products with which tributes were paid" (Gade 1975, 137), while "barley was undoubtedly an important commercial crop during colonial times for horse and mule feed" (Gade 1975, 134).[23] The bulk of tribute payments and incipient but still debile farming for markets thus did not directly finger the complexes of diverse crops.

Even in those uncommon cases when diverse crops comprised tribute and entered commerce, moreover, a modicum of variety was actually enlisted. For example, the hot capsicum, or aji chile pepper—a condiment cropped only in the lowlands and deepest Andean valleys—made up a frequent item of tribute lists. The spicy seasoning sustained a modest amount of diversity, however, relative to the major food plants (NRC 1989). Nor did much diversity enter the slowly growing stream of commerce in potatoes and maize. Although the fields of Indian peasants fed the expanding market flows of the colonial economy, they channeled a narrow variety of crop types. The potatoes and maize that traded widely in Cuzco, for example, enumerated but a few landraces that offered either exceptional culinary qualities or robust growing traits such as high yield (Glave and Remy 1980).

The colonial policy of *reducción,* or forced resettlement, posed a dire threat to the diverse crops, although one that eventually was met with some success by Indian resistance. In 1571 Viceroy Toledo declared that the dispersed populace of farmers and herders be "reduced" into nucleated settlements to aid the colonial regime with its collection of tribute, military suppression of uprisings, and religious conversion of Indians (Gade and Escobar 1982; Málaga Medina 1974; Wightman 1990). As many as 1,500,000 Indians were reported to be resettled during the next decade. In the Paucartambo Andes Toledo's edict consolidated the rural populace into the villages of Paucartambo as well as Colquepata and Challabamba (AAC 1595; Levillier 1925, 165; Villanueva Urteaga [1693] 1982). The detailed report on Colquepata village by a colonial official recounted a complex but not uncommon sequence: at first, the jolt of forced resettlement jeopardized the means of tilling the diverse crops, although, as discussed below, most Indians later regained this livelihood right through their effective opposition against the reducción.

Diego Maravier, the *Visitador y Juez Repartidor de Tierras* (Royal Inspector and Land Judge); his official scribe; and a surveyor reached Colquepata on December 15, 1594, and met with the region's chief Indian authorities, the kurakas García Pancorva and Gerónimo Compe.[24] Maravier's mission was to partition and privatize land in conjunction with the forced resettlement of the rural Indian populace into Colquepata village (AAC 1595). In the next days he reassigned fields to each of 355 Indian households among twelve extended kinship groups or local communities (*ayllus*). Most households were awarded a single parcel suitable for potato growing, a "topo to plant potato" (*topo de sembrar papa*), while roughly one third were also granted a maize parcel.[25] Maravier justified his partitioning of community lands on grounds that all Colquepata inhabitants would thereafter possess sufficient land for subsistence: "That all tribute-paying Indians possess the 'potato lands' that they requested so that they . . . have sufficient land for their fields and the subsistence of themselves, their wives, and their small children" (AAC 1595, 2).

By rupturing the existing pattern of ayllu land use, however, the partitioning designed by Maravier devalued the farm assets of Colquepata Indians. It fractured a commons style of land use that had efficiently paired individual usufruct rights with community territory by coordinating farming and herding into sectoral fallow (known in Colquepata at that time as *mañais* and elsewhere in Paucartambo and Cuzco as *suerte, laymi,* and *mañay* [Franquemont 1988; Gade and Escobar 1982; Orlove and Godoy 1986; Urton 1981; Zimmerer 1991e]). Sectoral fallow in the Colquepata commons had defined a number of so-called sectors. Individual families within an ayllu community coordinated their rotation of crops and fallow of fields through the several sectors. Because most Colquepata families held at least one field in each sector, Maravier's allocation of a single large field per household undercut the landholding basis of the sectoral fallow commons. It definitely made a bald fiction of his claim to guarantee subsistence.

As a result of the transformation of land tenure by the colonial official, a number of economic and ecological benefits gained in the sectoral fallow of Colquepata were imperiled. Maravier's radical transformation of land tenure upped the demands on Indian labor-time. Without sectoral fallow, their activities of livestock-grazing and crop-raising in the agropastoral economy were no longer so separated in spatial terms. Greater effort was required in successfully coordinating the two; any shortcutting of this exigency was likely to result in crop damage. The map of newly partitioned lands also inferred deteriorating soil fertility and worsened erosion in areas nearest villages due to the contracted periods of fallow and the loss of community control over cropping and livestock (Gade and Escobar 1982; Spalding 1984).[26] These added economic and ecological costs pressured the Indian farmers in Colquepata to allot less land and fewer labor resources for their diverse crops.

The impact of the early colonial reducción in Colquepata was, however, effectively countered by the diffuse opposition of local Indians. Over time Colquepata Indians reinhabited their dispersed dwellings and reestablished their previous means of territorial control including sectoral fallow. Of one thousand or more Indians residing in Colquepata in 1595, only three hundred remained in 1690 (Villanueva Urteaga 1982, 254), a loss of villagers that exceeded the probable toll of disease. Demographic changes in Colquepata suggested the reverse migration that followed forced resettlements in other Cuzco regions where peasants ultimately reoccupied their former habitations and tended anew to their old fields, pastures, and shrines (Gade and Escobar 1982; Glave and Remy 1983). Indians' opposition to partitioning in Colquepata held through the seventeenth century. In 1658 two ayllu communities adjoining the village, Hanansaya Ccollana Chocopía and Urinsaya Ccollana Kalla, managed to assert territorial claims with success while ignoring Maravier's survey and partitioning (LACH 1658). Not even later government administrators, it seemed, could find evidence of the ill-fated effort of Maravier and the others to reform land tenure.[27]

The triad of colonial policies consisting of the labor tax (mita), tribute, and forced resettlement (reducción) aided the Peruvian viceroyalty's collecting of revenues and its control over the masses of subjugated Indians; but the colonial institutions enforced in Paucartambo and other regions collided often with the capacity of Indian peasants to seed diverse crops. Nevertheless, this was plainly not the original or official intent of the viceroyalty's rulers. The viceroyalty of Peru announced various policies to protect the ability of the Indians to subsist, if only to bolster the supply of goods and labor buttressing the colonial economy (Fonseca 1988; Larson 1988; Spalding 1984; Stern 1982). Protecting Indian rights to food production and procurement was widely recognized as a pressing priority of viceregal policy and were intended to be put into practice: "Most important, the state validated their [Indians'] collective right to a minimum degree of subsistence security" (Larson 1988, 299).

Officials of the viceroyalty planned to protect Indian subsistence by enacting the colonial ideal of two coexisting societies in Peru: the "Republic of Spaniards" and the "Republic of Indians." The overarching intent of the twin republics was to govern a gamut of ecclesiastical, political, and fiscal affairs (Macera 1968; Mörner 1978, 1985; Roseberry 1993; Spalding 1974; Wightman 1990). At least a few policies germinated in the grand ideal did in fact protect the customs of Indian subsistence and encourage growing of the diverse crops during the colonial epoch. Tribute and tithes (*diezmos*) demanded of Indian peasants, for example, could be applied only to those crops brought by the Spaniards, such as wheat, barley, and alfalfa (Mörner 1978, 70; Spalding 1984, 182). This may well have caused Indian farmers to locate tribute-paying fields in areas separate from their subsistence plots (Collins 1988, 34; Stern 1982, 40–41). Looming large over the measures to protect subsistence and the ideal of twin republics, however, was a reality fraught with the internal contradictions of Spanish colonial rule and the chronic crises of its economy.

Far from the ideal of a secure subsistence, the diverse crops and the livelihoods of Indian peasants were devastated by the onslaught of catastrophic mortality known as "demographic collapse," desperate migration to other regions, the wholesale arrogation of Indian lands, and unsettling socioeconomic differentiation within their communities. Deadly attacks of disease epidemics inoculated unwittingly by the Spaniards—including smallpox, measles, typhus, and influenza outbreaks—decimated the Indians, who lacked immunological defenses against the foreign plagues. Epidemic outbreaks in the Cuzco region included measles (1531), typhus (1546), influenza (1558–59), smallpox (1585–91), another outbreak of smallpox (1687–92), and smallpox and influenza (1719–21) (Dobyns 1963; Wightman 1990, 46). By 1620, the Indian population of the highlands in southern Peru crashed to roughly one half of its total in 1570 (Cook 1981).

Indians in the Paucartambo Andes suffered rates of mortality even greater than in surrounding regions, due partly to the region's tragic loss of inhabitants in the coca mita (Cook 1981, 246).[28] The indigenous population of the region plummeted to a nadir in the mid-1600s, its lowest count of Indians languishing with little change during the long decades between 1628 (4,631 Indians) and 1690 (4,668 Indians). Their numbers then began to rise slowly until 1725 (4,819 Indians) and more rapidly afterward (7,141 Indians in 1754) (Wightman 1990, 65). Catastrophic mortality due to disease outbreaks nonetheless did recur episodically in Cuzco throughout the eighteenth century; a few of the widely reported epidemics erupted in 1701, 1720, 1726, and 1730 (Mörner 1978, 7). The epidemics thus continued to ravage the lives and livelihoods of Indian peasants throughout the colonial period.

Floods of "free" Indian immigrants poured into Paucartambo in the wake of its tragic depopulation. Numbers of the permanent immigrants, known as *forasteros,* or foreigners, swelled steadily during the late 1500s and the 1600s before cresting in the 1700s (Flores Galindo 1977; Mörner 1978; Wightman 1990).[29] By 1786, the immigrant population of 1,264 landless families (*foras-*

teros sin tierra) residing in the Paucartambo Andes outnumbered all other Cuzco provinces. Making up the overwhelming share of 90.5 percent of the region's populace, the density of landless migrants settling in the region far outweighed the rest of the southern Peruvian sierra where forasteros typically accounted for less than half of Indian populations (Flores Galindo 1977, 23; Mörner 1978, 51, 118).[30] Although some migrants to the labor-hungry montaña were lured via an unusual urban-to-rural flow from Cuzco, many relocated instead to the intervening highlands of Paucartambo (Wightman 1990). Their arrival in the farm country of Paucartambo was not, however, as colonizers or homesteaders on unclaimed land.

Some forasteros found themselves as second-class members of Indian communities, although they actually worked for the most part on nearby haciendas. Already in the early 1600s owners of private estates in the region had usurped sprawling territories from Indian individuals and communities. One of the first private estates claimed the territory of the "indigenous settlement" of Mollomarca (Gutiérrez 1984). In the northern Paucartambo Valley the estate of Challabamba displaced an ayllu community whose authority sold its lands to a Spaniard in 1639 (ADC 1639). Although the Spaniards who baptized Paucartambo Village in 1558 asserted the absence of Indians (Maurtúa [1558; 1572] 1906), this was a pious and self-serving fiction. By 1690, more than four dozen estates lined the Paucartambo Valley between Challabamba and Mollomarca. They consolidated an uninterrupted wall of haciendas that for centuries dominated the Paucartambo landscape (map 3).[31] The 106 haciendas set on the Paucartambo countryside in 1786 presented a bastion of estates unmatched in Cuzco. Mostly medium in size, the dominions of seignorial tenure at that time counted an average of seventy-nine Indian residents, many of them forasteros (Flores Galindo 1977, 30; Mörner 1978, 32; Villanueva Urteaga 1982).

As Spaniards and creoles usurped the Indian land in Paucartambo, some communities and individual leaders opposed them through legal recourse. This legal resistance of colonial Indians made reference to the ideal of a fit livelihood with its inference of diverse crops. In one claim the leaders Nicolas Callisana and Carlos Yapo—principal authorities of the ayllu community of Kalla in Colquepata—protested that the community lands grabbed by an adjacent estate were essential for the Indians in order to "maintain subsistence and tribute" (ADC 1794). Land for the twin activities, Callisana and Yapo asserted, was needed to ensure the accustomed livelihoods of themselves and their families. Hacienda owners, for their part, typically countered that such lands had stood unused and, therefore, presumably unowned. Heated legal disputes and violent skirmishes that broke out intermittently among Indians, villagers, and hacienda owners in the 1700s and 1800s were concentrated especially in one dozen or so ayllu communities near Colquepata (ADC 1639; ADC 1785;

ADC 1794). Community Indians there lodged a flurry of legal protests in order to thwart the land-grabbing ambitions of nearby haciendas such as Ccotatoclla (map 3).

Socioeconomic differentiation, or the growing gap in resources among Indian peasants, was a consequence of contradictions in the policies of colonial rule and the chronic crises of its economy. During the period from 1533 to 1776, a widening gulf separated the better-off from the most destitute "have-nots," perhaps especially in Cuzco and its surrounding regions (Wightman 1990). The nature of this gulf and the processes pulling it apart further—colonial policy, the indirect rule by local community authorities, land theft, migration, uncommuted tribute payments despite "demographic collapse"—remain a subject of much studied debate (Fonseca 1988; Larson 1988; Murra 1964; Spalding 1974, 1984; Stern 1982; Wightman 1990).[32] There is little disagreement, however, that sizeable differences in resource stocks polarized the Indian peasantry of the colonial period. As a means of gauging how critically differentiation in their ranks impinged on the diverse crops, we can turn to language common among the Quechua peasantry of that time. One source was a dictionary of Quechua words that were carefully catalogued and defined by Spanish priest and Quechua lexicographer Diego González Holguín, who lived in Cuzco and other Andean cities between 1581 and 1618.

Expressions analyzed in González Holguín's epic work reveal a key denominator of livelihood differences (González Holguín [1608] 1952). González Holguín defined a cauacachiqqueyoc (having one), as "one who possesses the means to produce and obtain a customary livelihood" (González Holguín [1608] 1952, 51).[33] By definition, "haves" among the Quechua held the rights to a fit livelihood. These persons he contrasted to the *huaccha,* or the "one who is poor or an orphan" (González Holguín [1608] 1952, 167). Orphan, in González Holguín's definition, figuratively meant "without resources," or by inference, without rights to a fit livelihood. Everyday language in the viceroyalty thus logged a polar split between the haves and have-nots within the Indian peasantry. Members of the have-not category could not meet their subsistence needs due to the lack of resources available for this effort. They could not, by definition, cultivate the full suite of diverse crops. In contrast the haves enjoyed the rights to pursue a fit livelihood.

Diverse crops thus assumed a highly ironic role within the Andean colonial world. On the one hand, most haves among Indian peasants continued to farm the living heirlooms of commoner cropping. Many have-nots in the Indian peasantry, on the other hand, lacked the requisite stocks of farmland, labor-time, or perhaps other resources that were required by the diverse crops. The diverse crops did not disappear widely, however. When the perspicacious Spanish botanists Hipólito Ruiz and Joseph Pavón ventured to the Andean fastness of the Peruvian viceroyalty in the 1700s, they found that many diverse crops, including the Andean tuber-bearer and nasturtium-relative massua (mashua, or

Tropaeolum tuberosum)—which they identified for modern scientific taxonomy—were being grown as "quotidian foods of the natives" (Ruiz and Pavón [1794] 1965, 77). Quotidian they may have been in terms of widespread cultivation, but the diverse crops were by no means a universal good among the viceroyalty's Indian peasants.

Oddly, this irony in the social role of diverse crops seemed to be lost on the dominant society, including for that matter the visiting Spanish botanists Ruiz and Pavón. The Spanish speakers commonly cast items such as potatoes, ulluco, and quinoa as the defining ingredients for food of the Indian (comida del indio). The diverse crops were, in other words, still unequivocal symbols of being "Indian," although the cultivation of these foodstuffs was not shared evenly in the ethnic masses. Notwithstanding its inaccuracy as a statement of resource availability, the expression *comida del indio* and its curiously dogged usage during the colonial period did readily confer a means of verbal denigration. Persons consuming these foods were clothed as Indian "Others," by association thought uncivilized, inferior, and degraded. The priest Bernabé Cobo, who lived and worked in Cuzco during the early 1600s, wrote the following of the term *Indian*: "The word Indians is used when we Spaniards speak to each other; but since its meaning is now *derogatory,* we do not use it when we speak with Indians" (Cobo [1653] 1979, 9; my emphasis).[34]

Unheard in the chronicles of colonial cropping was the verbal treatment of diverse plants in the segregated and less-visible spaces of Indian life. Continued cultivation itself suggests that the comida del indio and its base of diverse crops retained value and esteem in the undocumented venues of Indian peasants. In rural hamlets, familiar fields, and pastured recesses the subjects of Spanish colonialism still harbored their crop plants as sturdy vessels of the "having." In addition to consumption and cultural value the diverse crops lessened the specter of food shortages and thus improved Indian welfare. This was not an unfounded worry since shortages of food due to crop failure and poverty beleaguered colonial peasants more than their Inca predecessors. One reason was that the relatively weak and, in any case, less agricultural state of the Peruvian viceroyalty did not administer an insurance-like system of food redistribution akin to the Inca. Indian peasants of the colonial world, confronting an acute insecurity of food supply, found new reason for farming the diverse crops that were hardy and rarely failed completely.

Haciendas and Communities, 1776–1969

Two centuries of late colonial and republican rule ushered mostly quiet changes into the panorama of diverse crops and peasant livelihood in the Paucartambo Andes. Nonetheless, the cumulative effects of these small alterations prior to the Land Reform of 1969 resulted in diversity's first major curtailment. The undramatic decisions of the Quechua farmers to reduce plantings of the early *Solanum phureja* potatoes and the quinoa crop took place amid a setting of many manorial estates and a few Indian communities. Spread across the region's tortuous landrace, the haciendas and communities weathered a severe economic depression between 1776 and roughly 1900 and a slow recovery afterward. The farming of diverse crops continued in the context of market pursuits being paired with resilient and innovative customs of subsistence. The tensions between nonsubsistence and self-provisioning were not always resolved, however, in favor of diverse crops, and Quechua peasant life was made to change accordingly.

The regional economy of the Paucartambo Andes declined severely during the late 1700s. A Crown fiat in 1776 split the former viceroyalty of Peru. Severed from the Potosí mines and trade with the Atlantic port of Buenos Aires, Paucartambo and other regions of Cuzco suffered stiff tariffs and difficult communication. Coca production for Potosí miners shifted to the Yungas in the eastern Andean foothills of La Paz. Isolation replaced trade circuits that had once led urban Cuzco with 31,000 inhabitants in the mid-1700s to be the leading sierran city of the Peruvian viceroyalty behind Potosí (Brisseau 1978; Flores Galindo 1977; Mörner 1978). In Paucartambo Indian rebels waged bitter and bloody warfare with colonial royalists in the neo-Inca revolt launched by Tupac Amaru II in 1780–81. Finally, a crumbling economy caused the once rich coca estates of the region's montaña to declare bankruptcy between 1776 and 1810 (AAC 1746; ADC 1784–85a; ADC 1785–87; ADC 1807–8; Lyon 1984).

National independence of Peru in 1826 did not avert the economic collapse of either Cuzco or Paucartambo. By 1876, the population of Cuzco had fallen to 17,000, its lowest in more than a century. When the prospecting Italian geologist Antonio Raimondi visited the Paucartambo Andes in midcentury, he seemed startled that "Paucartambo, a growing population and center of economic activity during an earlier epoch, is now in a notable state of decline" (Raimondi [1894–1913] 1965, vol. 1:218).[35] While an international wool trade bankrolled by British finances enmeshed the southern Peruvian highlands, the economy of Paucartambo stayed mostly beyond its reach. Rugged terrain and the lack of a railhead precluded the expansion of wool production and the location of textile mills in the region (Flores Galindo 1977; Mörner 1978; Orlove 1977a).[36] Weakening Paucartambo ceded political territory later in the century when the Districts of Ccatca and Amparaes passed to neighboring provinces. Meanwhile the seignorial haciendas of Paucartambo stagnated; between 1690 and the early 1900s the workforce on a sample of seven haciendas barely grew (Mörner 1978, 55). Despite the stagnation, hacienda domination of land ownership in the region remained strong. In the core repartimientos, or districts, of Paucartambo, Challabamba, and Colquepata they outnumbered ayllus by thirty-one to seventeen in 1845 (ADC 1845).

The economies of the region's manorial estates began to revive by 1900 due to the purveying of potatoes, wheat, and cattle to the expanding populace of nearby Cuzco. Several textile mills and the growing presence of departmental government in greater Cuzco fueled its urban growth. In 1890 track of the Southern Peruvian Railroad connected to the city from the commercial centers of Puno and Arequipa. The population of greater Cuzco grew from 18,167 in 1906 to 20,000 in 1912, and then climbed to 40,657 in 1940 (Brisseau 1978, 21; Mörner 1978, 1990). Throughout this period, the difficulties and cost of transportation frustrated the ability of Paucartambo producers to meet the new demand of Cuzco markets. Nonetheless, the estate owners and Quechua peasants managed to augment at least slightly the flow of trade, transporting farm goods by horse and mule teams and drive livestock on-the-hoof from their hinterland. Amid economic expansion, the haciendas began to divide and grow while the ayllu communities struggled to keep their land.

Planning of a vehicular road to Paucartambo came on the heels of the Amazon Rubber Boom and a smaller export stream of cascarilla, or quinine bark. Speculation in the lowland economy led Swedish engineer Sven Ericsson to survey a Cuzco-Paucartambo highway route in 1911, and to lobby his plan with national authorities (Villasante Ortiz 1975). Beginning in 1921, a mandatory corvée of peasant labor for this project was ordered by President Augusto B. Leguía (1919–30) under the Ley de Conscripción Vial. It drafted more than 1,600 workers. The conscripts hammered with picks and shovels, chiseling a passable roadway that eked nearly fifty miles from the rail depot of Huambotío in the Urubamba Valley, curved alongside Paucartambo's westerly outpost of Caycay, and snaked uphill to the Kellkaykunka Pass. The road traversed the elevated basin between Sayllapata and Ninamarca before coasting downhill to reach Paucartambo in 1922 (see map 5). During subsequent decades the Paucartambo Road inched northeastward over the eastern Andean Ridge and in 1950, nearly half a century too late for the long-ended Amazon Rubber Boom, bridged the lowland settlement of Patria on the Amazon-feeding Pilcopata River.

A social landscape of farm estates greeted the first cars and trucks that pulled the steep grade from Huambotío and crossed the Kellkaykunka Pass into Paucartambo in the 1920s. Roughly one hundred and fifty manorial estates monopolized the best agricultural lands and the large majority of territory in the region, with the stark exception of roughly one dozen ayllu communities, most nestled around Colquepata. The grip of haciendas on Paucartambo during the twentieth century seemed to surpass that institution's hold on other provinces of Cuzco as well as most other sierran regions of Peru (Caballero 1981; Collins 1988; Gade 1975; Guillén Marroquín 1989; Guillet 1992; Watters

1994). By the mid-twentieth century, the majority of other highland regions could show a mixture of land tenure including peasant smallholding and sizeable numbers of ayllus and official Indigenous Communities. As a consequence of Paucartambo's distinctness, its estates framed the nearly exclusive context for the cultivation of diverse crops.

The Paucartambo haciendas offered a life of misery for Indian workers (Mörner 1978, 53). Hacienda abuses included violence and coercion aimed at recruiting workers to estate-owned zones of colonization in the montaña, a death sentence for many (Watters 1994, 250). Wealthy criollo families meanwhile put together portfolios of several haciendas. For example, the Yabar clan amassed more than ten *fundos,* or *quintas,* as the owners called them. The family's patriarchs posed for a fabled photograph with President Manuel A. Odría (1948–56) and one in-law, Juan Manuel Figueroa Aznar, became a well-known artist in Cuzco (Poole 1992). The Nuñez del Prado family held several other Paucartambo haciendas. Interestingly, members of both families were active in the "pro-Indian" *indigenismo* movement in Cuzco, hinting at the deep cross-purposes within its cultural and political motives.

Dual production, or the subsistence-commerce division between fields, held sway on the Paucartambo haciendas. Hacienda owners typically reserved prime field sites for the demesne while dispensing usufruct rights over other farmland, as well as rangeland, water, and firewood to their peasants. The spatial distinction between demesne cropping and the peasants' own farming did not much separate the economics of production. Indeed, cultivation of the two parties was bound by frequent flows of labor, land rights, and goods. The quasi-contractual obligations on Paucartambo haciendas were recorded by Gustavo Palacio Pimental, the son of a Paucartambo estate owner and a lawyer concerned about reforming the estates. He rode horseback through the region during 1955 and 1956 in order to study the economic arrangements between seignors and their peasant workers (Palacio Pimental 1957a, 1957b). Based on his study of twenty-two Paucartambo haciendas, Palacio Pimental summarized the general obligations of both parties: "The estate owner concedes the use in usufruct of a determined extension or area of arable land, the use of pasture, water, and firewood to a certain person, almost always an Indian, who, for an indeterminate time and as a non-transferrable lifetime right, provides the estate with a source of labor for a certain number of days each year or week" (Palacio Pimental 1957a, 188).

By the 1940s, the Paucartambo haciendas were mostly using this "Indian . . . source of labor" for commerce in the potato crop. The region gained a reputation for its potato agriculture throughout Cuzco and even in the distant market cities of Puno and Arequipa (Villasante Ortiz 1975). Some estate peasants were able to convert portions of their usufruct lands into commercial plots. The market-bound plantings, however, both of seignor and his serflike peasants, measured only a measly share of the diverse crops. The commercial growers preferred a single landrace named *qompis.* They could market its high yields and healthy seed tubers in major markets across the southern Peruvian sierra. Sauer noted this variety's widespread distribution in his brief visit to the southern Peruvian sierra, writing that "[qompis is] widely grown in Bolivia and southern Peru because of its heavy yield" (Sauer 1950, 514). It was Paucartambo more than any other region, however, that became synonymous with qompis production, since this single type thoroughly dominated the region's farm commerce (Vargas 1948, 1954).

Little variety likewise was seeded in the barley crop bound for market and in the smaller wheat planting. Wheat was once widely planted as the standard successor of potatoes in the prevailing sequence of crop rotation (AAC 1777; ADC 1845). At least one dozen water-powered gristmills, whose remains dot the present-day landscape, ground the wheat produced by estate owners, villagers, and peasants (map 3).[37] Paucartambo hacienda owners liked the image of their gristmills, seeing them as a sign of a bucolic and not unproductive rural life, and they chose often to have paintings and photographs set with a wheat-grinding mill in the background. By the mid-1950s, however, barley alone mattered much in Paucartambo agriculture. One reason for the abrupt changeover to the hardy barley crop was that wheat was being devastated by the black stem rust of the genus *Puccinia,* a European pathogen inadvertently transported to South America.[38] Wheat commerce from Paucartambo also sunk because price-deflating imports of inexpensive wheat were being imported from abroad and especially from the United States (Guillén Marroquín 1989).

By contrast, barley commerce in Paucartambo happened to prosper under soaring demand from the manufacturing plant of a national beer company located in Cuzco. The Cuzco factory of the Beer Company of Southern Peru paid top price for a two-row malting barley known as German, or *alemana.* Paucartambo farmers discovered that the attributes of this German barley, with respect to field rotation and location, so closely mimicked wheat that the latter crop was readily replaced. The plentifulness of horses among estate owners and even the well-to-do peasant workers, as well as the presence of countless rock-base threshing floors known as *eras,* facilitated the substitution of barley for wheat. When the observant Palacio Pimental surveyed Paucartambo estates in 1955 and 1956, the German barley occupied a place in farm commerce second only to potatoes (Palacio Pimental 1957a, 1957b).

Haciendas of the twentieth century sent a narrower spectrum of potato, barley, and wheat kinds to market than had their colonial-era predecessors. Earlier the Paucartambo estates had shipped at least somewhat diverse foodstuffs of potatoes, maize, ulluco, and quinoa, mostly to their coca-growing counterparts in the nearby foothills (ADC 1784–85a; ADC 1784–85b; ADC 1785–87; ADC 1800; ADC 1807–8; see also footnote 31).[39] At that earlier time, many hacienda owners possessed a large highland food-producing estate (*hacienda de pan llevar*), along with a coca hacienda and a smaller highland property near Challabamba that could be used for storing the coca, known as a *despacho*. In the mid-1900s, however, the fare of foodstuffs being produced on the Paucartambo haciendas was simplified. Qompis predominated among potato plantings. In the case of maize the Cuzco markets at midcentury mainly stocked Yellow Maize, or *maíz amarillo* (Glave and Remy 1980). Both the demesne cropping of Paucartambo estates and the market-growing of their peasants responded faithfully to the new market signals (Palacio Pimental 1957a, 1957b). Their market rationality contrasted more isolated settings such as the Kauri estate in Quispicanchis (twenty miles south of Colquepata), which apparently was guided by purely sumbolic incentives such as status and prestige (Mishkin 1947).

It was the subsistence farming of estate peasants that renewed most crop diversity between 1776 and 1900. In their self-provisioning the Quechua farmers expressed an ideal of customary diet and cuisine as part of a fit livelihood, using the term *kausay,* or *kawsay,* a modern-day equivalent of the concept defined centuries earlier by González Holguín. The twentieth-century meaning of this term was spelled out by Toribio Mejía Xesspe (Mejía Xesspe [1931] 1978). His account of Quechua peasant subsistence in the central Peruvian highlands described the customary cuisine of a kawsay style of fit livelihood. The kawsay cuisine was based on a *customary diversity* of crops and especially landraces. The landraces provided agroecological, culinary, and cultural resources to peasant farmers. In the case of maize more than fifteen culinary creations were derived from the ten or so most common landraces. Mejía Xesspe's account also accented the resilience of pre-European livelihood expectations among estate peasants during the twentieth century. In emphasizing their resilient customs he underscored the conservative meaning of kawsay.

In many cases, however, Quechua farmers had recast the contents of the kawsay concept since the Inca and early colonial periods. The Quechua peasants in Paucartambo, not unlike their neighbors in the Urubamba Valley, had adopted wheat and barley as key ingredients of subsistence at least as early as 1800 (Gade 1975, 137; 1993). New traditions of kawsay continued to be innovated in the Republican period after 1826. The Quechua estate peasants and their community counterparts in Paucartambo recast kawsay to incorporate the Old World broad bean or fava (*Vicia faba*) and peas (*Pisum sativa*) as important minor crops. They valued the agronomic traits of the broad bean and pea, which restored soil nutrients, and they liked the gustatory appeal of the new crops as fresh vegetables and dried seeds that could be boiled, parched, or even ground into flour (Palacio Pimental 1957a, 1957b; see also Gade 1975; Mejía Xesspe [1931] 1978). The willingness of Quechua farmers to adopt the valuable new crops signaled their readiness to test worthy additions. They definitely preferred, however, that the adoptions serve to complement existing crops rather than replace them.

Opposing pressure, however, especially undercut the quinoa crop. By the

1950s, enterprising peasants in Paucartambo mostly planted qompis potatoes and barley on the middle slopes below 12,500 feet (3,800 meters), which otherwise were suitable for landraces of valley quinoa (Palacio Pimental 1957a). Quinoa growing was also curtailed among the Quechua farmers on independent communities in Colquepata where Diego de Maravier had surveyed agriculture in 1595 (AAC 1595). At that time centuries earlier quinoa stood second behind only potatoes in overall area. Colquepata Indians of the early colonial era were following a widespread custom in sowing quinoa following potatoes. But a contraction of quinoa area began later in the colonial period when encomienda grantees and officials demanded wheat for the purpose of tribute. Wheat, like qompis potatoes and German barley later, excelled in the fields at middle elevations. First by force and then by choice, many Quechua farmers in Colquepata and other Paucartambo areas curtailed their quinoa fields at middle elevations, a trade-off also observed historically in the nearby Urubamba Valley and the Colca Valley in Arequipa (Gade 1975, 154; Guillet 1992, 98).

Worsening land shortages in the Colquepata communities also caused the Quechua farmers to question their commitment to quinoa cropping at the middle elevations. Their population density and person-to-land ratios swelled due to the influx of persons who were avoiding, and in some cases escaping, the onerous life on nearby haciendas (ADC 1845; ADC 1850; ADC 1890). The number of community Indians dwelling in the Colquepata interior climbed steadily after the Peruvian government abolished tribute obligations in 1896. The distinctive trend toward denser population in the communities was still unchecked after the Peruvian government titled the official Indigenous Communities in 1927. By the mid-1900s, the fourteen Indigenous Communities of Colquepata faced land shortages without par in Paucartambo. Farmers thus felt added incentive to convert their multipurpose fields at middle elevations to income-yielding qompis potatoes and German barley.

Labor acted as much as land to restrict the farming of diverse crops, maybe most notably on the haciendas. Throughout the colonial and republican eras, both peasants and estate owners in Paucartambo and other Andean regions complained of debilitating shortages of labor-time besetting their agricultural efforts (Bowman 1916; Denevan 1986; Fuenzalida 1970, 73; Martínez-Alier 1973). The seasonal scarcity of labor-time dealt a fickle fate to those estate peasants who tried to crop for commerce and kawsay-style foods, while meeting their tightly enforced labor obligations to the seignor (Martínez-Alier 1973). Most Paucartambo estates demanded five days of labor from their peasants each week, and some expected even more (Palacio Pimental 1957a, 1957b). One casualty of labor-time scarcity among estate peasants was the unique early potato of the species *Solanum phureja*. Known in the local vernacular as *chawcha,* literally precocious, fast-maturing landraces could turn out multiple crops each year only with the unending supply of labor.[40]

As late as 1940 small plots of the early chawcha potatoes were farmed by

Quechua peasants on various estates in Paucartambo. But the stiff labor and precious irrigation demands of this diverse crop no longer jibed so well with the livelihood pursuits of estate peasants. Short maturation and a lack of dormancy meant that the *S. phureja* landraces were planted in two or frequently three crops during each year, a cropping calendar that typically stretched to ten months. In the main growing area of the Paucartambo Valley, upstream of Challabamba, estate peasants could produce chawcha only by using scarce irrigation during the dry months of May, June, August, and September. Use of the scarce irrigation inferred competition with qompis farming, however, since farmers shored up its commercial production with timely watering. Another wave of *S. phureja* loss, perhaps for reasons similar to those in Paucartambo, took place in the region of Ayacucho during this century (Mitchell 1991, 73).

Social cleavages within the Paucartambo estates helped to determine who among the Quechua peasants benefited from the tillage of diverse crops (Palacio Pimental 1957a, 1957b). The lawyerly Palacio Pimental noted that a group of the more well-off hacienda peasants filled the ranks of tenants, or sector people (*mañayruna*) (appendix C.1). A second and middle category in terms of status and resources consisted of tenant sharecroppers, known as *yanapaq,* or helpers. Desperately poor servants, or *pongos,* rounded out the list of three main groups. Many full tenants of the estates tended to dispose of land and labor resources adequate for sowing the diverse crops.[41] But the other groups of estate workers suffered from resource shortages that frequently hindered the breadth of their planting. Tenant sharecroppers, for instance, often could muster little extra labor due to the combined demands of estate owners and their sharecropping with more well-to-do peasants. By the 1950s, it was obvious that the agroecological, culinary, and cultural advantages of diverse crops accrued unevenly to the Quechua peasants (plate 2).

Normative expectations grounding the kawsay ethic stayed strong during the republican period notwithstanding the noticeable curtailment of some diverse crops. The broad base of this concept could be heard in the legal disputes that the Quechua in Paucartambo waged over the ownership of territory, the autonomy of their labor, and their capacity to sustain a fit livelihood. Disputes about resource control especially flared in the Colquepata area, where the estate owners regularly preyed on peasant lands, even those that had been officially entitled as Indigenous Communities by the Peruvian government in 1927 (ADC 1793; ADC 1845; Villanueva Urteaga 1982; Zimmerer 1991d). The Indigenous Communities of Sayllapata, Viscachone, Micca, Chocopía, Colquepata, Sonqo, and Sipascancha became a geographical center of peasant protest (map 4). Their legal disputes illustrated various aspects of the kawsay ethic while deploying new arguments built on a curious concatenation of the axiom of property rights and the advocacy of the "pro-Indian" indigenismo movement (Deustua and Rénique 1984).[42]

While indigenista ideas and the Peruvian laws on private property were an-

Plate 2. Santusa, a Paucartambo farmer, in her Hill field of diverse floury potatoes in 1986. Under the dominion of a hacienda estate before 1969 she barely managed to cultivate this field. Her field, at about 12,800 feet (3,900 meters), is seeded with twenty-five landraces, including her favorite type, Llama Nose, or llama senqa.

choring new claims in their struggle, the Quechua of Colquepata still pleaded their cases on the basis of livelihood rights. A legal complaint filed by Isidoro Soncco of the Indigenous Community of Huaranca, for instance, reported that theft of his family's hogs jeopardized their "basis of subsistence" (ARA 1923). During the decades after 1920, Miguel Quispe, a renowned leader of the Quechua in Colquepata, blended traditional-style claims together with the new flavor of arguments. In accusing a Colquepata villager of stealing land and illegally forcing labor from Chocopía community, Quispe and a group of Chocopía peasants could invoke the complaint that they would no longer survive if such losses persisted. At the same time, they requested "a royal inspection or measurement or partitioning of lands such as were made during the viceroyalty," in order to guarantee their property rights (ARA 1935). Inferred in their arguments was that the land resources rightfully belonging to them would enable a fit livelihood.[43]

Diverse crops also played a role in demarcating the territory of livelihood-yielding farmland. An example of the territory-marking function surfaced when an overseer, or *majordomo,* of a bordering estate attempted to seize control of the remote Calzadapampa fields of Chocopía in the 1940s. That land theft was thwarted by the Quechua residents of Chocopía's Indigenous Community, who

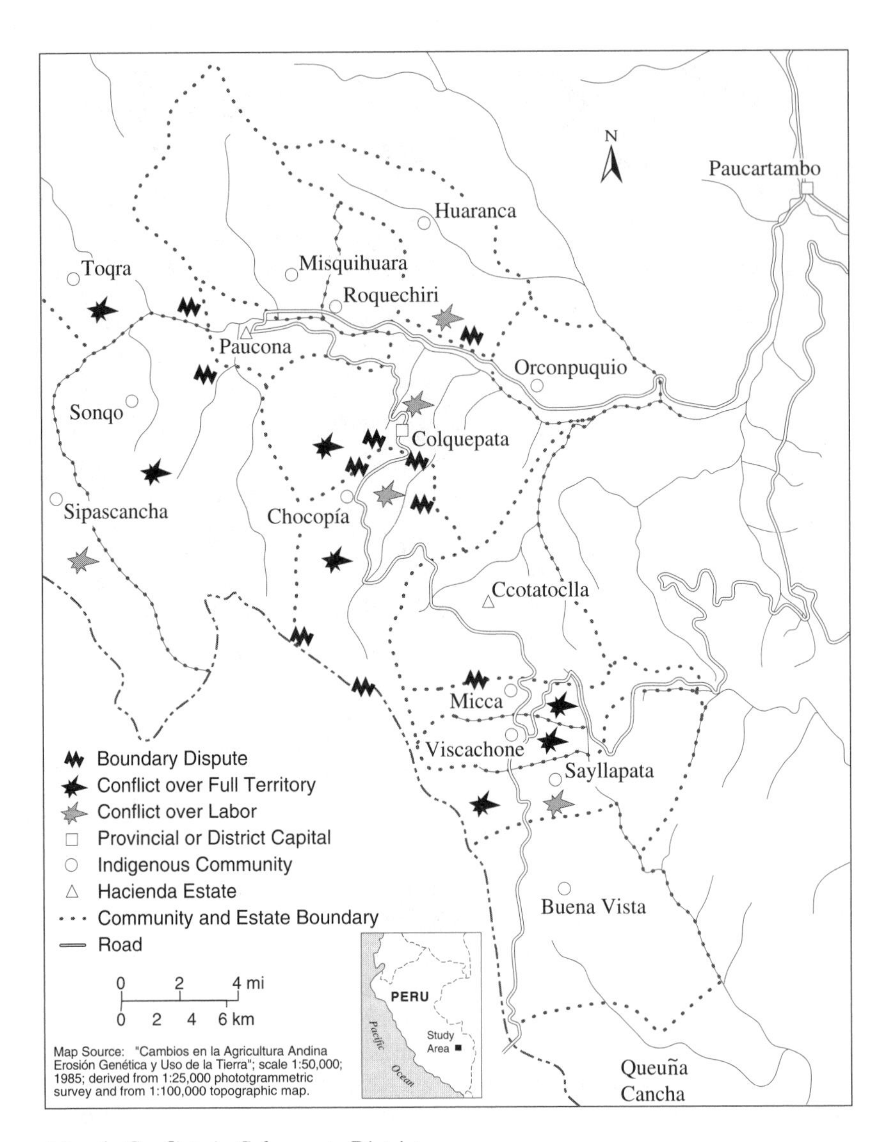

Map 4. Conflicts in Colquepata District.

protested that they customarily cropped floury potato landraces there in a sector of the commons typical in the upper reaches of peasant farming (LACH 1950). Their success closely resembled a similar protection of farm territory nearly two centuries earlier in 1658, when a survey of Chocopía territory defined its boundaries in terms of the potato fields sown in the sectors of their commons or as they were termed in the document *Mañais de Laymiña* (LACH

1658). Cropping of diverse landraces could thus support claims to the territorial requirements of a fit livelihood as well as vice versa (see also Larson 1988, 155–56).

Biodiversity and Long-Term History

De facto conservation of the diverse crops in Paucartambo and other central Andean regions was crafted during a great arch of agrarian history beginning more than seven thousand years ago. The successful conservation of biodiversity did not, however, imply an unchanging stasis of farming systems. It depended throughout its lengthy course on the capacity of farmers to adjust and redefine the role of their diverse crops. To be sure, Quechua farming and its traditions of a fit livelihood (cauacachiqqueyoc, kawsay) continued to yield a world-class cornucopia of the diverse crops ever since the fifteenth century. Even this feat, however, depended on farmers reshaping the use of diverse crops in ensuing epochs of the Inca empire (circa 1400–1533), Spanish colonial rule (1533–1826), and Republican government (1826–1969). As farmers adjusted and redefined the roles of diverse crops—both biological and social (broadly interpreted)—they also established patterns of the upmost importance to conservation.

Biological change has been key both to diversity and to the de facto conservation of the diverse crops over the long haul of evolutionary history. Exemplified by potatoes, maize, ulluco, and quinoa, the Andean crops evolved species and landraces suited to a great range of growing sites in the tropical mountains. Quechua and Aymara farmers and their mountain-dwelling predecessors skillfully chose types that would establish farming over habitats from the semiarid and subtropical to the humid and cold-temperate, and even to the near-tundra. A majority of the diverse domesticates evolved a number of key traits via the upslope extension of range, the divergence of maturation, and cycles of geographical isolation in the eastern Andes. Environments did not act alone, however, in the saga of biological evolution: rather than a one-stage play, the crops performed also in the formative settings of particular social and economic institutions.

First noticeable under Inca rule, the farming of Andean people for purposes other than household subsistence was perched on a narrow biological base. Farming for nonsubsistence in Paucartambo was designed to meet the specific demands of rulers, urban consumers, and still later new agroindustries who sought a modicum of diversity. Since at least the invasion of Inca Roca's troops in the 1400s, the region functioned as an agricultural hinterland that sent products and labor to outside rulers and extraregional markets. Yet neither the Inca economy nor the Spanish colonial institutions and the coca economy nor the market-directed cropping of haciendas called for the cultivation of many diverse crops. There was no essential or metaphysical reason why farming for non-subsistence purposes in Paucartambo remained poor in diversity for roughly six hundred years. The creation of taxes, tribute, and market trade through cultural, historical, and political economic mechanisms simply specified little biodiversity.

Subsistence growing furnished the main home for the diverse crops. Inspired by the cultural ethic of a fit livelihood (cauacachiqqueyoc, kawsay), subsistence-grown diversity and the concept of livelihood rights were the innovative, flexible products of peasant commoners. Indeed the Quechua peasants in Paucartambo altered the ethical concept of a fit livelihood in a flexible fashion to add new food plants—including European introductions such as wheat, barley, broad beans, and peas—while keeping the diverse crops at the core of their cultural tastes. Their food and farming habits did not persist in a perilous fragility. Changing subsistence did, however, at times result in the depletion of diverse crops. Orphan, or have-not, Quechua lacked the resources for cultivating or acquiring the accustomed spectrum of diverse crops, even though the colonial viceroyalty of Peru and the hacienda owners who controlled most of Paucartambo idealized a secure subsistence for their peasant subjects.

Another primary meaning of diversity-rich subsistence also changed during the long-term history of biodiversity. The diverse crops, labeled in the non-Quechua circles of colonial and later republican society as Indian food, were being named in an ironic fashion. In the Paucartambo Andes mainly the wealthy peasants still tilled their diverse crops of quinoa and *chawcha*, or *S. phureja* potatoes. Quechua estate workers and free peasants in Paucartambo still valued the two notable crops, but by the mid-twentieth century many, especially the worse-off, lacked the farmland and labor-time needed to grow them. Ironically, therefore, diverse crops were absent among some of the poorest in the so-called mass society of the region's twenty-thousand-odd peasantry. Conventionally thought of as cultural "markers," the diverse crops were neither produced nor consumed by the entire mass of Quechua peasants. The epithet *Indian food* was losing its claim to objectivity while still being used to denigrate the ethnic group in general.

Conservation was not solely an artifact of the alterations in plants, environments, and social milieu. Diverse crops and subsistence-growing played a key role in the changing course of development. In fact those Quechua in Paucartambo able to protect their subsistence customs were doing so not for the defense of an old custom but rather for the sake of using their resources to meet existing needs and opportunities. The crops most obviously provided the Quechua farmers with valued nourishment. A less obvious aspect of development derived from the ecological ability of the diverse crops to deter field losses. Quechua people in the Paucartambo Andes especially sought a hedge against crop failure because of decaying mutual aid pacts previously guaranteed by ethnic groups, or señoríos, and the Inca state. At no time during the period from 1400 to 1969 could they trust markets, moreover, to deliver the customary foodstuffs. Finally, the farmers could count on the diverse crops to aid in protecting their farmland from potential usurpers.

In sum the long arch of Andean biodiversity prior to 1969 witnessed evolution amid a multitude of adjustments in peasant farming. Since 1400, the Quechua peasants in Paucartambo redefined diversity-rich subsistence, we might even say *reinvented* it, in response to changing circumstances. After the radical juncture of the Land Reform of 1969, the de facto conservation of diverse crops became challenged by even more accelerated changes in their livelihoods.

3

Transitions in Farm Nature and Society, 1969–1990

Resource Paradoxes of the Land Reform of 1969

Late one June afternoon in 1969, the joking Eufemia had cajoled a then-youthful Faustino—who years later guided me to the Big Hill or *Hatun Loma* field—while she heaped platefuls of steaming floury potatoes with meaty hominy and ladled frothy maize beer into a glass that was being passed around. A tipsy Faustino was serving the homemade brew and the maize and potato plates to family and neighbors, who had finished stacking the adobe walls and thatching the roof of the newlyweds' home. The couple's jest hinted at their happiness with the new single-room hut. Its site even offered some space for future expansion of their quarters. They owed the site to Faustino's father, once a tenant worker of the Umamarca hacienda. His family was an updated version of Quechua haves, and both Faustino and Eufemia planned to pursue the current concept of a fit livelihood.

By 1990 the couple's home had grown to include a pair of adjoining huts. When Santusa, Faustino's now-aged mother, looked over the enlarged house compound that year, she pointed out the huts built since 1969, each a new room housing backpack cannisters for spraying insecticides, large piles of fresh seed potatoes, and bags of granular fertilizer.[1] The equipment and supplies stored in the corner huts in 1990 were not unknown twenty-one years earlier, but they have tended to be few. Santusa and her neighbors in Umamarca, literally the Head Place, had hoped their livelihoods would better after 1969, when the coup d'etat of a new Peruvian government—"the government" as the Quechua style it—declared a radical land reform. The hacienda of Umamarca was converted into the official Peasant Community of Umamarca; but their outlook of guarded optimism faded to troubling uncertainty, Umamarca's free peasants finding few hopes behind the hyperbole of the Land Reform of 1969.

Mixed fortunes also faced the diverse crops during the decades after 1969. Soon after the thatched roof covered Eufemia and Faustino's first hut, a small mound of multihued and many-shaped potatoes was heaped on its dirt floor. The modest-size mound of seed to be planted the following year was replete with tubers of the diverse landraces. Santusa had sacked the seed tubers from her larder so that, as she put it, a "decent subsistence" could be enjoyed by Faustino and his new wife. Recollecting her gift, Santusa referred to them with the vernacular floury potatoes, or *hak'u papa,* which aptly described the mealy or flaky texture of their pulp once cooked. The young couple was pleased to receive the seed tubers from Santusa, although the gift incurred yet another social debt. Indeed, the etiquette-conscious mother-in-law had lorded her gift by baptizing them with the precious epithet ancient potatoes, or *ñawpaq papa.*

A rainbow of types had colored the ancient or floury potatoes gifted by Santusa to the newlyweds. Two decades later, Santusa recalled a few landrace varieties present in the pile: Red Mother, or *puka mama,* a rose-skin tuber; Aborted Guinea Pig, or *qowisuyu,* with its mottled blood-red and purple skin; the lumpy Pig Manure, or *khuchi aka*; and the symmetrical Sunflower, or *sunch'u.* She also recounted having given them the variety called That Which Makes the Daughter-in-Law Weep, the unforgettable *qachum waqachi,* whose tortuously convoluted skin tried the peeling talents of the most adept wife. "Don't forget," she told me, "this is the one called *choqllos* [Corn-on-the-Cob] in other communities." The pile also held a few *saqma,* or Fist, the knuckled tuber that some farmers called Jaguar's Paw, or *puma maki.*

During the decades after 1969, floury potatoes like the ones gifted by Santusa were sprouting less widely in the farm landscape of the Paucartambo Andes. To be sure, some potatoes, like Red Mother, Aborted Guinea Pig, and That Which Makes the Daughter-in-Law Weep could still be said to flourish in 1990; however, the stocks of other varieties dwindled drastically or even vanished, which occurred in the case of the unique Fist type. One keen observer who noted the decline of saqma and other diverse potatoes in Paucartambo was the Cuzco-born and internationally renowned taxonomist Carlos Ochoa of the International Potato Center. Venturing to Paucartambo since the 1940s in order to collect potatoes, Ochoa first observed the local disappearance of precious landraces beginning in the mid-1970s:

> The variety Sajma [or saqma] from south Peru was first collected by us in 1948 at Humana [near Umamarca] in the important potato-growing zone of Paucartambo at 3600 m in Cuzco Department. Its chief distinguishing feature was the curious shape, similar to a clenched fist, very like a boxing glove. We have looked for it repeatedly during the last ten years on numerous occasions but without success. It appears to have become completely extinct. (Ochoa 1975, 170)

By 1990, more than one-third of Quechua farmers in Paucartambo did not sow the richly diverse floury potatoes. They also curtailed other diverse crops, withdrawing some to a greater extent than the floury potatoes. Less decisive outcomes were dealt to still others like ulluco and quinoa. During the two decades since young Eufemia and Faustino had thanked Santusa for her gift, farm families were refashioning both their livelihoods and their expectations about a kawsay-style subsistence. A number chose to make do without the diverse landrace-rich crops; however, the Quechua farmers who deserted them during the period from 1969 to 1990 were not culturally transformed. Although they were no doubt more worldly than their parents, their new experience and learning did not spell dramatic Western acculturation or massive out-migration. Instead, the process of curtailing landraces and undercutting the ethical norm of a fit livelihood was waged in the countryside itself, albeit inconspicuously, with thousands of Quechua peasants deciding to alter farming field by field.

Countless field-scale changes in the farming systems of the Paucartambo Andes were swept along in the wake of the land reform propelled across Peru by the national government in 1969. One of the most radical land-to-the-tiller fiats ever designed in Latin America, Peru's reform of 1969 abolished the rules of seignorial land tenure that for centuries had set the chief historical context for the diverse crops' fortunes (de Janvry 1981; de Janvry et al. 1989). The land reform did not, however, crest in a great transformation of society. Paucartambo peasants remained among the poorest in Peru, earning no more than their version of a few hundred dollars each year. Most of the countryside still lacked basic medical services, clean water supplies, and any electricity. Despite its personality as a backward hinterland, the region was being remade under upended circumstances. The Land Reform of 1969 was simply supplementing a series of powerful changes rather than dictating them.

The story of Eufemia and Faustino foretold a basic paradox of the Land Reform of 1969: How livelihoods of the Quechua in Paucartambo bettered somewhat, while their vaunted dressing of diverse food plants and other livelihood customs suffered. A perplexing set of contradictory processes was set in motion by the Land Reform. On the one hand, the Quechua farmers benefited from the nationwide overturn of seignorial land tenure in October of that year by the powerful, nine-month-old Revolutionary Government of the Armed Forces, a military junta headed by General Juan Velasco Alvarado (1969–75).[2] Within months, all Paucartambo haciendas were expropriated under the Supreme Edict 17716 on grounds that they operated "inefficiently," involved "anti-social forms of land tenure," and exploited "illegal labor conditions" (La Prensa 1970, 61). Almost overnight, the Umamarca hamlet and the other bastions of seignorial dominion ceased to exist as manorial estates. The Peruvian government ceded the Paucartambo countryside through usufruct to Santusa, Eufemia, Faustino, and more than twenty thousand other Quechua farmers.

Peru's national government and the regional Agrarian Reform Offices headquartered in cities like Cuzco created three categories of reform units to replace the former haciendas and Indigenous Communities. The three categories were Peasant Communities, Peasant Groups, and Peasant Cooperatives (Fonseca and Mayer 1988; Mayer 1988). Reform officials defined Peasant Communities to be "slightly capitalized peasant agricultural enterprises." In hinterland regions like Paucartambo the newly designated Peasant Communities far outnumbered the other reform units. The uncommon Peasant Cooperatives were described as "small or medium-size, moderately capitalized peasant agricultural enterprises." Peasant Groups formed a residual category of units awaiting official title from the government. By the mid-1980s, the numbers of each reform category in the core area of Paucartambo were as follows: Peasant Communities (forty-seven), Peasant Groups (thirty-one), and Peasant Cooperatives (three). Paucartambo's social landscape had become a complex of peasant institutions (map 5).

Resource availability in the period between 1969 and 1990 did not vary much among the different government-designated reform units that were designated in Paucartambo (Fonseca and Mayer 1988; Mayer 1988; see also Flores Galindo 1988). In all three categories, the Quechua farmers found that their access to resources rested with their families and with the direct control of their community institutions (such as the sectoral fallow commons). Farm resources adequate for the diverse crops appeared to be ensured by the main reform decrees, the Agrarian Reform Law of 1969 and the follow-up Special Statute of Peasant Communities issued in 1970 (La Dirección y Promoción de la Reforma Agraria n.d.; La Prensa 1970). The law of 1969 seemed to seal the fate of labor resources, since former estate peasants were immediately freed of their labor obligations in order to pursue "completely autonomous economic ends." Diverse plants, it could have been wagered, were assured farmers' labor-time due to the proclamations, although that appraisal would later seem naive in light of ensuing transitions.

Farmland for the diverse plants also seemed assured by the laws of 1969 and 1970 (La Dirección y Promoción de la Reforma Agraria n.d.; La Prensa 1970). Former estate peasants gained the expropriated land minus up to twelve acres, or five hectares, that could be retained by the ex-owner. Since the official Peasant Communities of Paucartambo were deeded the boundaries of previous haciendas and Indigenous Communities, most new territories encircled the various farm habitats needed by the diverse crops (map 5). In fact, the majority of the region's newly recognized communities, like their prereform predecessors, held lands from the river channels near 8,850 feet (2,700 meters) to rolling grasslands above 13,450 feet (4,100 meters). They characteristically combined a range of generalized land use activities based on agriculture and livestock-raising in various elevation-related, or "compressed" environments (Brush 1976; Brush and Guillet 1985; Guillet 1981b; Rhoades and Thompson 1975). Most Paucartambo communities thus exemplified the contiguous

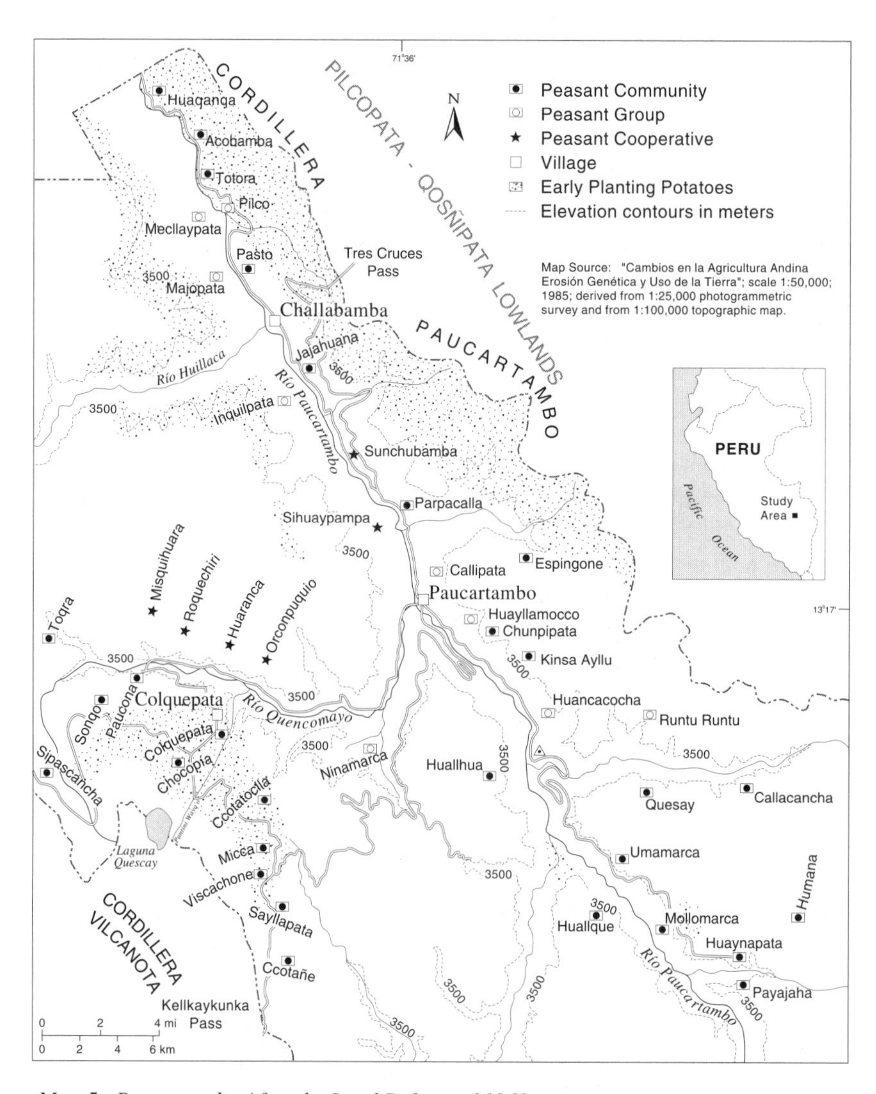

Map 5. Paucartambo After the Land Reform of 1969.

elevation-related compressed model of verticality, still widely utilized in the eastern Andes of Peru and Bolivia.

While the legal measures of Peru's Land Reform of 1969 appeared to grant ample grounds for tilling the still-esteemed diverse plants, the inferred promise was not fulfilled for many Quechua farming in Paucartambo. In fact, a broad series of social and economic transitions powerfully undermined the apparent assurance of autonomy in the government's legal provisions. An irreconcilable contradiction was riven between the Velasco junta's much-publicized goal of protecting Peru's peasant economy through its Land Reform of 1969 and its less-publicized but more forceful macroscale policies of economic development, which were biased strongly toward urban growth, industrial expansion, and mechanized, large-scale agriculture in the coastal valleys and lowlands of the country. The development model of General Velasco's "revolution" enforced a strong bias against the economies of Andean regions, contrary to its rhetoric (Alvarez 1983; de Janvry 1981; Figueroa 1981, 1984; Guillet 1992; Mitchell 1991; Wilson and Wise 1986).[3]

While the policies of the *velasquista* Land Reform and development model of 1969 filtered into Paucartambo, many of the region's twenty-thousand-odd Quechua peasants were already embroiled in contentious issues of control over their farmlands and resources. Protests by estate workers had begun erupting by the early 1960s. The peasants openly challenged strictures on a number of haciendas, including Sipascancha, Majopata, and Mollomarca. Organizers from the Departmental Federation of Cuzco Peasants (*Federación Departmental de Campesinos de Cusco*) helped plan many of their protests. One FDCC rally in the central plaza of Paucartambo village was described in the diary of peasant leader Domingo Cruz:

> At this time, around 9 A.M., it was full of groups of *campesinos* who came from high altitudes playing cornets and drums of pig skins. [Peasants had come] from the southwest, others from the north, also the high lands of Tahuanpata, and also from the east, the *campesinos* of Llaycho. It looked like an army of *campesinos* arriving, commanded by Melquiades Huaman, director of the FDCC and a native of Urubamba. . . . Approximately 2,000 *campesinos* [arrived]. . . . The list showed 110 comunidades participated and later at 12 o'clock everyone assembled in the Plaza de Armas and each group expressed their protests. (Watters 1994, 255–56)

Their protests renewed attention to land rights and to the role of farmlands as a sine qua non of fit and customary livelihood. In 1964 peasant squatters occupied sparsely used lands belonging to the large 7,207-acre (2,918 hectare) estate of Majopata across from Challabamba in the northern Paucartambo Valley (ADC 1962a, 1962b). They rallied behind their right to land for "subsistence and other purposes," thus dramatizing the claim to farmland in both their livelihoods and politics. Upstream on the hacienda of Mollomarca, the tenants and tenant sharecroppers of the estate rejected an offer by its modernizing owner to substitute wage payments in exchange for his full control of farmland (Cotler 1975). In the actions of estate workers it was obvious that the influence of organizers and ideology from Cuzco's peasant movement did not detract from the local issues of land resources and fit livelihoods but rather reinforced them. Livelihood rights were still a vital umbrella concept in the stormy world of the Quechua peasants in Paucartambo.

Contention over the rights of estate peasants to their own labor-time was also waged in various protests. The tenants on several Paucartambo haciendas including the large Majopata estate staged a full-scale strike at the height of the planting season from October through December in 1963, a season when hacienda owners tended to be notoriously violent and coercive with their workers (Martínez Alier 1973). Edgar Vizcarra Rojas and Abraham Vizcarra Rojas, owners of Majopata, along with the other hacienda owners, conceded to improvements of pay and working conditions in a negotiated "Act of Reconciliation" in order to bring the peasants' strike to an end. Perhaps most significant for the diverse crops was a legal cap of fifteen days per month placed on the work obligations owed by estate peasants. It seemed to promise them the labor-time needed to ensure their livelihoods with a kawsay-style cuisine, protecting their labor-time reserves even in seasons of peak demand (Cotler 1975, 148–49, Whyte and Alberti 1976).

The owners, however, callously disregarded their agreement to curb the labor-time forced from estate workers. The unresolved issue of the claim of estates on peasant labor persisted at the center of bitter conflicts until the appropriation of Paucartambo haciendas. Although the protests of Quechua peasants would eventually cease after the Land Reform of 1969, a series of related albeit less vocal struggles for resources replaced them. The estate peasants and owners—about to become community members and ex-owners respectively—vied to find favor in local implementation of the reform. Both groups discreetly pressured some officials, quietly evaded others, unobtrusively disobeyed laws, created unannounced alliances; in short, they deployed on-the-ground tactics of discreet "everyday resistance." Furthermore, both groups met goodly success in evading the legal edicts of the Land Reform of 1969. Their accomplishments resulted in mixed consequences for the ecology of farming. Three instances of their de facto practice taking precedence over de jure policy foretold the fate of diverse crops: the continued de facto custom of semiprivate land ownership, new immigration to Peasant Communities, and the decapitalization of estates prior to expropriation.

Long-term usufruct rights to farmland had traditionally fastened the livelihoods of the Quechua peasants on estates as well as the Indigenous Communities to their diverse crops. Individual families had controlled the right to farm parcels, although their claim to sole use typically complemented forms of common property access, such as sectoral fallow and gleaning, at times when cropping did not occur. Yet Article 119 of the Agrarian Reform Law of 1969 declared the Peruvian government's intent to remake the holdings of individual families into community cooperatives. Although the scenario of land tenure transformation spelled out in Article 119 was vague, it nonetheless held the potential to undermine the cultivation of diverse crops. Soon after declaration of the 1969 law it became clear, however, that cooperative landholding was doomed due to the resourceful, unabated resistance of peasants. One tiny leaf in this voluminous resistance against forced cooperatives was turned in Umamarca's Peasant Community by the youthful Faustino, then eighteen and newly married, who had purposefully misled a visiting reform official: "He [the inspecting official from the Agrarian Reform Office] was crazy to ask openly about my fields. I would never tell him what I own. That which grows here is what we eat. Other production from these fields is what we sell. Every community member has that right."[4]

The edict for cooperative landholding was rescinded, not surprisingly, in the Special Statute of Peasant Communities of 1970, issued six months after the Agrarian Reform Law. The statute of 1970 legalized the lease and inheritability of land by families within communities. In its words "The Peasant Community is an aggregation of families that possess and identify with a certain territory" (La Dirección y Promoción de la Reforma Agraria n.d.). Through their own dogged resistance and the Special Statute of 1970, the Quechua farmers in Paucartambo solidified their grip on field parcels as quasi-private property. Contrary to the reform's prohibitions, they even rented them to noncommunity members. While de facto field ownership was planted neither in outright favor of the diverse crops nor in direct opposition to them, it was a foundation of land tenure that could not help but shape the scenario of ultimately influential social and economic transitions.

De facto norms of property rights also prevailed over de jure statutes in the case of measures to change the territorial jurisdiction of communities. For example, the ill-fated cooperative named Peasants' Union, or Unión Campesinos, was ordained by the reform officials as a collectivized conglomerate of small prereform Indigenous Communities in Misquihuara, Roquechiri, Huaranca, Orconpuquio, and Roquepata (bordering Orconpuquio) (map 5). Each community had commanded a slice of territory in the middle section of the Quencomayo watershed that officials sought to sandwich together and thus gain an advantageous economy of scale.[5] In the wake of the government's titling, however, residents of the five communities stonewalled this unpopular unification into the Peasants' Union Cooperative so effectively that reform officials eventually withdrew their mandate. By the mid-1980s the residents had successfully petitioned for official dissolution of the Peasants' Union into separate communities, each derived from its prereform predecessor (Mayer 1988).

Immigration of numerous persons into peasant communities after the Land Reform of 1969 counted as yet another instance of de facto practices redoing the Quechua farmers' access to resources. Most of that immigration was illegal according to reform law.[6] A great flow of immigration occurred during the prolonged lapse while estates were expropriated and communities entitled (1969–72). Well-disposed residents of communities coaxed the immigration of family members and friends, usually winning over or bypassing their reluctant neighbors. Although much of the massive immigration of Quechua peasants violated rules of the reform of 1969, the weak and disinterested local government did not deter it. In fact, reform officials in Paucartambo unwittingly acknowledged the extra-community origins of peasants in census lists recorded for each Peasant Community. These immigrants arrived from various departments of southern Peru (Puno, Apurímac, Ayacucho) and most Cuzco provinces as well (appendix C.2). Networks of newcomers arriving from the same and neighboring locales peopled many Paucartambo communities in a peasant version of geographical chain migration.[7]

The weight of this underground immigration pushed a nearly twenty percent climb of population in Paucartambo between 1961 and 1981 (appendix C.3). Force of the reform-era surge of migrants pressed unevenly on different places within the region. In some peasant communities the rate of population growth far outstripped the regionwide average. Population of Pasto Grande near Challabamba in the northern Paucartambo Valley, for instance, grew from zero families in 1955 to twelve by the late 1960s and no less than sixty-three in 1975 (ARA 1972; Palacio Pimental 1957b). Other communities near Pasto Grande also counted major increments of population growth. Surreptitious migration in the wake of the 1969 reform eventually weighed against the landraces by adding to factors that shrunk the availability of land to individual families. Access of families to the prime parcels for farm commerce in particular was whittled away by the flow of new immigrants.

Further sapping of the resources for landrace-growing took place when a number of Paucartambo hacienda owners illegally transferred or destroyed the productive assets of their estates in the years prior to expropriation. Forewarned by nearly one decade of rural unrest, few owners failed to anticipate the Land Reform of 1969. In advance many liquidated their valuable farm implements and livestock; the owners of Majopata hacienda, for instance, sold livestock and destroyed irrigation works and fences prior to expropriation. Such removal of estate capital and infrastructure deflated the working assets later granted to the Peasant Communities. This easy evasion of reform law by the hacienda owners again proved the plain weakness and obvious disinterest of numerous government officials in Paucartambo and Cuzco. With this tacit approval "from above," the ex-estate owners exiting from Paucartambo's countryside stamped one last mark on the landscape to be inherited by the Quechua residents.

Soils and vegetation were also depleted on the eve of estate expropriation. In some cases chronic overcropping drained the fertility of fields or flooded them with buildups of insect pests and fungal pathogens. Some owners shortened fallow periods and planted fewer fertility-enhancing crops such as leguminous broad beans and tarwi. In one instance tenants of Pasto Grande near Challabamba filed a legal protest with evidence of how the hacienda owners, Plácido Corrales Barrionuevo and Martha Yabar de Corrales, had mined soil resources and degraded field environments during the years preceding expropriation (ARA 1972). Incensed, the tenants wrote "it's clear to see that there's no rotation of crops so that later we will not be able to sustain ourselves." The same legal brief attested that the owners had also destroyed vegetation at the brink of the reform's takeover, indiscriminately cutting full-grown trees of fruit-yielding Bird Cherry, known as *capulí,* or *Prunus serotina.* Indiscriminate tree felling deprived the peasants of convenient food and fuel items and sent them on the uncertain road of postreform livelihood change minus an important resource (ARA 1972).

The Land Reform of 1969 clearly helped to catapult the farm resources and diverse crops of a desired livelihood in Paucartambo into a new postreform version of the region's peasant economy. Dismantling of seigniorial estates promised that demands on peasant resources would lessen. It turned out to be true that some reform-era pledges were adhered to and that the lot of Quechua peasants was improved as a result. The Land Reform of 1969 did not, however, improve their livelihoods either significantly or pervasively. Nor did it alone determine the future of the diverse food plants but rather it fixed major new settings in the economy of Paucartambo. Like unsettling aftershocks, the jolts of economic diversification and social differentiation, dietary change, and the recasting of ethnicity and personal power repositioned the recent role of the diverse crops in peasant life.

Diversification and the Postreform Political Economy

The U-shaped compound of farmery occupied by Eufemia and Faustino in 1990 was in contrast to the image of their dwellings twenty years before. In the scene of 1990 the center hut served as a kitchen-bedroom-living quarter for the couple and their three children. A few yards away one new hut sheltered several large mounds of seed tubers composed of the improved high-yielding varieties of *mariva* and *yungay,* what they often called money potatoes (*qolqe papa*). Another new hut quartered an assortment of farm tools and field inputs: five well-used picks, a worn oxen-drawn plow, fertilizer sacks, half-used pesticide tins and fertilizer bags, three lanterns for the nighttime barley harvest, and new hose and pipe fittings for an irrigation sprinkler.

Their new farm equipment was set within the framework of a restyled peasant economy. The most palpable change after the Land Reform of 1969 reverberated from the surge and diversification of farm commerce. The Quechua in Paucartambo angled widening flows of their farm goods and labor toward either Cuzco or eastward to the upper Amazon reaches of Pilcopata and Qosñipata. Their growing streams of commerce were directed influentially, albeit at a distance, by the national economic policies emanating from Lima, first of

Velasco's military government and then its successors—the military regime of General Francisco Morales Burmúdez (1975–80), the conservative civilian government of President Fernando Terry Belaúnde (1980–85), and the populist platform of President Alan García Pérez (1985–90). Despite their differences, the development models of all the governments promoted urban and lowland growth at the expense of highland agriculture (Alvarez 1983; Caballero 1981; Hopkins 1981; Thorp and Bertram 1978; Wilson and Wise 1981). Peasant producers throughout the Peruvian sierra forced themselves to diversify commerce in a quickening search for income in the post-1969 political economy (Collins 1988; Deere 1990; Figueroa 1981, 1984; Guillet 1992; Mitchell 1991; Watters 1994). At home, the Quechua in Paucartambo sought to profit from the new flows of commerce.

Dramatic transitions in the farming of Quechua peasants in Paucartambo were witnessed firsthand by anthropologist Catherine Allen, who conducted field studies in the Peasant Community of Sonqo in the heart of Colquepata beginning in the mid-1970s:

> Since 1975 the amount of land in Sonqo devoted to cash crops has increased dramatically. Most of Sonqo's barley is now destined for sale to the Cuzco brewery, and, while the oats grown in Sonqo are fine for animals, they are not fit for human consumption. . . . For the first time, the people of Sonqo are putting major amounts of land and energy into crops they cannot eat. . . . Rapidly, Sonqo is being transformed (Allen 1988, 30).

Impetus for the Sonqo farmers and their counterparts in other Paucartambo communities to diversify commerce after 1969 was driven by an admixture of new consumption needs, government policies, chronic inflation, the changing demand for farm goods, and environmental deterioration. Demand for consumer wares greased their market integration. Enamelware bowls and tin spoons, plates and pots, plastic buckets, soap and matches, and hoe blades were taken up as standard possessions during the years from 1969 to 1990 (Figueroa 1981, 1984). Families also purchased school supplies, treadle sewing machines, kerosene for lanterns, and, in a few cases, kitchen stoves. Bicycles, sold in village stores in large numbers by the 1980s, allowed much faster travel than walking, at least downhill. With leftover savings, the peasants bought transistor radios that broadcast daily news, popular music, and personal messages. Their acquisition of the durable consumer goods grew severalfold during the 1970s and the 1980s, and diversifying commerce was concomitant with making the new purchases.

A treadmill propelled the commercial endeavors of the Quechua farmers. Buying and selling linked them to the national economy of Peru and to the needs of its urban centers. Lubricated by the consumer demands of Quechua farmers, the treadmill was turned by so-called terms of trade, which compared the changing value of farm goods relative to nonfarm goods. Throughout most of the postreform period, the terms of trade motored in a direction counter to the interests of producers in Paucartambo and other Andean regions, since the prices received for their farm goods rose less rapidly than the costs of other commodities (Alvarez 1983; Collins 1988; Figueroa 1981, 1984; González de Olarte 1987; Guillet 1992; Hopkins 1981; Mitchell 1991; Thorp and Bertram 1978).[8] Since the Quechua farmers in Paucartambo purchased a variety of consumer goods including farm inputs like manufactured pesticides, fertilizers, and insecticides, the terms of trade treadmill was not easily sidestepped. Rampant inflation in much of the period from 1969 to 1990 further accelerated the treadmill effect of the terms of trade.

Accelerating commerce in the peasant economy could not, however, merely slide along the familiar tracks of previous market endeavors. During the decades prior to 1969, wheat and potato farming by the Quechua in Paucartambo and other Andean regions had successfully gained a sizeable share of sierran markets (Baca Tupayachi 1985, 25; Caballero 1981; Lehman 1982a, 1982b). Wheat and potato markets plummeted after 1969, however, due to government "cheap food" policies designed to supply the staple foodstuffs at rock bottom costs to the masses of low-paid urban workers in Peru's teeming cities. Wheat markets collapsed under government-subsidized imports, most from the United States, that rose more than fifty percent during the 1960s and the 1970s. Potato prices during the same period became subject to below-market caps that were imposed by government policy (Alvarez 1983; Guillén Marroquín 1989; Hopkins 1981).[9] While the Land Reform of 1969 was ending economic competition from estate owners, the Quechua farmers faced a steep drop-off in returns from their mainstay products.

By contrast, market demand for two other major crops—beer-making barley and rice—kept pace with Peru's rapid industrialization and urbanization during the postreform decades. The beverage preferences and food tastes of the urban populace elevated barley and rice prices. Unstinting government subsidies for agroindustries and lowland agriculture lifted them still further. Malting barley made up the principal ingredient of beer, which claimed almost four percent of the country's industrial output in the late 1970s (Hopkins 1978, 5). In the Paucartambo Andes, Cuzco, and other regions south of Lima the Beer Company of Southern Peru, known as Companía Cervecería del Sur del Perú, nearly monopolized the brewing industry.[10] With a large factory located on a main street in Cuzco, the company processed a vast tonnage of malting barley raised in nearby regions. Prescient of what occurred in Paucartambo after 1969, the neighboring Urubamba Valley had witnessed more agricultural change during the mid-1960s due to demand from the Cuzco brewery than from any other driving force (Gade 1975, 67).

The Paucartambo Andes bypassed the Urubamba Valley and other Cuzco regions as the cultivation center of malting barley during the decades after 1969. Barley cropping blanketed a major share of Paucartambo's rugged terrain, increasing severalfold by 1990. It built on an already substantial business that Parcartambo estates and their peasant workers had conducted with the Beer Company of Southern Peru as early as the mid-1950s (Palacio Pimental 1957a). By the 1976–77 season, the yearly cultivation of more than 6,000 acres (2,500 hectares) of malting barley in Paucartambo surpassed all its other sources of raw material in the southern Peruvian sierra (Hopkins 1978, 26). By the 1970s, a high-yielding six-row barley, bred by the brewery and known as *griñon,* had replaced the older two-row German type. Yields in Paucartambo averaged almost one ton per acre (1,400–2,000 kilograms per hectare), enough for the brewery to post one company agent—a field "technician"—permanently in the region.

The improved *griñon* variety, required by the Beer Company of Southern Peru, seeded an exceedingly narrow biological base (table 3). Its biological uniformity did not, however, deter widespread adoption in the region. Quechua farmers perceived a regular market and income source in the company's crop. Via legally binding contracts, or so-called bailment, the Quechua farmers and a slew of subcontracting villagers were credited seed and fertilizer from the brewery's Paucartambo-based agent. The brewery contracted the right to purchase all harvested barley. By fixing the prices of seed, fertilizer, and barley, not surprisingly, it realized comfortable profits. Price setting more than the bailment contracts themselves—which put the full burden of economic risk on barley growers—had on occasion unloosed peasant anger and even led to an ill-fated boycott of Paucartambo growers in the mid-1970s.[11] Nonetheless, facing few alternatives, plentiful numbers of Quechua farmers opted for the malting barley as a chief route of their new commerce.

Rice production also skyrocketed during the post-1969 period, outpacing the other major food crops and embedding firmly as the principal staple of urban diets and a new addition to many rural ones (Alvarez 1983, 39; Figueroa 1981, 1984; Hopkins 1981; Mitchell 1991; Weismantel 1988). In the Andean foothills, or montaña, fields of upland rice sprawled over former tracts of tropical rain forest. Rice farms added to the day-wage work available in logging, placer gold mining, tropical fruit production, and coca fields that boomed in the Pilcopata and Qosñipata lowlands east of the Cordillera Paucartambo (map 5).[12] The frontier economies paid a day-wage three times greater than sierran farms. This wage gap drew hordes of short-term or seasonal migrants who could arrive in less than eight hours on the many trucks that careened down the Tres Cruces Pass. During intervals of one week or more, the Paucartambo migrants searched for wage-earning work in the thriving extractive and plantation enterprises. Usually successful, they could not, at the same time, escape new stresses and strains in the familiar framework of farm production at home.

Short-term migration after the Land Reform of 1969 was substantial. An

Table 3. Commercial Crops in Paucartambo after 1969

Commercial Crop	*Principal Variety*
Barley	*griñon* (improved variety)
Off-season potato crop	*mariva* (improved variety)
Off-season ulluco	*wawa yuki* (Cradled Baby)
Rainfed potato crop	*mariva, mi Perú, yungay, papa blanca, renacimiento, revolución, mantaro* (improved varieties)
Maize	*k'ellu sara* (Yellow Maize)

upswing in seasonal migrations was due to the accentuated forces of "pull" from the prosperous lowland economies and "push" from the stagnant or deteriorating economy of Paucartambo (Baca Tupayachi 1985; Figueroa 1981, 1984; Radcliffe 1986; Skeldon 1985). Still, the region's frequency of short-term migration registered as moderate compared to the neighboring sierra, where it was frequently the major form of "capitalist penetration." Only thirteen percent of families in the Ninamarca community near Colquepata, for example, supplied a migrant during the 1976 to 1979 interval, while the average in other Cuzco provinces exceeded sixty percent (Figueroa 1984, 69). Nonetheless, some communities in Paucartambo, such as Callacancha (estimated twenty-five percent) and Huaynapata (estimated thirty-four percent), saw migration rates approaching the Cuzco average. There was, in any case, an unprecedented laboring for off-farm wages among the Paucartambo peasants.

Breaking the agricultural calendar with off-season crops of potatoes and ulluco also became a common route for the Quechua in Paucartambo seeking to escape from the market traps of wheat and conventional rainfed potatoes. They sowed off-season fields in a variety of humid sites (bogs, irrigation, floodplain, cloud forest) during the nearly rainless "winter" season in June and July and harvested them as early as January and February (map 5). Their off-season crops included an early planting of potatoes, known as *papa maway* or *papa maguey,* and an early planting of ulluco or *maway lisas*. Farmers reaping the two off-season crops prospered from the seasonality of price swings in the rapidly growing Cuzco markets (Zimmerer 1991d). The prices for potatoes from the early plantings harvested in January fully doubled those depressed by the glut of the rainfed crop in July. An even more extreme peak in ulluco prices due to its perishability led the Paucartambo farmers to convert their fields into a major source of this off-season crop as well. By the mid-1980s, more than one half of the region's farmers grew off-season plantings of potatoes and ulluco.

*Plate 3. Cultivators filling gunny sacks with their Early Planting ulluco known as Cradled Baby (*wawa yuki*), which will be sold at the Cuzco market. Behind them are the slopes of Majopata, now a Peasant Community but formerly one of the largest haciendas of Paucartambo.*

A single improved high-yielding potato variety known as mariva and a single ulluco landrace—the Cradled Baby, or *wawa yuki*—were seeded in the majority of their off-season fields (table 3; plate 3). Farmers preferred the mariva potato, a member of *S. tuberosum* subsp. *tuberosum,* since its yield rose sharply in response to fertilizer additions. Mariva, popularized as one of the money potatoes or new potatoes, excelled in the off-season planting. Ripening in six months or slightly less, the mariva crop could be shipped early to market. Loans from the national Agrarian Bank, increased under the ill-fated populist program of President Alan García Pérez in 1985 and 1986, helped to swell the wave of off-season cropping. The bank's regional branch in Paucartambo counted 270 peasant borrowers in 1983, 430 in 1984, and 800 in 1985. In Colquepata district a nongovernmental organization funded in the Netherlands ran a credit program for farmers that also supplied much-needed capital.

The peasant borrowers, or *prestatarios* as they called one another, invested in seed tubers for their off-season crops as well as good-size amounts of various commercial inputs. They regularly applied commercial fertilizers (ammonium nitrate, potassium chloride, phosphorous, and a 12-12-12 mix of N-P-K), pesticides, and insecticides (including the toxins aldrin and parathion). In some areas of Paucartambo farmers found that irrigation technology was a requisite for successful off-season cropping. Their new Green Revolution package of money potatoes and inputs was depicted faithfully in the scene from Eufemia and Faustino's storeroom circa 1990, stocked with its sprinkler apparatus, piles of improved seed, sacks of Cachimayo fertilizer, and half-used agrochemical tins.

Potato cropping during the rainfed growing season was also a standard-bearer of postreform commerce. Cropping in the high-sun rainy period between November and April was already widespread before 1969, earning the tag big planting, or *hatun tarpuy.* Given the impetus of greater Cuzco's market, many Quechua in Paucartambo found the pursuit of commerce in rainfed cropping more accessible than its off-season analogue, since the big planting did not depend on irrigation. Their big planting did, however, call for the moderate use of agrochemicals and high-yielding varieties (My Peru, or *mi Perú*; the mariva; Mantaro [a valley in central Peru], or *mantaro*; and White Potato, or *papa blanca*) (table 3).[13] It thus parlayed meager diversity and in this resembled the other forms of expanded farm commerce. Although thin profits scarcely measured up to the hefty gains of the off-season crop, the big planting remained true to its name, forming the largest cropping regime in Paucartambo (Mayer and Glave 1990, 1992).

Nor did even a modicum of landraces line the lengthening rows of commercial maize. Maize commerce in Paucartambo seeded a single landrace type known as Yellow Maize, or *k'ellu sara* (table 3). Market demand for the region's Yellow Maize emanated from scores of popular *chichería* cantinas in Cuzco that poured endless pitchers of homestyle beer brewed from the maize that had been germinated and ground. Many Cuzco entrepreneurs turned to a handful of Paucartambo and Challabamba villagers who negotiated maize inexpensively from peasant farmers. The middlemen mainly bought from growers with fields at the prime sites downstream of Paucartambo and below 9,850 feet (3,000 meters). The Quechua farmers that supplied the Yellow Maize planted it solo in single fields and apart from other landraces in order to prevent the diluting and devaluing effects of hybridization. Their commercial rationale to confine maize diversity in the market-bound plots was thus reinforced by a biological imperative.

Sheep and cattle raising supported other avenues of postreform commerce. Farm animals were long combined and closely coordinated with cropping in Paucartambo and other Andean regions (Brush 1977; Brush and Guillet 1985; Guillet 1981a, 1992; Mayer 1985; Orlove 1977a). In Paucartambo the post-1969 phase of livestock-raising meshed well with other pursuits because of its small labor requirements and its by-product of fertile manure. Applications of livestock manure to potato fields totaled as high as 1,600 pounds per acre (1,800 kilograms per hectare; Mayer and Glave 1992, 127). Between the early 1970s and 1987, the farmer-herders of the region nearly doubled the average size of sheep flocks and cattle herds that were made up of rangy mixed breeds (appendix C.4). Stocks of pigs, horses, llamas, and alpacas—although smaller in number—grew at a similar pace. The Quechua in Paucartambo multiplied the size of their livestock herds more after the Land Reform of 1969 than did their counterparts in other Cuzco regions (Figueroa 1984). This unbridled emphasis on livestock-raising in Paucartambo accented the relatively land-rich and labor-poor nature of the region's peasant economy.

Marketing in particular and the tradition of nonsubsistence production more generally were deeply rooted among the Quechua farmers. Their postreform commerce differed little from earlier trends in designating a small handful of crop varieties. They chose griñon barley, mariva and other improved potatoes, Cradled Baby ulluco, and Yellow Maize to fill the swelling streams of commerce after 1969 (table 3). Meanwhile, similar to earlier periods, the fate of the diverse crops was enacted in their farming slated for self-consumption, that of subsistence. In their self-provisioning the protagonists such as Eufemia and Faustino confronted a mounting challenge: could they bring together the resources to husband their diverse crops and livelihood expectations while wedding farm assets to the new demands of expanded and diversified commerce? Fresh challenges of the postreform economy did not provoke bland responses; the Quechua in Paucartambo innovated piquantly, but not always with the desired result for diversity.

Synopsis: Biodiversity's Fate

By 1987, more than thirty-five percent of Quechua farmers in Paucartambo did not cultivate one or more of the potato, maize, ulluco, and quinoa crops (appendix C.5). Many fell from the ranks of diversity's tenders when, like Eufemia and Faustino, they failed to renew their diverse plantings. Asked why they had abandoned the potato landraces gifted years ago by Santusa in the memorable pile, Faustino once replied there was an obvious reason: they lacked seed tubers.[14] The answers to why Faustino, Eufemia, and the others forfeited seed of their former crops were also less obvious and more deliberate. Fuller answers hung on the rationales related to farmers' shrinking stocks of resource that were prescribed by the field systems and agroecology of the plants themselves.

In a sketch of the role of resources a line could be drawn between the limiting resources that often inhibited farmers from growing their diverse plants and the sufficient resources that typically enabled them in the years from 1969 to 1990 (Zimmerer 1991c, 1992a, 1992b). Limiting resources were headed by the quantity or quality of farmland and labor resources whose status many farmers were finding inadequate. Typical sufficient resources encumbered a different class of crucial inputs—farm technologies, techniques, and knowledge—accessible in adequate amounts and thus offering little impediment to the cultivation of diversity. A first-order estimate of diversity's fortunes then rested on a farm family's access to the resources that either enabled or conversely constrained its means to couple the kawsay-style foodstuffs—and the agroecological desideratum prescribed by their farming systems—with the no less desired network of new commerce.

This first-order estimate introduced below of how diverse crops fared based on farmers' access to resources does not imply that input status alone determined the course and quality of diversity's fortunes. Indeed the Quechua in Paucartambo proved able and sometimes ingenious innovators in adjusting a number of their kawsay-yielding traditions that nurtured the diverse crops. Although the skillful innovations and artful adjustments of the farmers filled a key chapter after the Land Reform of 1969, it was one that followed from the pressures and capacities sketched out first in the concrete terms of limiting and sufficient resources. The two categories were like markers fixed on by Quechua families as they blazed their trails of agricultural change.

The sets of limiting resources and sufficient resources hoisted markers near the trailhead that could be gauged in terms of both social and environmental units. The main social units were the farm household, house-family, or *wasi familia* in the vernacular, and its woman head. Most house-families among the Quechua in Paucartambo were conjugal groups of parents and children, although some spread across three generations and included close social kin, such as widows, orphans, and medium-term visitors (an unwed cousin or a displaced friend). While the households were patriarchal, women wielded authority and ruled work in various farm tasks. The Peasant Community, recognized by national government, sanctioned the resource use of its house-families but it did not exert much direct control. Groups of farm households frequently influenced one another's use of resources informally, but there, too, the power to choose rested in large measure with the family and its individual members (Collins 1986; Deere 1982; Guillet 1981b; Mayer 1977; Mayer and de la Cadena 1989; Orlove and Custred 1980; Radcliffe 1986).

The individual field was to the space of Quechua farming what the household and woman head were to its social sphere. Farmers endowed each field as an elemental unit in terms of agronomic inputs, crop types, and techniques and technologies. A plot of high-yielding, Green Revolution mariva potatoes was tended in a fashion quite unlike the mixed landraces of floury potatoes. Rich cultural relevance reinforced the economic salience of individual fields. Conspicuous meanings were conveyed in the names assigned to each field, such as the Big Hill field of Líbano, and in farmers' easy recollections of each field's past use. Nearly all fields were defined as outfields, or *chacras,* located at a distance from the farmhouse.

Most farm families also planted a small dooryard garden or infield—known as *huerta* or *kanchón*—though their garden plots lacked main crops, such as potatoes, maize, ulluco, and quinoa. Dooryard gardens nonetheless flourished with a wide variety of temperate fruit trees (such as Bird Cherry, or capulí; the Andean elderberry, or *sauco* [*Sambucus peruviana*]; native lucuma [*Lucuma obovata*]; medlar, or *níspero* [*Mespilus germanica*]; apples; and peaches), flowers, spice plants like *rokotu* chile peppers, and vegetables, all closed off with rock walls, hedgerows, and spiny "living fences" (Zimmerer 1989).[15]

Limiting Resources

The number and quality of chacra outfields foretold much of the fate of potatoes, maize, quinoa, and ulluco. The median number of parcels cultivated annually by a Quechua household in Paucartambo was eight, half sown with potatoes (table 4). Each family held an average of five acres, or two hectares, including fallow sites. Their modest number of fields was typically scattered across community territory; some, such as floury potato parcels, were located as far as one mile or more from the farmhouse, while others, like early potatoes and maize, tended to be closer. The combination of quasi-paritable inheritance, a desire to hold a range of growing options and reduce the risk of crop failure, and community institutions like the sectoral fallow commons together conspired to enforce the scattering of field parcels, which was common in the central Andes (Brush 1977; Brush and Guillet 1985; Goland 1992; Guillet 1981a, 1981b, 1992; Mayer 1979; Mitchell 1991; Orlove 1977b). As a result of the moderate quantity of fields, typical farm families enlisted a limited capacity to diversify their venues of commerce without reducing their repertoire of landrace-rich plots.[16] If they wished to add parcels of barley or off-season early potatoes, they likely faced difficult decisions with respect to the sowing of a diverse crop.

Deteriorating quality of their fields added to the constraints apparent in the sheer smallness of holdings. When farmers intensified cropping and livestock-raising in the wake of the Land Reform of 1969, they often depleted the fertility of field soils. In order to harvest more crops many were shortening fallow and eliminating the rotation of lower-yielding, nitrogen-fixing leguminous crops, such as broad beans and tarwi. Negative consequences often ensued for the diverse potatoes, maize, ulluco, and quinoa, which all depended on fields with at least moderate nutrient status. The vulnerability of diverse crops to declining soil fertility was compounded because they augmented yields only slightly in response to mineral and chemical fertilizers (Evans 1980; Wilkes 1983). Thus purchased amendments did little to remedy the effects of falling fertility, quite unlike the fertilizer-responsive cropping of improved high-yielding varieties.

Access to field sites was not fixed. Quechua farmers in Paucartambo could supplement their holdings by sharecropping, rental, or purchase. Sharecropping, termed *in parts* (*a partir*), provided use of a field in exchange for the provisioning of labor and often times inputs, the parties agreeing to split harvest

Table 4. Cultivated Fields per Family in the 1986–1987 Growing Season[1]

Crop Type	*Number of Fields*
Maize-Quinoa	1
Barley	1
Other nontubers (fava beans, tarwi)	1
Rainfed Improved potatoes	2
Early Planting Improved potatoes	1
Floury Potato landraces	1
Ulluco, oca, mashua	1

1. Values are the median of sixty families.

equally. Sharecropping, however, afforded the Quechua families meager help in finding farmland for diverse crops. While some may have wished to sharecrop for that purpose, they were usually lacking the labor that was demanded. Their shortfalls of labor-time derived from the demands dictated by bustling commerce in the years from 1969 to 1990 (Zimmerer 1991c).[17] New activities geared toward the openings offered by markets and agroindustry—such as short-term migration, off-season crops, and contracted barley-growing—pressed heavily for the reallotment of labor-time.

Tasks in the new work regimes fell to both men and women, although it was the latter who shouldered a heavier share of responsibility for the families' diverse crops. Women customarily managed seed, playing primary roles in harvest, postharvest processing, selection, and storage. Their work was especially intense in crops like maize, where a barrage of duties started at harvest: cutting canes and drying them in shocks; pulling and husking ears; drying ears; and selecting, storing, and shelling ears.[18] Although the gender-based division of labor in Quechua households was flexible, women's duties seemed to multiply disproportionately after the Land Reform of 1969. They weeded and monitored the finicky off-season plantings, herded livestock, and sometimes migrated with their husbands. In addition they labored in the unrelenting and familiar routines of cooking and housekeeping, raising children, spinning wool, gathering firewood, and, occasionally, marketing farm goods (Collins 1986, 1988; Radcliffe 1986; Weismantel 1988).

Quechua families relied on workers from extra-household sources, especially the reciprocal exchange of labor (ayni) and the payment of a daily wage (*jornal*), to supplement their own workforce. Their recruiting labor was, however, far from effortless. Its well-known difficulties led most people to recruit from among social kin—also known as fictive kin and ritual kin—established in the widespread custom of godparenthood, or *compadrazgo*.[19] Farmers also devised variations on the major forms of labor recruitment to make them more flexible; *mink'a* or *minga,* for instance, referred to the sending of another person in one's stead in the process of labor swapping. Only through a combination of resourceful recruiting could field tasks be finished. Potato growing, the cornerstone of Paucartambo cropping, typically mixed fifty-seven percent of labor from the family, twenty-four percent from labor exchange, and nineteen percent from wage payments (Mayer and Glave 1992, 131; see also Deere and de Janvry 1981). Peasant Communities qua communities, it should be pointed out, did not source much labor for farming; they used the community corvée, or *faena,* mainly for collective projects such as school building and repair, roadwork, and waterworks.

In sum the Quechua farming in Paucartambo suffered a constricting supply of labor-time for diverse crops after 1969, when they greatly expanded and diversified commerce (Zimmerer 1991c, 1992a). Labor shortages had long plagued sierran farming, where low yields and marginal environments seemed to demand a standing army of farm hands (Golte 1980, 11). The farming term *topo* and its etymological history illustrated this point. During the Inca and early colonial periods, topo had referred to a land unit, one topo defining the field area suitable for the subsistence of a household (Cobo [1653] 1979; Rowe 1947a). Its later definition, however, denoted a unit of labor-time expenditure. The Quechua in Paucartambo took it to mean the field area that a single skilled ploughman could till in one day with a team of oxen.[20] While commonly calibrating farm work by vernacular measures of labor-time, the latter was a resource that for their diverse crops the peasants found in shrinking supply during the years from 1969 to 1990.

Sufficient Resources

A host of other resources afforded most Quechua in Paucartambo with sufficient supplies to farm their diverse crops. Key albeit inconspicuous inputs included field tools and technologies, farm knowledge, and culinary preferences and know-how. The Quechua farmers worked the landraces with a versatile array of rustic hand tools: from various plows and hoes to mattocks and husking-and-shelling and threshing devices (appendix C.6). In handpicking their farm tools through home manufacture or purchase they matched technologies to growing conditions and to their own labor and capital resources. They tilled thick sod and dense clay with the levered foot-plow, for instance, while preferring the hand pick for use with sandy soils on steep slopes. Meanwhile, the oxen-drawn plow worked well in sandy but less-steep sites.

Standard tools for landrace cropping were affordable to most Quechua farmers in Paucartambo. Farmers purchased iron and steel components such as blades, plowshares, nails, and fasteners at village stores, while they gathered other materials from sites and sources on the farm or nearby. Cowhide sinew for lashing the traction plows, for instance, was plentiful. Wooden handles and pegs could be carved from the stout limbs and slow-growing trunks of preferred hardwoods like *q'euña* (*Polylepis racemosa*), *kishwar* (*Buddleia incana*), *chachakoma* (*Escallonia resinosa*), and *llok'e* (*Kageneckia lanceolata*). None of the trees formed dense or even full stands due in part to overcutting for fuel, but they remained adequate in number to furnish the farm tools. Eucalyptus trees, most common of all, also rendered usable wood for farm implements. Although the Australian gum tree had naturalized in the Andes during the 1800s, it was less desirable than stronger native hardwoods.

Farm knowledge equipped the Quechua in Paucartambo to produce their landrace crops. Farmers rarely betrayed their diversity-filled fields due to inadequate or lacking skills. Their rising interest in commerce after 1969 did not, in other words, devalue the relevant stock of local knowledge for tending to the diverse plants. Local farm knowledge consisted of technical skills for work routines, social acumen for such tasks as recruiting workers, and cultural beliefs such as ritual calendars that helped time those tasks such as planting and harvest. Farmers gained the know-how necessary for diverse crops through a wide variety of experiences beginning in childhood. Younger Quechua in the postreform period clearly cobbled farm skills in a number of new settings. Nonetheless, even the more complex procedures, such as the selection and storage of seed, exceeded the practical skills of no more than a few persons.[21]

Culinary preferences and the cultural commitment to a kawsay-style cuisine stayed planted firmly in favor of the diverse crops. Farmers' liking of their traditional crops did not, however, translate into an aversion toward new foods; many Quechua in Paucartambo welcomed greater rice-eating, for example, during the postreform decades (see also Mitchell 1991; Weismantel 1988). But they also partook eagerly of long-familiar fare for everyday consumption, such as *wayk'u papa* (boiled potatoes); hominy-like boiled maize kernels, or *mot'e*; and parched *hank'a* maize, which all derived directly from diverse landraces. Quinoa and ulluco were basic foodstuffs for hearty soups and stews. More than gustatory appeal whetted their appetites for these traditional crops; lightweight hank'a maize, for instance, could be easily packed to distant fields or pastures as a snack food, a genre of parched grains and beans that the Quechua called *qaqaw* (Zimmerer 1992b).

Religious uses affirmed the formidable status of diverse crops. The Paucartambo farmers proudly embellished their chief religious fiestas with particular dishes made from the tasty staples. The temptable *thimp'u* dish served during Carnival, for instance, was prized for its meaty maize kernels shelled from the native landraces. The farmers also celebrated numerous other ceremonial events with landrace-containing special occasion foods (Isbell 1978;

Mitchell 1991; Weismantel 1988). They served them at minor holidays, family gatherings, and work parties like the one Eufemia and Faustino sponsored in 1969. Special occasion foods were drawn from an eclectic palette: the diverse potatoes served alongside rice; roast guinea pig, or *qowi,* with bottled beer from the Cuzco brewery; quinoa soup with maize thickener and wheat breads bought in one of the villages. Customs of special cuisine among the Quechua in Paucartambo clearly did not rule out new foods but they also placed high regard on diverse crops.

Farmers also rated highly the agronomic qualities of the diverse crops. Quechua peasants held keen regard for the tolerance to climatic stress, which buffered the risks of failure that many farmers were amplifying by adopting the especially risk-prone commerce that was based on off-season production. Potato landraces withstood drought and frost better than the improved varieties, quinoa and ulluco yielded dependably in a gamut of environments, and maize, although subject to damage from frost and drought, was protected with special cultivation and irrigation techniques. The peasant tillers typically planted mixtures of landraces in all four crops. Their intercropping of diverse varieties lessened the threat of crop loss and thus subsistence failure, adding an increment of insurance that they would taste the fruits of their desired food plants.

Resource assets and debits notched after the Land Reform of 1969 bore a resemblance to some landmarks inscribed in the history of diverse crops under Spanish colonialism beginning in the 1500s. Farmland and labor-time shortages stressed Quechua cultivators in both periods and pressured them to surrender their preferred food plants. In both epochs the availability of other resources, such as farm tools and technologies, knowledge, and outright preferences, did not detract from viability of the varied crops. Skillful cultivators at both times tried to meet the challenge of dwindling resources by innovating new farm practices that could replenish their diverse crops under an altered agriculture. The continued creativity of kawsay customs and cultivator innovation could not, however, curtail fully the consequences of recent resource shortfalls.

Socioeconomic Differences and Dietary Change

The diverse crops did not shed evenly among the Quechua in Paucartambo who tilled the rugged countryside during the post-1969 period. A peasant house-family deciding whether to sow the landraces sighted a pair of circumstances on their economic horizons. In one line of sight the household eyed the prospect of combining the diverse crops in some fields with the farming of high-yielding varieties in others. Every farmer in the region sought to couple the perceived advantages of subsistence with commerce. In another corner of the decision-making horizon, the Quechua cultivators took stock of their personal resource endowments and consumption customs. There, all families did not view the same landscape, for their production capacities and consumption patterns differed markedly among one another.

Differences in resource stocks surfaced as the individual families sought to forge ahead with their commerce while embossing a range of diverse crops on their livelihood styles. Disparate effects led the relatively well-to-do peasants to number disproportionately among the tenders of diversity by the 1980s (table 5). Many extremely poor families, however, suffered input shortfalls that prompted them to disown the diverse crops. Socioeconomic differences in the years from 1969 to 1990 tended to follow the fissures formed earlier among estate peasants. The descendents of full tenants or the sector persons (*mañay runa*) on Paucartambo haciendas held up to 28 acres (12 hectares) when the Land Reform of 1969 was actually enacted in the early 1970s.[22] Former tenant-sharecroppers held medium-size areas half that size or a bit more. A number of the poorest families, former servants, clung to less than seven acres. Although the land areas appeared substantial, only between one third and one fifth was typically arable.

Socioeconomic differences within the Paucartambo peasantry were not altered much by the land reform. Although Article 102 of the Special Statute of 1970 had sought to moderate landholding differences, those inequalities persisted and were plainly documented in the reform's census reports registered shortly after 1969 (appendix C.7). Since reform officials filed the census reports along with community titles, they legalized the transfer of the preexisting disparities of landholding into the postreform period. Expropriation of the forty to sixty percent of an estate's land that typically had belonged to the demesne did not lessen these disparities and may have worsened them (Mayer 1988; Martínez Alier 1983). While reform officials sometimes awarded a share of ex-demesne land to a community's poorest peasants, they often granted the expropriated parcels to better-off persons; or, as was still rumored decades later, the powerful and more wealthy Quechua in the community seized many former demesne lands, often the very finest fields.

Differentiation resulted in the socioeconomic ranks of rich, middle, and poor peasants being evident among the Paucartambo people during the postreform decades (Mayer and Glave 1992, 76–77). The hierarchy of economic status compared a family's wealth, particularly the quantity and quality of its land, livestock, labor, and nonfarm business activities (like running a small dry goods store). Skewed differences existed in the size, number, and location of fields, the kind and size of livestock herds, and the age and gender composition of household members. Names assigned to the three categories stressed the salience of contrasts, although none of the peasants were especially prosperous

Table 5. Growers of the Diverse Crops by Socioeconomic Group[1]

Economic Category	*Potatoes*	*Maize*	*Ulluco*	*Quinoa*
Well-to-do peasants	75%	100%	38%	38%
Middle peasants	43%	71%	14%	19%
Poorest peasants	61%	48%	22%	6%

1. Based on a sample of eight well-to-do peasants, twenty-one in the middle of the socioeconomic spectrum, and thirty-one of the poorest peasants (on categories, see Brush and Taylor 1992; Mayer and Glave 1992).

by nonlocal standards. A similar degree of socioeconomic differentiation in the period between 1969 and 1990 was noted in other Cuzco communities and ones elsewhere in Peru, Bolivia, and other Latin American countries, such as Mexico (Baca Tupayachi 1985; Fonseca 1988; Godoy 1990; Guillet 1979; Isbell 1978; Mitchell 1991; Orlove 1977a; Sheridan 1988; Watters 1994).

The Quechua themselves noted the salient economic contrasts within their communities. Paucartambo people termed the well-to-do peasant a *qhapaqruna* (wealthy person), while the poorest person was cast as a *wakcharuna,* or orphan. A family in the middle of the socioeconomic spectrum did not bear a particular label, although they were often distinguished as neither qhapaqruna nor wakcharuna. In some cases they were described as "being like qhapaqruna without really being qhapaqruna." Agile ability of the perceptive Quechua peasants in Paucartambo to categorize one another's wealth was not surprising; they shared a well-seasoned and sometimes envious familiarity with their neighbors' endowments of land, livestock, and family composition.[23] A community member could easily spot a wealthy person, or qhapaqruna, typically someone with lots of farmland and large herds including oxen for plowing. With similar ease, he or she could distinguish a poor orphan, or wakcharuna, that held scant land and few livestock.

One indicator of a household's wealth was whether its head, usually a male, regularly held a role in community government. By law the Peasant Communities in Paucartambo periodically elected a leader known as either the president (*Presidente*) or lieutenant governor (*Teniente Gobernador*), as well as a several member governing body known as the *Junta Directiva* or *Concejo de Administración.* Community members expected that their leaders would come from the ranks of the more well-to-do. A wealthy person, it was thought, would wield more clout in the enforcement of community institutions like the corvée, or faena, which although sanctioned by a moderate fine for absence was unpopular at times and thus in need of powerful leadership. An orphan, by contrast, was often noted as someone too weak and too poor to lead. The powerful roles of post-1969 community government were inherited in part from hacienda life; tenants or sector persons had dominated the intrapeasant politics of estates, often both enforcing the discipline of owners and heading the resistance of peasant workers.

Economic divisions within the Paucartambo peasantry could be reckoned readily with access to the diverse crops. Better-off farmers were more likely than their poorer counterparts to seed the diverse landraces of potatoes, maize, ulluco, and quinoa (table 5). Nonetheless the poorest peasants, the wakcha orphans, proved in general to be more probable planters of diversity than peasants in the middle of the socioeconomic spectrum. (The reverse, it should be noted, held with respect to the diverse maize crop.) Although both the better-off and the most destitute sowed the diverse crops, the two groups of socioeconomic opposites created consumption roles for them that were dissimilar. On the one hand, for the well-to-do, the diverse crops harnessed prestige and reinforced status, frequently helping the leaders host a community or religious feast (see also Mitchell 1991, 84). For the poorest farmers who could not afford anything else, on the other hand, the diverse crop plants were more an unchosen item in their hardscrabble diets.

Production-related rationales also differed among the three groups of peasant farmers. Wealthy planters who could command upward of forty fields—as many as twenty cultivated in a season—abounded in lands that were apt for the varied landraces of each diverse crop. They also could convert many fields to commerce while keeping ample others for the diverse crops. Labor-time likewise impeded them less than the other socioeconomic groups. They recruited workers through wages, sharecropping, labor exchange, and payment-in-kind. In the case of labor exchange they could benefit from the local custom of mink'a that permitted paying another person as a substitute in the exchange. Payment-in-kind also helped assure the wealthy a ready supply of workers for their diverse crops. They regularly gained the helping hand of their less well-off neighbors by offering them either a fixed amount or a portion of the work effort in harvest—meted out by their harvested rows and piles known as *q'ama, qhaña,* and *phiña.*

Many of the poorest peasants in Paucartambo, inhabiting the other end of the socioeconomic spectrum, also cropped the diverse plants. As a result of the Land Reform of 1969, the poorest farmers tended to hold a modicum of land; however, their production-related rationales for the diverse crops were not the same as those of their well-to-do neighbors. Field holdings of the poorer Quechua clustered in marginal sites and, especially, at the high elevations unsuited to farm commerce. Their poor farmland pressed them to grow potato and ulluco landraces in order to furnish the nearly sole source of everyday subsistence. Their capacity to cultivate maize and quinoa was, however, much less. A majority of the poorest peasants lacked access to the fields valued located at elevations low enough for maize and quinoa parcels (below 11,650 feet, or 3,550

Plate 4. Faustino, a Paucartambo farmer of middling means, threshes the stalks of freshly cut quinoa. The quinoa belongs to his well-to-do neighbor, who will pay Faustino with a small sack of quinoa seed.

meters). In general their measly means of farm commerce left them with little choice but to till the diverse tuber crops.

Diversity's greatest absence from agriculture turned out to be in the middle of the socioeconomic spectrum (plate 4). Peasants in this middle group confronted critical shortfalls of one resource or another due to the channeling of sizeable assets to growing commerce. Labor-time, for example, strained them sensibly when they exhausted much of their farming effort on new activities and, in doing so, were less able to recruit extra-household workers than their better-off neighbors. Many families being squeezed for resources failed to muster the workers needed for the diverse crops. Central to their predicament was that the middle group could adopt a fair share of farm commerce during the postreform period; they were, for instance, among the chief adopters of improved potato varieties (Brush and Taylor 1992; Zimmerer 1991b). They overcame the obstacles that otherwise impeded the pursuit of commerce, but they often did so without the means to continue cropping their diverse plants.

That Eufemia and Faustino had deserted the multicolored floury potatoes gifted by Santusa in 1969 was not unexpected in light of the couple's hopeful but only half-founded foray into new commerce. By 1980 the couple, with the help of their young children, was forging ahead with four parcels of high-yielding potatoes—mostly mariva—destined for Cuzco markets and two medium-size barley plots contracted by the Beer Company of Southern Peru. In some years Faustino traveled to the gold mines at Quince Mil, not far from the Paucartambo montaña, where he earned piecemeal pay for the glittering specks he uncovered by hand-washing and screening placer deposits. Still other endeavors dated to 1985 when they became prestatarios, borrowing a low-interest loan from the Agrarian Bank in order to buy irrigation equipment for off-season potatoes. Strapped for field parcels, their family decided that tending the potato landraces was no longer affordable.

It was in order to obtain a small supply of the savory spuds that Faustino religiously sharecropped with his neighbor Don Líbano in the Big Hill field of upper Umamarca. While sharecropping gained him and his family the desired foodstuff, it nonetheless cost them labor-time. Although he and Eufemia viewed the arrangement to be convenient, it was not what they hoped for. Barter arrangements offered a similar trade-off for those persons angling to obtain the diverse crops. The most common form of barter involved the swapping of maize and potatoes. They were traded according to a standard unit of volume known as the *chimpu* for the shawl that farmers used to carry them: one chimpu of shelled kernels could be exchanged for an equal amount of freeze-dried chuño or moraya. Alternatively, one chimpu-worth of unshelled maize traded for an equal volume of fresh potatoes. Although farmers could, therefore, turn to barter to obtain diverse crops, this mechanism did not make up for the supplies lost due to cultivation changes.

Consumption habits of the Quechua in Paucartambo shifted swiftly in the years from 1969 to 1990. A number of major changes delved deeply into their dietary customs at the expense of diverse crops. High-yielding potatoes grown at home were filling their blackened cooking pots and worn enamelware with a spiralling frequency. Although most high-yielding types ended up in market stalls, the Green Revolution ingredients also served to sustain many farm families. High-yielding improved varieties, such as *renacimiento* (Rebirth) and later mariva, could be cooked like the diverse floury potatoes, that is, by boiling to serve in soups or with a requisite spicy relish known as ají, which was ground from the hot capsicum, or aji chile peppers. Good-size harvests of the high-yielding varieties could be substituted for the less-prolific landraces. The inconspicuous substitution of high-yielding varieties thus subsidized the growth of farm commerce. Quechua farmers also shifted their dietary fare by adopting more barley-based dishes into their cuisine. By discreetly retaining unnoticed portions of the company-contracted crop or tending their own, they consumed more barley soups (*lawa, chupe*), gruel (*mashka, hak'u pichisqa*), and homemade beer (*teqte*).

After the Land Reform of 1969 consumption by the Quechua in Paucartambo also shifted toward the purchase of marketed foodstuffs. A detailed study of their consumption habits revealed that large percentages of inhabitants were buying the following foodstuffs by the 1980s: onions (ninety-four percent), rice (ninety-four percent), carrots (eighty-eight percent), sugar (eighty-six percent), macaroni and spaghetti noodles (eighty-six percent), cabbage (sixty-seven percent), and cooking oil (forty-five percent) (Fano and Benavides 1992). Commonplace purchases of noodles and rice signaled a region-wide shift that was displacing the older eating customs of many farmers. Since the macaroni and spaghetti noodles sold at the lowest prices, countless Quechua families realized that the store-bought fare cost less than sowing some of their diverse crops. Their cost-saving substitution of noodles for a landrace-based cuisine did not, however, usher in a radical upheaval of taste preferences; consumers lamented and joked derisively about the abject lack of flavor and "cardboard" texture of the noodles and other cheap foodstuffs given as aid (for example, canned meat from the European Economic Community).

Diverse crops and the corresponding culinary dishes went on garnering unique esteem for their taste, nutrition, and ceremonial value in the post-1969 decades. Rather than relinquish renown to the new foodstuffs, the diverse landraces of potatoes, maize, ulluco, and quinoa actually attracted quite a bit of prestige in the course of accelerated economic and dietary change. Potato landraces such as the Red Mother, Aborted Guinea Pig, Village Plain, One Who Cries for Her Inca, and Fist once grown by Eufemia, Faustino, and their neighbors in Umamarca were heaped with renewed feeling when they became rarer. The diverse crops, much like the finely woven and increasingly rare garments of homespun wool that gained local status, were becoming a sort of traditional luxury. Even Eufemia and Faustino, who no longer cultivated the floury potatoes, agreed without hesitation that a larger supply in their larder would bring a welcome renewal of kawsay-style eating.[24] They agreed that maize, ulluco, and quinoa similarly grew in memorability even while, ironically, they made up ever-smaller portions of the family diet.

The social fissure that cut through cultivation of the diverse crops widened since the ready supply of prestigious foodstuffs helped the better-off farmers to recruit crucial labor. Whether a Quechua family in Paucartambo prospered in expanding postreform commerce turned on its recruiting extra-household labor. In addition to direct payment with the diverse crops a well-to-do family used them to sweeten the culinary prospects of the midday meal that was served to workers. The midday meal, referred to as *mesa puesta* (provided meal, or simply as food [*comida*]), appeared at first glance to be a minor fare compared to a worker's basic return in his or her wage, ayni labor exchange, or payment-in-kind. The laborers took serious stock of the meal, however, when deciding whether to toil for a certain field owner. Together with coca leaves or cigarettes, a robust midday meal could decisively sweeten the terms of a labor arrangement. Prominence of the meal also affirmed the pivotal role of household women not only as skillful cooks but also as agents of labor recruitment, a function often overlooked.

As a result, more wealthy farmers like Don Líbano of Umamarca and Don Pedro seen in Nova's *Seeds of Tomorrow* could use their quotient of diverse foodstuffs to edge economic advantage still further in their direction. The unexpected twist of diversity's usage within the peasantry was illustrated by the household headed by Líbano and his wife, Doña Natividad. The well-to-do landrace-growers regularly deployed their prestigious crops to help recruit extra-household labor for their suite of commercial endeavors. By serving a renowned midday meal heaped with steaming floury potatoes and succulent maize kernels, flavorful soups and stews stocked with quinoa and ulluco, and generous helpings of maize-brewed chicha beer, Don Líbano and Doña Natividad held a reputation among the poorer Umamarca peasants as the most rewarding field owners for whom to work.

Outcomes of the manifold transitions in farm nature and society after the Land Reform of 1969 took on definitive shape in the small fields and out-of-the-way storehouses of the Quechua, who tilled the scores of Peasant Communities in Paucartambo. Yet the full story of farming in the mountainous countryside and of diverse crops was unavoidably set across the broader arena of the Paucartambo Andes, where the peasant tillers resided together with villagers and merchants, government officials, agribusiness agents, and the personnel who staffed the national government's Agrarian Bank and nongovernmental aid programs. By exerting economic and political power as well as pull on people's feelings of ethnicity, the various groups in regional society put in motion several dynamics that moved the post-1969 saga of diverse crops in more complex and, at times, surprising ways.

Ethnicity, Power, and Biodiversity

The Land Reform of 1969 led to the relocation of many ex-estate owners to villages in the Paucartambo Andes—Paucartambo (1981 population: 1,928), Challabamba (1981 population: 201), Colquepata (1981 population: 484). The erstwhile seignors joined a few thousand other non-Quechua residents already living in the country towns. There, the new villagers craftily redid their economic portfolios. Their influence as economic agents, and especially their interaction with the Quechua farmers, compounded the effects of new commerce and socioeconomic differences, thereby adding new complexities to the already varied social roles of diverse crops. An anecdote of the well-to-do Don Líbano in 1986 illustrated a common scene amid the regional-scale forces of power and ethnicity that pressed on diverse plants:

> One reason I grow the floury potatoes is so that I can host my village friends when they come to visit and relax and to do a little business here in the countryside. Just last week Señor Juancito was here in Umamarca. You know that he used to be the owner of the Huaynapata estate near here. Together we ate three big platefuls of floury potatoes with aji chile-pepper relish and fresh cheese that Natividad made last week, and that was after he finished two bowls of quinoa soup. Anyway, with the election victory of the APRA party, Señor Juancito is now the assistant tax collector for motor vehicles, which I know is good for me. [Líbano was part owner of the Darwin truck at the time.][25]

Peru's Land Reform of 1969 mandated that Líbano's friend Señor Juan—the diminutive Juancito bespoke an etiquette of deferential affection—and the other hacienda owners immediately cede most estate land.[26] The majority of ex-estate owners relocated to one of the country towns of Paucartambo, Challabamba, and Colquepata, while a handful resettled in Lima and Cuzco. At least a few of the latter benefited from the enlarged apparatus of the Peruvian state; one, for instance, eventually became entrenched as a middle-level bureaucrat in the government's national oil company PETROPERU. Former estate owners who took up permanent residence in the Paucartambo villages still managed to extract their livelihoods from the region's agriculture. With plentiful social power and economic aspirations, but without farmland of note, they immersed their assets and energy in the commercial flows that followed the Land Reform of 1969.

The ex-estate owners flexed a surprising muscle in regional farming, their economic power reconstituted but not much reduced if at all. Refracted through the prisms of resettlement and entangled development, domination by the new class of villagers must have been unforeseen, or at least unstated, by the planners who designed Peru's most radical land reform. An invisible mesh of economic entanglements bound the powerful but landless villagers to the Quechua farmers of the Paucartambo countryside. Described as patron-client bonds and asymmetrical reciprocity, the power relations between villagers and peasant farmers were "a response of individuals to grossly unequal distributions of wealth and power and to situations of risk and uncertainty" (Orlove 1979, 84; see also Collins 1988; Fonseca 1988; Guillet 1992; Mayer 1988; Mitchell 1991; Orlove and Custred 1980; Thurner 1993; Weismantel 1988).[27]

In the Paucartambo Andes the "grossly unequal distributions of wealth and power" after the Land Reform of 1969 translated geographically into a flaring-out of the village economies like funnels that diverted quickening flows of goods, labor, and capital being swept loose in the expansion and diversification of the peasant economy. Although the Quechua in Paucartambo possessed the vast majority of land and composed nearly the whole of the rural labor force, they resorted out of necessity to social ties with their village patrons, often fictional kin. Their village patrons helped to ensure the off-farm supplies and services crucial to farm commerce in the period from 1969 to 1990. Four resource types—field inputs, money, transportation, and brokering with government officials—levered the farm economies of the Quechua into new positions of commercial advantage, but led directly to the tightening of patron-client ties with villagers in Paucartambo, Challabamba, and Colquepata.

The chief field inputs transferred through the villager-peasant mesh were seed, fertilizer, and pesticide. According to the terms of one common pact between the landless villagers and the Quechua farmers, the villagers contributed seed and other inputs in a sharecropping or a partir deal that stipulated an even division of harvest. Under that sharecropping, the farmland and labor-time were proportioned entirely by input-poor peasants. Such sharecropping was widespread and crucial in the spread of much input-intensive off-season cropping after the Land Reform of 1969. Powerful villagers also managed access to productive fields and peasant labor through small cash sums exchanged for land mortgages and land rental, transactions that prevailed in the Peasant Communities of Paucartambo despite prohibition by laws of the Peruvian land reform.

Money advanced as credit knotted a mass of patron-client ties between the powerful villagers and many Quechua farmers seeking to adopt new commerce. Formal credit that was dispensed by the Paucartambo branch of Peru's Agrarian Bank ran out before reaching more than a small fraction of farmers in the region. Even at the height of its lending in the mid-1980s the bank forwarded loans to less than fifteen percent of the region's credit-hungry cultivators. Seizing economic opportunity, numerous villagers financed the major share of farm credit for the Quechua, who committed to repayment either in cash with interest or in farm goods at harvest. In setting terms for harvest-season sales the villagers manipulated indebtedness in order to fix prices that were seen as either invitingly favorable or scandalously exploitative contingent on one's perspective. This exploitative practice was already familiar, in fact, for it had been common under the haciendas prior to 1969. Credit, either in money or goods, thus continued to entrap those peasants who fell into debt in the lean preharvest months or in years when revenues, harvests, or both, failed outright.

Without any external sign of a twin function, the retail stores in the country towns of Paucartambo discretely doubled as credit agencies and buyers of farm goods for thousands of Quechua cultivators. In the mid-1980s the cobbled streets of Paucartambo were lined with more than sixty such dry goods outlets, which tended to occupy a single storefront room, venturing in sale items from candy to kerosene to loudspeakers (Mayer 1988). Stocking also a disarming array of out-of-date bric-a-brac, many stores were properties of the ex-estate owners. These "new *hacendados,*" as more than one frustrated peasant farmer called them, used their stores to forge a potent alchemy that combined liquid money reserves for credit and the political power of local government offices for patronage.

One index of the far-reaching power and influence of storeowning villagers was the absence in Paucartambo and Challabamba of an open-air marketplace. Lack of a periodic marketplace in the two villages contrasted nearly all the analogous-size villages throughout the southern Peruvian sierra, including nearby Colquepata.[28] In effect the powerful village patrons completely captured the wholesale trade of rural clients, leaving farm markets cornered and marketplaces closed without ever opening. Efforts to develop a marketplace—and there were some attempts by the Paucartambo branch of the nationwide Ministry of Agriculture, as well as discussion by visiting nongovernmental organizations—failed in not being able to sever the social mesh binding peasants and villagers. A farmer trying to sell in a new marketplace risked the wrath of his or her patron along with the withdrawal of inputs or other supports. In defense of their rural ties the provincial villagers of Paucartambo liked to say in the late 1980s that the channels of villager-peasant communication were what deterred the Shining Path from locating in the region.[29]

Control of transport likewise rewarded the powerful albeit landless villagers. With few exceptions, such as Líbano's ill-fated stint owning Darwin, it was the villagers who held the fleet of workhorse flatbeds that trucked goods across Paucartambo's decrepit dirt roads. Truck owners shipped farm goods from the hinterland to markets in Cuzco and Sicuani and occasionally Arequipa, with a smaller fraction headed eastward to the sprawling towns of Pilcopata and Puerto Maldonado in the upper Amazon. As many as eighteen trucks with names as colorful as Noble Merchant (*Noble Comerciante*) departed daily from the villages of Challabamba, Paucartambo, and Colquepata at the peak of harvest in May and June. The Quechua farmers, for their part, depended more than ever on timely shipping. Success of their commercial ventures, such as off-season plantings and the brewery-bound crop of malting barley, rested on prompt delivery to extraregional markets. Faustino's abrupt concern about the upcoming departure of the trundling Darwin when we checked out the Big Hill field, described in the opening chapter, attested to transport's everyday presence in the Paucartambo countryside.

Constraints in the transportation network of Paucartambo truly bottlenecked farm commerce and benefited the powerful villagers. Villagers could use their lumbering vehicles as a convenient platform to loan money and lock up the purchase of farm goods. The vehicle owners operated trucking oligarchies, moreover, that were empowered by their small number and provincial location. They fearlessly inflated freight costs. One sign of oligarchic pricing was that per unit shipping costs from Challabamba to Cuzco averaged twenty-five percent more per unit than those from Paucartambo to Cuzco, notwithstanding the close proximity of Challabamba and Paucartambo. While the Quechua farmers resented and occasionally decried the blatant exploitation by the truckers, they dutifully aided at regular intervals in repairing the maintenance-hungry mountain roads that carted the fruits of their burgeoning commerce. Due to its many branches into the farm economy, the obstructionist role of transport was not strictly analogous to a simple roadblock.

Newfound functions of regional government seated in the hopeful country towns also added glue to the patron-client bonds between villagers and peasant farmers. The Quechua cultivators increasingly had reason to ply the bureaucracy of Peru's Ministry of Agriculture, for instance, which ran a regional office in Paucartambo and branches in the other villages. The local outposts of the country's farm bureau administered tasks varying from the arbitration of land conflicts to the approval of loans from the national Agrarian Bank and the institutional coordination of food aid and development projects, of which there were few. For the farmer beset by a pending problem or an unresolved request, the issue's outcome stood a better chance of success with the assistance of a well-connected, influence-peddling villager. The enterprising and circumspect Don Líbano, wealthy and well-known in the village, wisely consulted his social kin such as Señor Juan on even minor affairs, such as the disliked truck tax, as well as on more potentially upending matters.

The favor-brokering villagers garnered further patronage from farmers due to the worrisome uncertainty of reform measures that remained unresolved long after 1969. In fact the Ministry of Agriculture fanned the deep-seated insecurity of the peasants in 1986 when it reverted the ownership of prize farmland on a community near the Paucartambo-Challabamba roadway to a powerful villager and former estate owner. In response to urging by the ex-owner and his dispossessed but influential clan the ministry's headquarters had ruled that the peasant community demonstrated "inefficient use." This manipulation of widely recognized national laws heightened the chronic anxiety of Quechua tillers throughout the region about the vulnerability of their benefits from the Land Reform of 1969. Still further anxiety was instilled by the slow pace of the government's issuance of titles: even in the late 1980s twenty-two Peasant Groups anxiously awaited the minor procedural adjustments that would award them their legal title as Peasant Communities (Mayer 1988).[30]

The patron-client bonds that lashed down regional commerce and reached into the farming of diverse crops were merely retied rather than permanently undone following the Land Reform of 1969. A comparison of scholarly notes on the functions of the many bric-a-brac stores run by patronage-wielding villagers in Paucartambo illustrated the grim scenario. Perceptive commentators had recognized that those storefront rooms counted primarily for the purpose of patron-client exchanges rather than retailing merchandise. Prior to the Land Reform of 1969, the sixty-plus stores run by villagers—store people, or *tendayoq,* as the Quechua called them at times—were thought to lift a hopeful sign of multiple options and, therefore, a seller's market for the Quechua farmers in need of patronage. That was a perceptive rendering of the town's social landscape at the time (Cotler 1975, 151). By the mid-1980s, however, the similar presence of dry goods stores in Paucartambo had come to symbolize the entrenchment of economic domination and inequality. By the later date this was an equally accurate but decidedly more pessimistic verdict, well-versed in the economic reality of the postreform decades (Mayer 1988).[31]

The diverse crops lay inextricably entwined in the web of patron-client ties spun after the Land Reform of 1969. While the facility of farmers to seed the floury potatoes, maize, and other plants did owe to their economic assets, their access to the resources took shape through the economy and its actors, both powerful and weak. Peasants wishing to strengthen their economic status through patronage managed to make good use of the diverse crops in a new assertion of cultural identity amid the many strands of patron-client webs. Villagers, for their part, were remaking their expression of a distinct "townsperson" culture—one cemented via opposition to the countryside—that prided itself on public parlance of the Spanish language, albeit in a provincial dialect. Village people most commonly referred to one another as neighbor (*vecino*). They mostly used the derogative Indian when referring to a Quechua person, with the notable exception of public addresses, which since the Land Reform of 1969 were heavily dosed with the term *peasant,* or campesino.

The new ethnic role of diverse crops in the years from 1969 to 1990 was evident in the identity practices of villagers—with their provincial culture and plentiful ties to the countryside—and the analogous expressions of the Quechua farmers. Like the villagers, the postreform Quechua cast their personal identities in opposition to the other social group (Allen 1988). In frequently referring to themselves as either person (*runa*), or peasant (campesino), or community member (*comunero*) they set their identities apart from the villagers that they cast as whites, or *mistis.* That difference was exaggerated further in face-to-face settings where the latter were deferred to as "sir" or "madam," in keeping with an "etiquette of inequality" (van den Berghe and Primov 1977).[32] The linguistic artifices behind this ethnic barrier between runa peasants and vecino villagers were accurate reflections of pronounced difference; however, they also belied a small but definite permeability and changeability in the ethnic categories.

Greater permeability in the social categories was made plain by the growing number of Quechua peasants, especially the more well-to-do, who outfitted a village residence as well as their main dwelling in a Peasant Community. The four settlement quarters of Paucartambo Village, for instance, showed a growing number of residents that bridged neighborhoods in the capital directly to the countryside: Carpapampa to the southern Paucartambo Valley, Callispuquio and Virgen del Rosario to the central and northern valley, and Quencomayo to the interior. The villages of Challabamba and Colquepata also housed neighborhoods that, although no more than a few blocks in size, were inhabited mainly by Quechua peasants rather than "white" villagers. In Challabamba and Colquepata some peasant villagers followed the dual residence pattern of Paucartambo people, while others resided full time in town and still retained membership and access to lands in Peasant Communities.

The coexistence of deeply felt difference and subtle sentiments of similarity gained a peculiar expression in the ethnic status of the diverse crops. Irreconcilable difference was indeed one meaning of the landraces and their farming after the Land Reform of 1969. Cast as cultural icons of mutual opposition, they typified the identity of the Quechua in Paucartambo, both to themselves and to villagers. The farming of landraces, like the chewing of coca leaves, stood for an integral part of their widely shared idea of being Quechua. By contrast, the provincial townspeople would no sooner nurture the diverse landraces of potatoes, maize, ulluco, and quinoa than they would publicly masticate the mildly stimulating and sacred leaf. Growing landraces was inimical to mestizo identity. The Quechua names of the scores of peculiar landraces and their myriad dishes did indeed express exclusively the cultural realm of the Quechua. Such strong symbolism of the diverse crops suggested that its place rested solely on one side of the bipolar barrier, much like the purported allegiance of Indian food, or comida del indio, which had begun in the colonial period.

Nevertheless, the unity of diverse crops as ethnic expressions after the Land Reform of 1969 was in some ways more apparent than real. One reason for this ambiguousness resembled the persistent fallacy of the term *Indian food,* which was purported to, but did not, furnish an all-inclusive cultural marker. After the Land Reform of 1969 many Quechua tillers and the commerce-strained middle peasants in particular no longer farmed the diverse crops, even though they remained as rural and Quechua as any group in Paucartambo.[33] The ever more striking presence of the Quechua farmers who grew little diversity did not detract from the continued vitality of the cultural meanings owed to diverse crops. It did indicate, however, that use of them was not uniform among the ethnically indigenous.

The inflated value of diverse crops as cultural icons for the Quechua was due partly to scarcity's effect in making the familiar dear; however, a major new source of cultural value also emerged. Rural clients and village patrons regularly relied on the diverse crops as means to renew their bonds of social kinship. In one diversity-rich custom that flourished after the reform of 1969, the diverse crops furnished food gifts that the Quechua farmers lavished on their social kin in the village. Many Quechua sacrificed sacks replete with hak'u papa (floury potatoes), fresh corn, and recently butchered beef, pork, or even *qowi* (guinea pig), hauling them down country paths and discreetly into the back doors of the villagers' houses. At other times, the wealthy Quechua farmers hosted feasts at home in their Peasant Communities where they served diverse crops and other foods to the visiting villagers.

Farmers also paraded the diverse crops in performing public and religious offices, or *cargos,* of the civil religious hierarchy. An office-holder, or *carguyoq,* was expected to sponsor a fiesta specified in his or her office, usually in celebration of a Saint's Day (appendices D.4 and E.5). As part of sponsorship, maize-brewed chicha, food, and often one or more bands and dance troupes were secured. Food was served in large feast meals consisting of certain prescribed dishes and customary cuisine. Some Saints' Days specified special soups and platters, such as thimp'u at Carnival, *merienda* at Cruz Velacuy (Saint Elena's Day), and *chiriuchu* at Corpus Christi. Diverse crops supplied the basis for the unique feast dishes as well as the special foods served in other celebrations. During the years between 1969 and 1990, well-off Quechua farmers frequently invited some village friends and patrons to their fiestas, lubricating their economic relations with liberal portions of cherished food, not to mention chicha.

Some fruits of the Paucartambo countryside, with potatoes and maize foremost, plainly did not present anathema to white villagers. The element of cultural commensurability grew due to the myriad patron-client ties in regular need of reinforcement after the Land Reform of 1969. Boiled floury potatoes served with the savory chile-pepper relish of spicy ají furnished one example of a shared cuisine built on the diverse crops. Villagers frequently referred to them as native potatoes, or *papa nativa,* but they unhesitantly embraced the flaky texture and delicate tastes of the sumptuous country-style meals hosted by their fictive kin. A few other types of diverse crops eagerly consumed in public meals by feasting villagers and Quechua farmers had in fact quietly been an ingredient of non-Quechua cuisine for centuries. Ulluco and maize in particular, which enjoyed great popularity during the postreform decades, had already been absorbed into non-Indian food habits during the colonial epoch.

The widespread popularity of diverse crops took a new turn after 1969 in further cleaving economic differences among the Quechua themselves. Wealthier Quechua farmers dined powerful village patrons to their mutual advantage. Don Líbano—the well-to-do Quechua farmer of Umamarca—regaled in telling of the hearty quinoa soup and relished floury potatoes that he and Doña Natividad served as both symbol and sustenance to the visiting Señor Juan, tax collector and loyal member of the ruling APRA party.[34] (He knew that I, too, was an avid eater of these foodstuffs.) Líbano's poorer neighbors, many who did not seed the landraces, apprehended the shrewed investment of his surfeit. The diverse crops were not being cast as simple symbols of social status, in other words, but were being used to help regenerate that status. Advantages of a fit or kawsay-style livelihood were advancing the future prospects of those who already benefited.

Biodiversity and Recent History

Diverse crops of the Paucartambo Andes were not eclipsed in the remaking of culture and ethnicity, town and countryside, and subsistence and commerce after the Land Reform of 1969. The landraces supplying floury potatoes and parched maize epitomized a double usefulness of the diverse crops: they afforded the valued staples of a fit livelihood as well as the chief ingredients of delectable feasts and food gifts. The double rewards of everyday value and special purpose testified to a re-created vitality of the diverse crops as cultural matter during the years from 1969 to 1990. Unassuming organisms were being reinvented as expressions vital to a broad-based popular culture. In post-reform Paucartambo, unlike premodern Europe, accustomed potatoes were not becoming a poverty food of last resort but rather a surprisingly elusive foodstuff akin to a traditional-style luxury item.

Quechua farmers in Paucartambo did not surrender their diverse crops as a matter of first choice. Yet families like Eufemia and Faustino labored mightily to expand and diversify their commerce of the off-season high-yielding potatoes, plots of barley contracted with the Beer Company of Southern Peru, and market-bound ulluco. In their budding endeavors they glimpsed an incongruous pairing of the modernizing urban markets—even global outlets for their farm goods, since the brewery exported its Cuzco Beer abroad—and the intensifying sweat of gritty work routines at their mountain-based field sites. Much like the nonsubsistence economies of earlier epochs, moreover, specialized market farming from 1969 to 1990 called for a sparse range of diversity in the way of crops and landraces. As before, therefore, diversity was reaped almost entirely in a family's cultivation for its own larder.

New dilemmas beset the self-provisioning share of farming both before and after the Land Reform of 1969. The Quechua peasants in Paucartambo practiced a micropolitics that seized on subsistence rights and the claim to diverse crops in their recurrent efforts to secure better livelihoods in the late 1950s and the 1960s. The radical reform of 1969 cast the diverse crops into a medley of political roles, since environmental resources in general struck a high profile among the contending parties of anxious peasants and exiting estate owners. The reform also introduced paradoxical effects in the region via peasant migration and soil degradation that set in motion some resource scarcities that would worsen sharply after 1969. The macroscale economic policies of Peruvian governments further worsened the scarcity of resources. Overall, the Quechua in Paucartambo were the losers in the post-1969 models of development that exerted strong biases in favor of economic growth among other geographical regions, producing sectors, and social groups.

Many Quechua farmers in Paucartambo nonetheless succeeded in coupling the diverse crops for self-provisioning with the pell-mell demand of their new commerce. The successful cultivators managed to squeeze them into a smaller albeit still vital corner of their farming. Other families, however, were unable to adjust and innovate adequately. On the one hand, numerous families in the middle of the socioeconomic spectrum found that their struggling engagement with new markets did not permit the diverse crops. Shortfalls of land, labor-time, or capital at the farm level pressed their unwelcome decisions. The farmers curtailing diversity often shifted to less expensive diets, including home-grown improved potato varieties and cheap noodles and other food aid. The well-to-do peasants, on the other hand, reaped growing rewards from the diverse crops in recruiting the labor and loyalty of other peasants as well as capital and patronage from villagers. Enterprising use by the better-off Quechua peasants reflected new social and ideological roles for the diverse crops.

The Quechua in Paucartambo transformed not only the peasant economy from 1969 to 1990; they also reshaped the geography of farm space. They were especially successful at recasting the spatial order of dual production—the coupling of commerce with kitchen-bound farming. Their most recent reconfiguring in the region capped a long series of spatial changes. As early as the Inca period, discussed in chapter two, the major nonsubsistence crops such as coca and maize derived from well-demarcated lands set in environments and districts distinct from the self-provisioning fields of commoners. Under Spanish colonialism, that spatial distinction was an intra-regional one, since Indian peasants sent tribute such as wheat, barley, and chuño from their own fields to the rulers. After 1969, however, unprecedented marketing drove farmers once again to reorder their use of territory. When the crux of conservation took shape during the 1969 to 1990 era—that is, landrace loss versus diversity's revival in continued cropping—it became clear that the environmental outcomes were a matter of farm spaces and farming places.[35]

4

Innovation and the Spaces of Biodiversity

Seeding Landraces

Outside the cluttered compound of adobe brick huts in Umamarca, Líbano heaved the bulky sack of floury potato seed onto their strongest burro, while Natividad steadied the load and roped it snugly. They exchanged greetings with their neighbor Eufemia, who was striding toward the nearby spring with empty pails for water. On this October day in 1986 Líbano, Natividad, and I were headed for their Big Hill parcel, and the dimpled sack of floury potatoes needed to be well balanced and surely lashed. Although frequently aided by Eufemia's husband, Faustino, Don Líbano and Doña Natividad nevertheless did toil some in the distant Big Hill, and planting was one such task. From their home with its pleasant hedge of shrubby q'euña trees, the three of us climbed steadily. We suddenly gained a panoramic vista of the rolling grassland plateau that was perched atop the plunging slope of the Paucartambo Valley.

More than one-half mile above his distant huts, Líbano peered again at the expansive upland and the dark, overturned rows of tilthy soil:

> I can plant Big Hill and my other fields here with the seeds of floury potatoes. Big Hill is also a good one for growing fine seeds—it's been fallow more than five years. This field always yields lots of good-size tubers: not too big nor too small. In any case I have another field like Big Hill on the other side of those rocks. If one field fails me, I'll get seed tubers from the other one. Look down there [he pivots motioning in the opposite direction toward the canyon], can you see my maize fields? My *Rumiriyoq* [Rock Place] looks like an alpaca with its long neck below the cactus patch. We'll go down there tomorrow. I need to plant some maize—you know, mot'e—for an early harvest so that we have fresh ears for

> Carnival. Look how the stone fence around the neck of my Rock Place was damaged by a couple of Vicente's cows. They ate half my field last year and nearly left me without maize seed.[1]

Farm spaces in the Paucartambo landscape were created separate and unequal with respect to diverse crops after the Land Reform of 1969. Some spaces at varied elevations pooled reservoirs of great diversity while others harbored little.[2] References to them signposted even a casual conversation. Talkative Líbano's planting of potato landraces, for instance, took place in the loma, or Hill. Beyond Líbano's Big Hill field and a cluster of his neighbors' plots bearing floury potatoes, ulluco, oca, and mashua, a floral carpet of greening grasses, and a few ground-hugging herbs spread to the neighboring community of Mollomarca. The kheshwar, or Valley field, a mere dot from our vantage at one-half mile (about one kilometer), marked the space where Natividad would tend the diverse maize types amid a dense packing of similar fields that descended to the Río Paucartambo. She planned to sow quinoa into the gaps inevitably left by ungerminated maize in the Rock Place field, and she was also thinking of putting in crookneck squash (*lakawiti*) and common beans (*purutu*), ideal ingredients for soups and stews, respectively.

The biological riches peppering Hill and Valley fields clashed with the sparse diversity cared for in their other farm spaces, the yunlla Oxen Area and the maway Early Planting. Líbano, Natividad, and their industrious neighbors enlarged the conspicuous lacunae of diversity with new fervor in the economically eventful decades after 1969. The sprawling Oxen Area, named for their yoked beasts whose brute force of traction pulled a single-share plow, was marked by its distinct style of tillage and its crops. Uniform, high-yielding potatoes and barley cloaked most Oxen Area fields, while wheat and broad beans ranked next. Its outstretched area was interdigitated with both Hill and Valley parcels extending downhill into the inner gorge of the Paucartambo and uphill near to the grassland moor. The Early Planting was the commercial, off-season crop that held only a few genetically identical varieties. Cobbling the farm landscape from their four units—loma (Hill), kheshwar (Valley), yunlla (Oxen Area), maway (Early Planting)—the Quechua in Paucartambo split farm space diametrically in terms of the diverse crops.

Capping the bald biological contrast, the quartet of farm spaces also mediated the course of agricultural change, setting parameters for the decision-making farmers. Land use spaces, in other words, acted reflexively as both artifact and organizer (Pred 1984; Sack 1980, 1986). The dual function of the farm spaces was itself rooted in another two-sided relation also at work, that of crop seed. Germ-bearing propagules were born in farming and, at the same time, they renewed its economic viability. Crafted by nature and cultivators, seed regenerated the diverse plants each year while it also resulted from them. When Faustino and Eufemia, reviewing their farm endeavors after the reform of 1969, surmised that it was the lack of seed that led them to forego their Red Mother (puka mama) and other floury potatoes, the couple relied on a standard style of reasoning: Quechua farmers often figured seed to be a prime mover of cause-and-effect.

Those Quechua in Paucartambo growing the diverse crops trusted in a suite of innovative farm tasks, some exceptionally skillful, in order to care for their seed. Tasks in seed management were like scripts or "routinized behaviors" (Alcorn 1984; Guillet 1992, 79). The seeds of landrace crops, like the proverbial pauper princes, were crowned with extraordinary treatment in a minor but defining moment. In that brief interval the Quechua farmers removed seed, or *muhu,* which they then titled it, apart from their remaining stores. That brief albeit exceptional treatment called on special knowledge and innovative techniques that enabled them to cope with the economic transitions and ecological changes between 1969 and 1991. While the farmers construed some seed-related knowledge and techniques as direct guides for the diversity of their crops, many diversity-defining skills were to meet the more general aim of ensuring the viability of propagules.

The overarching aim of assuring and enhancing seed viability ranked first in the minds of cultivators and it was, of course, a key to the evolution of vigorous diversity. Female farmers among the Quechua in Paucartambo held the greatest share of knowledge about seed quality, a gendered division of technical expertise documented in their routines of selecting and storing. In choosing seed the women farmers surveyed a deceptively simple assortment of visual clues that flagged the viability and type of propagules. Whether a potential seed would yield well—its viability—framed the basic question behind their scrutiny of visual clues. With deceptive ease, cultivators would inspect the recessed eyes of their potato and ulluco tubers where the plant-producing buds emerge and where the deformities of disease would appear. Similar concerns about viability in the grain-yielding maize turned their focus to the health and intactness of kernel tips that lodge the life-giving endosperm.

After the Land Reform of 1969, the women farmers altered a number of procedures dealing with the cumbersome selection and storage of seed. Farmers drew from their technical know-how—including their specialized insights into seed viability—in order to streamline the allotment of labor-time in these tasks. They managed to cull seed from the bulky harvest through a crucial one-step task, both diagnosing seed health and disease and estimating seed size. By selecting their future seed on the basis of health and size, the busy cultivators put into practice a series of well-defined and widely shared mental categories that also saved labor-time in other stages of their seed work. Such small modifications, quietly incorporated into their work routines, wisely helped to make the production of diverse crops more efficient, and thus more rewarding and viable in postreform farming.

One prime example of time-saving efficiency pioneered by the Quechua women in Paucartambo was the sequestering of size-based categories—medium-size, or second class tubers, and small to medium, or third class tubers—in order to select potato and ulluco seed. In sequestering the pair of size-based classes the cultivators noted the health and disease status of prospective seed, ridding those tubers with deformed eyes. Tubers set aside on the grounds of size or insalubrity, in potatoes known as spent potatoes or *gasto papa,* were used differently but not discarded. Extremely small, bruised, or diseased tubers were fed to farm animals or, alternatively, they were freeze-dried for the sake of future consumption. Extra-large tubers could be saved for impending use as special gifts or *wañlla.* The less-hale and hearty but nonetheless sound tubers of the second class and third class categories were segregated from seed to serve in subsistence and commerce. By shrewdly heeding the twin criteria of size and morphology in a single step, the farmers could quickly stockpile the seed that was biologically viable and economically apt.

Analogously, the Quechua farmer chose her maize seed from intermediate-size, healthy ears. While she handpicked whole ears, the farmer also surveyed quickly the quality of kernels. Selecting seed of quinoa expended still less labor because the farmers merely culled large numbers of its minute offspring from one or more sound-looking plants. With rapid culling, the selection and storage of seeds for the diverse crops encumbered no more than a few additional days of labor-time each year. The size categories and quality criteria designated by the Quechua in Paucartambo to route the seed of all their crops—minus quinoa—illustrated how the innovation of mental categories in farm work could impose a key filtering effect on diversity. Only those tubers and grains that passed through these comprehensive screens were replanted. The cultural sieves sifting seed on the basis of size and health did not delimit the exact sum of landraces but they did exact the first and most crucial cut in the material—diverse and otherwise—that subsequently was sown.

Innovating a still more basic set of categories known as landrace groups also streamlined seed work and filtered diversity. The Quechua in Paucartambo collapsed the vast numbers of landraces into a small handful of groups throughout their sequence of seed selection, storage, and planting (table 6). They carried out this simple grouping of landraces prior to even the joint size-based categorization and health-focused check discussed above. So defining was the role of landrace groups that each main category occupied its own separate class of fields and distinct storage niche. Scores of the diverse potato landraces, for example, were aggregated throughout selection, storage, and planting into a mere triad of core categories: the hak'u papa floury potatoes, precocious chawcha potatoes, and bitter *ruk'i* potatoes for freeze-drying. In similar style the cultivators clustered their maize landraces into three supracategories on the basis of ripening period and culinary use; they were *huch'uy muhu* (small seed), *chawpi muhu* (medium seed), and *hatun muhu* (big seed). By contrast, each of the minor crops such as ulluco and quinoa comprised only a single group of

Table 6. Landrace Groups of the Diverse Crops

Landrace Group	*Main Range of Elevation*	*Areal Distribution in Region*	*Production Space*
Floury potatoes	12,100–13,300 feet (3,700–4,050 meters)	endemic (moderate)	Hill
Bitter potatoes	12,800–13,450 feet (3,900–4,100 meters)	widespread (moderate)	Hill
Early potatoes	9,500–10,850 feet (2,900–3,300 meters)	endemic (strong)	Oxen Area
Big Seed maize	8,850–9,500 feet (2,700–2,900 meters)	endemic (strong)	Valley
Medium Seed maize	9,200–10,850 feet (2,800–3,300 meters)	widespread (strong)	Valley
Small Seed maize	9,850–11,650 feet (3,000–3,550 meters)	widespread (strong)	Valley
Quinoa	9,200–12,950 feet (2,800–3,950 meters)	widespread (strong)	Valley
Ulluco	11,500–12,800 feet (3,500–3,900 meters)	widespread (weak)	Hill

landraces. In their lumping of landraces into a few large groups, the Paucartambo growers saved on the separate handling of individual types and thereby economized their labor-time.

One consequence of grouping landraces into a few large categories was to cluster the local ranges of each group within the alpine landscape of the Paucartambo Andes (table 6). Potato landraces of the floury potatoes, the precocious chawcha, and the bitter ruk'i, for instance, each covered a distinct spectrum of environments within Hill and Oxen Area farming: the floury potatoes at 12,100–13,300 feet (3,700–4,050 meters), freeze-drying ruk'i at 12,800–13,450 feet (3,900–4,100 meters), and precocious chawcha at 9,500–10,850 feet (2,900–3,300 meters). Floury potato landraces, by far the most numerous of the three classes, clustered together in the lower section of Hill farming. Ensembles of the landraces shunted into the other main landrace groups similarly spread over shared, albeit substantial, segments of farm space. This innovative grouping of landraces not only framed the basic outlines of local biogeographic properties but in effect cast them as groups into one of a small set of agroecological brackets.[3]

To grow the future seed of their diverse crops, the Quechua families in Paucartambo planted their standard field sites, with the exception of the potato crop discussed below. A farmer culled most of her future seed for a certain landrace group from whatever fields of this sort that the family had sown. Due to crop rotation and field fallow, the cultivator's chosen seeds were certain to have been grown at earlier times in field sites unlike the present source. Over time, this breadth of seed-yielding fields tended to match the agroecological brackets of the landrace group. A broad swath of field environments, rather than narrowly defined or particular ones, thus equipped her family with its crop seed. Damaging frost, drought, and disease did nonetheless sometimes destroy harvests, thus shifting seed-provisioning onto better sites. On these occasions, the farmers sometimes gathered seed from surviving plants, thereby peppering their seed supply with hardy landraces. As a result they noted, for instance, that the floury potatoes known as Red Mother, or puka mama, and Broken Knife, or *kuchillu p'aki,* withstood frost better than most other landraces.[4]

Storage tasks supported the efforts of selection work, ensuring the viability of landrace seed. In this arena too the innovation of special knowledge belied a reliance on commonplace inputs. The woven upright cylinder called a *taqe,* piles tapped with straw, and elevated troughs of wooden slats all permitted the Quechua in Paucartambo to slow the desiccation of tubers and block bacteria-induced rot and the penetration of direct sunlight (appendix D.1). Their actual choice of caching technique for the tuber crops hinged mostly on climate; year-round humidity of the northern Paucartambo Valley, for example, recommended the elevated, well-ventilated trough, or *marka.* Farmers also protected their tuber harvests perforce from an army of pests including the boring Andean Weevil, or *Premnotrypes sp.* They defended their deposits with aromatic sprigs of a pest-repelling mint dubbed *muña* (*Minthostachyus* aff. *tomentosa*). By the 1980s, they also shielded them with broad-spectrum insecticides, although mainly these were applied to the more susceptible, market-bound seed of high-yielding potatoes rather than the multifarious floury potatoes. Fieldside storage pits for potatoes, once common, were no longer used.

Few insecticides were applied to maize and quinoa in keeping with the hard seed coats granted by nature and the lesser degree of commerce and the lack of high-yielding varieties that were governed by farmers. Growers of the grain crops did not waver, however, in devoting much care to the imperatives of storage and especially the regulation of humidity through which rot and a host of potential pests were restricted. Maize growers of semiarid climates in the southern valley and the Quencomayo watershed managed to regulate the humidity of their hard-earned caches via the customary use of taqe cylinders—in effect bottomless baskets—woven of local grass, similar to the ones guarding potato and ulluco tubers (appendix D.1). In the humid northern valley, by contrast, the Quechua preferred to cure their stored maize ears with ceiling-level marka platforms of wooden slats. Quinoa harvests kept far more effortlessly than maize or the other crops, a field's worth of its tiny spherical seeds could be packed handily into a large tin or small sack.

Due both to the agroecology of diverse crops and the region's peasant economy, seeds were not generated purely through self-provisioning. The twin facts assured a number of dispersal flows that bounded the most basic patterns of diverse crop biogeography. In the case of seeding floury potatoes the farmers periodically procured their seed tubers, rather than producing them, because of the crop's chronic vulnerability to soil pathogens such as nematodes and viruses that infested even those fields with the healthiest soils (Brush et al. 1981; P. Harris 1978; Horton 1987). Although a farmer's potato fields located at high elevations spurned the heaviest pathogen loads and thus allowed her to "freshen" seed periodically, even the colder climates were not free of infestations. As a result, the cultivator sometimes sought the seed of floury potatoes from sources other than her family's own plots or those of nearby neighbors. In the case of maize landraces episodes of drought, hail, and frost in Valley cropping likewise led the farmers on occasion to obtain seed from outside sources. In the period from 1969 to 1991 bartered exchanges and farmer-to-farmer sales sufficed in supplying the sought-after seed of both potato and maize landraces.

Both regular replanting and intermittent exchange helped to reinforce the coherent character of the four farm spaces in Paucartambo. Each year the farmers in effect reestablished the environmental range of biophysical limits of these spaces, when they harvested their seed from one field in order to replant it the following year, often times at a quite different elevation yet still within the same spatial unit. This cohering effect of seed self-provisioning on farm space was reinforced further on the occasions when families procured their seed from other sources. Landraces of the floury potatoes, for instance, tended to be traded from one Hill field for use in another. Such patterns of seed exchange—as well as the rotation of planting sites by individual farmers—resulted in the potatoes once sown in the more temperate climate of lower Hill areas near 12,500 feet (3,800 meters) sometimes being planted during the following season in a cold climate at 13,300 feet (4,050 meters), or vice versa. Periodic exchanges of seed in the other diverse crops also spread them across the range of either Hill or Valley farming and this reinforced the area of each space.

Seed exchange thus sculpted patterns of areal distribution that united the same farm spaces of different places. In the case of floury potatoes most farmers in Paucartambo sought new, disease-free seed from the potato specialists dwelling in nearby uplands, thus establishing a custom of seed exchange that strongly shaped the outcome of spatial patterning (Zimmerer 1991a). Planters typically acquired seed from their contacts among upland cultivators at intervals of five to ten years. Often familiar as family relatives, friends, and social kin, many upland farmers tilled five or more parcels of floury potatoes each year. Their characteristic exchange of floury potato seeds acted to interlace the well-defined biogeographic units known as "cultivar regions" or "landrace regions" around high-elevation cores (map 6).

In the mid-1980s at least four cultivar regions could be charted in this crop:

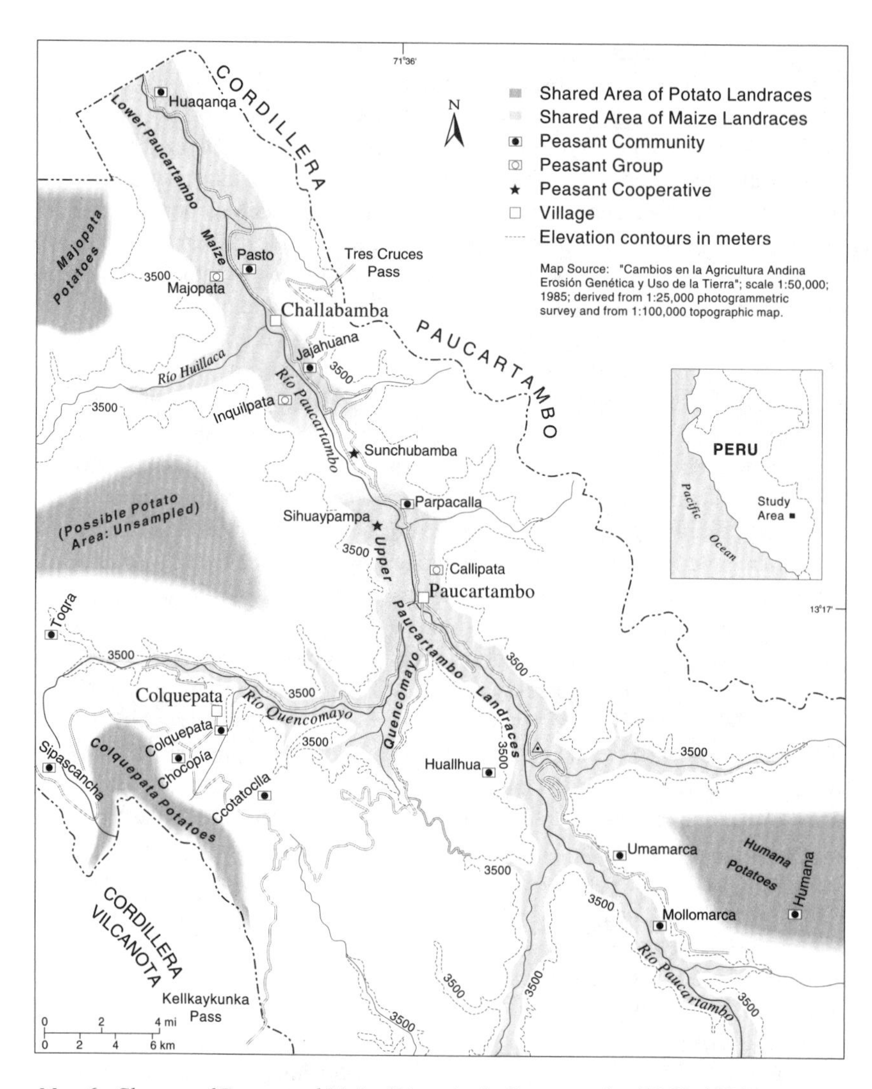

Map 6. Clusters of Potato and Maize Diversity in Paucartambo (1969–1990).

Majopata, Colquepata, Humana, and the unsampled upland of Sihuaypampa. Many potato landraces united within the reticulated network of a cultivar region were moderately endemic in biogeographic terms, that is confined to this intermediate-size area. The types known as Andean Gull, or *leqechu,* and the untranslatable *charqanwalla* and *kharwis* were restricted to the Majopata cultivar region. Llama Nose, or *llama senqa,* and *trumbus* characterized Colquepata. The cultivar region of Humana-Mollomarca uniquely fielded the landraces of Louse, or *ch'iya k'utu*; Jaguar's Paw, or *puma maki*; and wakan kayllu. Roughly one quarter of the floury potatoes, such as pitikiña and alqay warmi, showed broader distributions that reached across the borders of Paucartambo into the neighboring regions of Calca (for example, Patallaqta), Urubamba (for example, Cuyo Grande), and Quispicanchis (for example, Ccapana).

Maize seed also moved through a distinct circuitry of exchange and into resulting clusters of valley farming. Quechua farmers in the upper Paucartambo and Quencomayo Valleys planted the two groups of landraces known as small seed and medium seed. While both plantings suited a short season and farmland marginal for maize-growing, they were nonetheless damaged and even destroyed entirely at times by drought, frost, and hail. Full-blown crop failure on occasion caused the cultivators to secure new seed supplies from maize growers in the subtropical-tending climate between Callipata and Parpacalla in the central Paucartambo Valley near the capital. A distended cultivar region thus fingered along the two river valleys and joined at the confluence (map 6). Only the exceptionally hardy and early *huch'uy kullu,* a landrace whose name resembled Small Moon, or *huch'uy killu,* was pocketed as an endemic in the upper valley, being confined to the southern valley at Huaynapata and Payajana. Exchange circuits likewise created a second cultivar region of maize downstream in the northern Paucartambo Valley.

Naming of maize and potato landraces signposted the geography of this seed exchange, showing how deeply it contoured biogeographical distributions. The distributions of names in a few far-flung or cosmopolitan types coincided closely with the boundaries of cultivar regions. One ubiquitous maize landrace, for instance, was labeled Yellow Maize, or k'ellu sara, in the southern and central Paucartambo Valley and in the Quencomayo Valley, the pair of fingerlike areas joined by seed exchange. Growers of the northern valley, by contrast, knew the same landrace by the name Eight Row or *pusaq wachu.* In similar fashion various widespread landraces of floury potatoes carried a certain tag within one cultivar region and another one where seed exchange circumscribed a separate area; for example, the untranslatable *maktillu* in the Mollomarca cultivar region was known by the name Narrow One, or *suyt'u,* in the vicinity of Colquepata.

Standard means of producing the landrace seeds stood out vis-à-vis the new post-1969 economies of high-yielding potatoes and contracted barley. Rather than supply seeds from their own farms and through local trade, the Quechua farmers stoked the new commerce with seeds bought in regional markets—often Cuzco—and in agribusiness contracts with the Beer Company of Southern Peru. The new seed sources complemented changing production processes. Quechua farmers of improved varieties like the mariva cultivar sowed them with a potent arsenal of fertilizers, pesticides, and insecticides in order to maximize yields. Because mariva and the other high-yielding potatoes did not bear adequately in the chilly climate at higher elevations, the farmers entertained no choice but to buy from regional markets or to barter from extra-local sources.

In the case of griñon barley for beer-making the contracts of the brewery stipulated the use of its own certified seed in order to raise productivity and enhance profit.

Adoption of acquired high-yielding potatoes and contract-grown barley propelled a strong wave of biological uniformity across the Paucartambo Andes during the post-1969 decades. Out of necessity the region's Quechua cultivators adjusted the mechanics of their seed economy. Farmers came to rely less on that seed obtained from their own yields and those acquired from their neighbors. Their adjustments in seeding customs inferred a retuning of their experience of farm nature. In everyday language the common verb form *muhusqa,* meaning "to sow for seed," lost ground to the noun *seed,* or *muhu.* On the landscape, the farmers' eager adoption of the new mariva potatoes and griñon barley translated into the blanketing of single varieties over unprecedented portions of the region. In their own ways language and landscape each expressed the potency of the new seed economies.

Improved seeds in the postreform economy did not, however, simply supplant the seed economy of the diverse crops. Indeed, a number of Quechua farmers in Paucartambo adjusted the tradition of cultivating landraces in order to pursue more marketing. This innovation called for reapportioning the fractions of yield among a triad of purposes: seed, subsistence, and commerce. Innovation in the apportioning of yield aided the farmers of diverse crops by adding flexibility according to a family's needs. They could couple goals, in other words, rather being locked into a single strategy. Inclusion of commerce, it must be emphasized, did not necessarily antagonize the richly diverse landraces. In fact, the farming of diverse crops during the mid-1980s coupled commerce with seed and subsistence growing in no less than fifteen percent of floury potato fields and as many as fifty percent of the diversity-rich maize parcels (appendix D.2). Although the farmers reserved the term *money potatoes,* or *qolqe papa,* for the new improved varieties, this catchy moniker masked a more complex reality.

At times, however, the new flexibility given to management of floury potato yields did, in fact, proceed at the expense of diversity. Many Quechua cultivators seeking to stiffen commerce chose to market a major fraction of yield from the diverse plants. Although little advertised, the price of diverse floury potatoes regularly topped that of improved varieties. Floury potatoes in the Paucartambo and Cuzco markets sold for twenty-five percent more than the scientifically bred high-yielders in the mid-1980s. A higher selling price apparently prevailed at various times in other regions of the Peruvian highlands as well (Brush et al. 1981; Brush and Taylor 1992; Horton 1987; Rhoades 1982). Biodiversity did not prosper, however, in this state of relative advantage. Farmers who most pursued the potentially rewarding route of commerce tended to metamorphose their style of seeding.

Rather than plant diverse mixtures of multiple landraces, most entrepreneurs of floury potatoes in Paucartambo adopted a single cultivar per field. When fortifying commerce they in effect crossed over a threshold simplifying the degree of biological diversity. The phenomenon of reduced diversity—that is, a single landrace—occurred mostly in those fields where substantial yield was being caught up in commerce (appendix D.2). This predilection of commercializing farmers toward single-landrace simplification could be found in the diverse maize and ulluco crops as well. Their rationale for the single-type plots was to lower the costs of labor-time and seed capital otherwise incurred in the growing of landrace mixtures. By sowing one of the floury potato landraces, such as qompis, pitikiña, or olones, each known to yield well and reliably, the single-landrace farmers could boost their returns from the market.

The diversity-reducing threshold did not imply that their marketing and commercialization of landraces was intrinsically inimical to biological variety. The dominance and particular character of market rationales, rather than mere existence of them, led the single-landrace farmers to adopt a uniform biological base. Numerous other farmers meanwhile succeeded in initiating and expanding the commerce and exchange of their diverse crops without resorting to this single-landrace logic. By apportioning harvests flexibly to the goals of seed, subsistence, and commerce, they could more ably respond to the fluctuating demands of markets and to their changing food supply and resource availability. Such flexibility was paramount for general economic success during the years from 1969 to 1990, since the expansion and diversification of commerce often caused farmers to alter livelihood strategies from year to year.

The trade-offs among seed, subsistence, and commerce fostered dissent within households when members disagreed about the use of harvest. Intrahousehold dissent mostly plagued those households short of resources. In resource-poor families the female heads-of-household, who typically seeded and cooked the diverse crops, tended to be notoriously vigilant in alloting yield for seed and subsistence. By contrast, their husbands who labored more in the farming primed for commerce were prone to emphasize marketing. The divergent aims of heads-of-households sometimes collided, although as of the late 1980s most families could still agree on suitable splits of their harvest. Resolution of difficulties was eased since the women who labored in landrace work did not shun commerce and its opportunities. Eufemia once remarked that she, rather than Faustino, had directed them to forsake Sunflower, Pig Manure, Fist, and their other floury potatoes that were becoming incompatible with commerce.

Farming the Landscape

A quartet of farm spaces in the Paucartambo Andes—the Hill, Valley, Oxen Area, and Early Planting—was the creation of the Quechua peasants' laboring.[5] Like the seeds, these production systems and territories were both a result and a conditioner of their farm livelihoods. An anecdote on landrace growing by Don Líbano suggested how the geographical dynamics of farm territory wrought concrete influences on his family's handling of their diverse crops. While Líbano was trekking toward the Big Hill field soon to be sown with the twin sacks of seed, he remarked that the plot would ripen in conjunction with a few dozen other parcels: "It's advantageous that our floury potatoes are close to one another in the Hill. Otherwise whole fields could be eaten by the herds of cattle and sheep that graze up here."[6] In other words, his and Natividad's resolve to seed a wealth of potato landraces in their distant field may have been weakened or even upended if not for the community-scale cohesion of farm space.

Each of the four farm spaces—Hill, Valley, Oxen Area, and Early Planting—was a territory that together formed the agricultural landscape of Paucartambo. Except for ecologically specialized communities hemmed into the high-elevation upland, the full quartet occurred in communities throughout the region.[7] Many families in generalized communities that were environmentally diverse like Umamarca prided over a portfolio of plots representing each of the farm territories. Don Líbano and Doña Natividad, for instance, supplemented their landrace-rich Hill and Valley parcels with plentiful Oxen Area and Early Planting fields. Less wealthy neighbors, it should be noted, often lacked fields in one or more of the four units. Faustino and Eufemia, for instance, did not own a Valley field apt for maize growing. Boundaries of the peasant community thus set limits on whether the adoption of a farm territory—like the Valley—was possible for its members, although individual families differed widely in their own access to each space.

Characteristic systems of production reliably distinguished each of the farm spaces cultivated during the postreform period (table 7).[8] By managing the fields within each of the four units differently, the Quechua in Paucartambo defined a series of production-related distinctions. Via this battery of culturally drawn distinctions, they steered a differential supply of resources to the diverse crops and to their other farming. One resource obviously available in unequal amounts among the foursome of farm territories was that of crop type. Cultivators created the convention of sowing the floury potatoes and ulluco in Hill farming, while they reserved maize and quinoa for the Valley. Other resource inputs—labor-time, fertilization, plow technology—followed suit in being handled distinctly in each of the four pieces of the farm landscape.

The quality and quantity of farmland also authored contrasts among the four production spaces. Nature's signatures such as climate and soils differed among them. The elevation range of field locations within each production space could be used to estimate its typical spectrum of biological and physical conditions. The greatest contrast set the comfortable subtropical Valley between roughly 9,200 feet (2,800 meters) and 11,500 feet (3,500 meters), well

Table 7. The Agricultural Production Spaces of Paucartambo

Production Space	*Disposition of Yield*	*Main Range*	*Spatial Cohesiveness*
Valley	consumption-seed-commerce	9,200–11,500 feet (2,800–3,500 meters)	low
Oxen Area	commerce-seed-consumption	9,200–12,800 feet (2,800–3,900 meters)	moderate
Early Planting	commerce	9,200–12,100 feet (2,800–3,700 meters)	low–moderate
Hill	consumption-seed-commerce	12,100–13,450 feet (3,700–4,100 meters)	high

apart from the cold climate of Hill farming above 12,100 feet or so (3,700 meters). Based on their idea of each space, the Quechua farmers prescribed a particular medley of cultivation techniques for its production, such as plowing, crop rotation, and fallow. They designed many of the farm techniques specifically for managing soil fertility. In addition the farmers marked their production spaces by allocating labor-time in distinct quantities and according to disparate schedules attuned to the peculiarities of growing seasons and cropping calendars in each unit.

Contrasting features of Hill, Valley, Oxen Area, and Early Planting territories arose in part from the same processes and rationales that caused the units to coalesce into discernible units. The Paucartambo landscape itself suggested one clue for decoding the dynamics of farm territory. The strikingly small number of units seemed at odds with the extreme variation of environments and the social complexity of the region. This straightforward but still surprising observation revealed that Paucartambo farming and the surrounding landscape were not unique, for the organization of production space in peasant farming in the other Andean regions of Peru has displayed the same incongruent simplicity during recent decades (Basile 1974; Brush 1976, 1977; Mayer 1979; Mayer and Fonseca 1979). Even the yawning Cañete Valley—an environmentally diverse western Andean flank south of Lima that is more than one dozen times larger than the Paucartambo Andes—totaled only ten production spaces during the mid-1970s.

One force behind the surprising fewness of spatial units on Paucartambo farms was the standardization of work routines in the rural economy. Frequent reliance on extrafamilial labor pressured the owners to keep small the number of production systems as a means of regularizing the core tasks. Farmhands among the Quechua peasants typically toiled in groups without a supervisor overseeing the application of specific techniques. Workers were expected instead to perform their duties in keeping with local norms and expectations, without tailoring or customizing them to individual fields or much less to portions within them (Zimmerer 1994b).[9] One evidence of this pressure came from the rigid but widely shared parlance for work roles such as Plowman (*wachuq* or *chakitakllero*), Sod Turner (*rapachoq*), Fertilizer Person (*wanuq*), and Planter (*husk'adero* or *husk'aq*). Since most field owners commonly recruited labor from outside the household, their practical concerns about the transfer of techniques limited the array of production spaces that could be created.[10]

Another force consolidating land use was the weight of two perceived contrasts. Cultivators beheld a fundamental cleft between Valley maize farming and Hill potato agriculture, a dualistic concept of farm space shared by many Andean people (O. Harris 1985; Platt 1986; Salomon 1985). Numerous contrasts ranging from climate to cropping techniques coincided with the symbolic distinction between Valley and Hill spaces. In their minds' eye the Paucartambo farmers also severed the units designated primarily as subsistence providers—the Hill and Valley—and the ones destined principally for commerce—the Oxen Area and Early Planting (table 7). Although their mental map charting an antipodal alignment between subsistence and commerce was, in fact, frequently a fiction due to the mixed disposition of harvests, it was a salient feature perceived by the Quechua farmers in their everyday delineation of the landscape.

Impetus to cleave a mental division between the production spaces of subsistence-destined and market-bound cropping was reinforced by the powerful momentum of commerce in the period from 1969 to 1990.[11] While the Quechua in Paucartambo channeled increasing flows of resources into the widening commerce of the Oxen Area and Early Plantings, it was the symbolic significance of the Hill and Valley that inspired their more abstract images of the landscape. When discussing their impressions of the farm landscape, the Quechua farmers sometimes referred to the Valley as a domesticated space and the Hill as a space in need of domestication, both references in keeping with the role of subsistence-provisioning (see also Allen 1988; Isbell 1978; O. Harris 1985).[12] Conversely, the farmers held much vaguer ideas of the symbols evoked by the Oxen Area or Early Planting. Indeed, the crispest image of the Early Planting was that of its high-yielding money potatoes, or qolqe papa, which, although not explicitly a spatial metaphor, was revealing of a new image of farm space.

A purposeful orderliness thus pervaded the farming of Paucartambo, at least in the minds of its cultivators. The Quechua farmers perceived at least some of the landscape's order in terms of an axis between "upper" and "lower" halves of land use, settlement, local kin groups, and trading partners (O. Harris 1985; Isbell 1978; Platt 1982; Salomon 1985, 1986a; Urton 1984; Weismantel 1988). Their concerns about territorial rights also led them to emphasize order. By casting land use into widely known categories, the Quechua cultivators could better define and thus defend their rightful claims to an accustomed array of farmland. When the cultivators of a community countered threats to their territory, a frequent need, they could claim their rights to Valley or Hill fields as being crucial for subsistence and thus for survival. Peasant farmers of the Chocopía community near Colquepata twice demonstrated this nearly two centuries apart, each time making claims that referred explicitly to the names of territory, for example, *mañais de laymi* (the sectoral fallow commons of Hill farming) (LACH 1658, 1950).

Messier than the neat categories of the symbolic mindscape, farm spaces on the ground were, nonetheless, somewhat ordered. A degree of spatial cohesion owed to unequal environmental tolerances of the crops: maize produced best in the lowest range of farm habitats, potatoes in higher environments, ulluco and quinoa in still others. Each crop yielded most bountifully in an optimal setting. Agronomic adaptations alone did not, however, instill the sole reason for the cohesive character of production spaces. Indeed, if that were so, the distributions of so many crops and production spaces would not overlap as they did.

A key addition of cohesion came from the strong incentive to coordinate cropping and livestock-raising in the peasant economy. By sowing the same crop, owners of adjacent fields lessened the risk of crop damage that otherwise might be caused by straying livestock, especially those grazing nearby stubble and crop residue. Synchronized cropping made sense especially in those areas where there was sizeable overlap in the elevation limits of Valley, Oxen Area, and Early Planting territories. Interspersion of production spaces did in fact raise the specter of worsened crop damage abetted by the riskily incongruent cultivation calendars of adjacent parcels. Farmers responded by informally coordinating their choice of crops. On the lower slopes of the Majopata community, for example, the interspersion of fields belonging to the Valley (maize, broad beans, quinoa, tarwi), Oxen Area (barley, wheat), and Early Planting (potatoes) led farmers to create informal clusters of the parcels of each production unit (map 7).

This coordination of land use with their neighbors was prosaic but geographically potent. The force itself was not unfamiliar, of course, since the imperative to wed farming with livestock-raising had long distinguished Andean peoples' agropastoralism with its host of social regulations and special techniques (Brush and Guillet 1985; Guillet 1981b, 1992). Peasant Communities in Paucartambo continued after 1969 to renew one key tradition by fining herd owners for crop damage incurred by their livestock. The joint pursuit of farming and livestock-raising was facilitated also by experienced peasant herders who skillfully minded the movements of their animals by calling, whistling, and slinging rocks. Tethering cattle to a stake, or pinning, was designed to keep stubble feeders from stepping outside field boundaries. The existing techniques

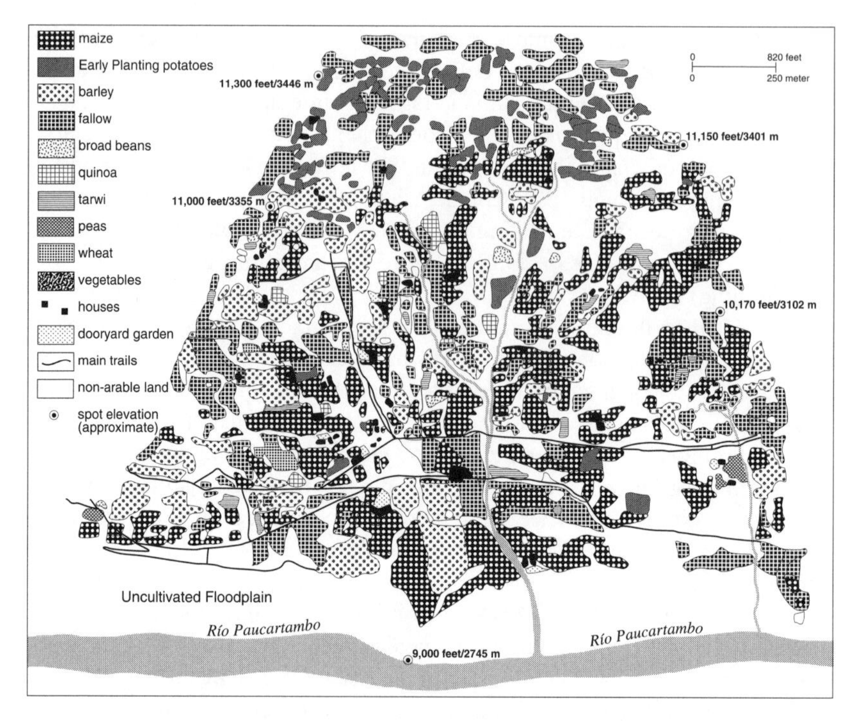

Map 7. Lower Slope Fields of Majopata Community (Paucartambo) in the Mid-1980s.

did not, however, obviate the acute risk of crop loss during the years from 1969 to 1990. Doubling of livestock numbers in most Paucartambo communities as well as the widespread adoption of off-season cropping provided plenty of added incentive to synchronize field calendars where possible.

Irrigation also crafted a cohesiveness of farming units in Paucartambo, but only slightly. Adequate precipitation for farming, common in the eastern Andes, kept the incentive for irrigation weak. Paucartambo's precipitous topography and weak sedimentary surface rocks of shale and sandstone also discouraged the building of canals. Not surprisingly, the earthen ditches dug by the Quechua farmers in the region paled in comparison to the elaborate stone structures and intricate social regulation and cultural ritual typical of irrigation in the more arid western and central Andes (Denevan 1986; Donkin 1979; Guillet 1992; Knapp 1991; Mitchell 1991; Winterhalder 1994). The rustic water control works in Paucartambo were, nonetheless, essential in supplying much-needed soil moisture early in the growing season for fields of the Valley, Oxen Area, and Early Planting. The simple irrigation systems also motivated farmers to adjust their hand-digging, grading, and use of ditches in order to coincide with the efforts of neighboring cultivators and thus to cut back on labor costs.

Hill farming created the most cohesive farm space. Large blocks of dozens of Hill fields were either cultivated—typically with the same crop—or fallowed together and grazed as a common property that was accessible to all community herders. This cohesiveness derived from the system of sectoral fallow. More than one half of the region's Peasant Communities—especially those in the southern and middle Paucartambo Valley—implemented this form of land use during the postreform decades. The complete number of sectors in a cross-section of Paucartambo communities varied as follows: Carpapampa (six), Espingone (six), Huallhua (six), Huaynapata (seven), Humana (five), Inca Paucar (seven), Majopata (six), Mashuay (eight), Mollomarca (six), Payajana (six), Parpacalla (five), and Umamarca (seven) (Zimmerer 1991e).

The Quechua in Paucartambo rotated cropping and livestock-raising among the sectors, each known among the farmers as a *suerte.* While a suerte was being cultivated, individual families planted their own parcels in that sector, although the entire community typically agreed to sow the same crop throughout its area. While grazed, the sector became converted to a rangeland commons where any community member could graze his or her livestock. In terms of economic function this coordinated land use lightened the labor devoted to Hill farming since the distant fields could otherwise be protected from the possible damage of livestock herds only with a sizeable effort. Indeed, as labor-time scarcity exerted a new pressure on the Quechua in Paucartambo, as well as on their counterparts in neighboring regions, the economic benefits of labor-efficient sectoral fallow remained notably strong (Kervyn 1989; Orlove and Godoy 1986; Zimmerer 1991e).

An irregular mosaic of land use, exemplified by the lower slopes of Majopata community (map 7), predominated over the majority of Paucartambo since the sectoral fallow commons were restricted to Hill farming at upper elevations. Evidence of the distinctive mosaic could be grasped in marked overlapping of the elevation limits that were notched by each production space recorded in a detailed survey of thirteen Peasant Communities widely scattered throughout the region (appendix D.3). That highly visible overlapping and its irregular extent indicated an incomplete cohesion of production spaces. It could be seen especially on the valley slopes below 11,500 feet (3,500 meters) like that of Majopata, where all farm spaces save the high-elevation Hill sunk to riverside floodplains nearing 9,200 feet (2,800 meters). Pictured at an intermediate scale, much of the cultivated landscape of the Paucartambo Andes was thus sewn into a seeming "crazy quilt" of fields that belonged to a variety of production spaces.

This uneven mosaic meant that the environmental properties of planting sites played an enabling rather than a controlling role. The adequacy of yield within limits, in other words, mattered to the Paucartambo farmers more than its optimality. In some production spaces adequate yields could be obtained across a wide array of elevation-related environments. The Oxen Area, for instance, traversed the thorn and shrub savannas and reached even into the lower fringe of the grassland moor (table 7). A similar range, albeit slightly smaller, was covered by the Early Planting, which carpeted a variety of perhumid patches permitting off-season cropping. In essence, the Quechua in Paucartambo contoured the geography of their production spaces according to conditions that were of both nature's making and human modification. As a consequence, the farmers created a land use mosaic that was stitched irregularly across the landscape.

The distinctive, irregular mosaic of farm spaces in Paucartambo disagreed conspicuously with the model of precise zones forming stacks of nonoverlapping, elevation-conforming tiers (Brush 1976; 1977; Dollfus 1981, 1991; Milstead 1928; Mitchell 1991; Murra 1972; Orlove 1977b; Pulgar Vidal 1946; Troll 1968).[13] Its imbrication of land use into the irregular mosaics that were readily observable disallowed the concept of a precise "zone" that, according to the model, resulted from the optimal adaptation of farm space and cropping system to climate-generated environment.[14] This messy reality of the region's production spaces in the years from 1969 to 1990 inferred that the spatial reordering of land use, both at present and in the past, may be unlike the patterns and processes deduced from the adaptationist model. Some historical evidence suggests that the irregular mosaics of suboptimal agronomic patterns were in fact not uncommon at least as early as five hundred years ago.

Spanish colonial policy had forced Indian peasants in the Peruvian viceroyalty to carve new Oxen Area fields in order to sow wheat and barley for tribute (Crosby 1986; Gade 1975; Larson 1988). Many Cuzco Indians, including the Paucartambo people, also pioneered an Early Planting of alfalfa in irrigated parcels and on river floodplains to meet colonial demands for the thousands of mules, horses, and donkeys common by 1650. Historical records of Paucartambo and the nearby Urubamba-Vilcanota Valley noted that many Indians possessed suites of fields—all in close proximity to one another—belonging to the tribute-yielding Oxen Area, the Early Planting of alfalfa, and the maize-containing Valley (ADC 1784–85b; ADC 1807–8; Gade 1975, 1992). This history of spatial overlapping inscribed a saga of past farm space that was at odds with the just-so picture of precise zones held to be optimally adapted to fine-grained differences of elevation. Farm territories, it is safe to conclude, tended to be neither fine-grained nor optimally adaptive in a strict ecological sense.

Imbricated mosaics of territory implied that agroecological factors have acted as the coarse outlines of distributions rather than as microscale determinants of a fine-grain ordering. General boundaries have permitted the Quechua farmers in Paucartambo to enlarge some production spaces at the expense of others. They encountered few environmental obstacles, for example, in planting the Oxen Area and Early Planting regimes at elevations between 9,200 feet (2,800 meters) and 11,500 feet (3,500 meters) that otherwise sited landrace-rich Valley fields (plate 5). Received wisdom of the adaptationist model of

Plate 5. A wide spectrum of growing environments in the southern Paucartambo Valley ranges from 9,850 feet (3,000 meters) to above 13,100 feet (4,000 meters). Farmers seed diversity in their maize-containing Valley fields on lower slopes and potato and tuber fields at the higher elevations. Their success hinges on piecing them together with brewery-bound barley (Oxen Area) and off-season potatoes (Early Planting) on the slopes below 12,500 feet (3,800 meters).

zonal space, by contrast, ignored the potential of this historical dynamic, assuming that farmers of nonoptimal crops would confront insurmountable obstacles. It was thought, for instance, that the Oxen Area crops such as wheat, barley, and potatoes would *not* be extended into the lower elevations of maize-growing (Brush 1976, 159; 1977, 82). Underlying this adaptationist error was the certitude that farm space could be fathomed simply by probing the abstract "interaction" between environmental properties and crop tolerances (see also Camino et al. 1981).

Recognition of farm spaces as overlapping and nonoptimal leads us to rethink whether good-size variation in resource management may have existed within single farm territories. The zonal model and its assumption of optimal land use had led to a logical emphasis on the homogeneity of each zone. In the Paucartambo Andes, however, major features of resource use such as crop types, field technologies, and growing calendars sometimes varied significantly within each of the different spaces. A few Oxen Area fields planted in the decades after 1969, for instance, were sown with floury potato landraces otherwise typical of Hill farming. Agriculture of the Valley meanwhile displayed even more noteworthy variation in the management of its resources. Such variation within a single unit of farm territory suggested that the Paucartambo cultivators themselves were not subscribing to an adaptationist perspective when they hastened farm change in the sequel to the Land Reform of 1969.

Spaces of Biodiversity: Reinventing Flexibility

Hill Agriculture

Hill agriculture was the fabled farming of the mountainous Andean rooftop. The Quechua in Paucartambo sowed their highest fields with a bounty of hardy tuber-bearing crops—potatoes, ulluco, oca, mashua. Although a cold climate stymied their adoption of improved potatoes and cereals, the planters of Hill plots could be confident in a secure harvest from the landraces of these potatoes and minor crops. Floury potatoes towered as preeminent among the diverse crops of Hill agriculture. Even partial sampling of only thirty parcels—two hundred plants in each field—revealed a Protean diversity of seventy-nine distinct landraces among the floury potatoes (table 8; plate 6). A still smaller sample of the ulluco crop turned up several landraces. Notwithstanding the large numbers of landraces, each type was named and known individually by farmers.

The growing environments of Hill farming cloaked the grassland-moor that was popularly referred to as the loma. This sobriquet served in Paucartambo for the more common label of *puna* heard in most of the southern Peruvian sierra and northern Bolivia (Gade 1975; Montes de Oca 1989; Pulgar Vidal 1946).[15] Hill fields sloped gently amid an open undulating terrain, which was rounded by alpine glaciers of the late Pleistocene about ten thousand years ago. Scattered boulders, poorly drained bogs, and U-shaped valleys bore witness to massive landform-sculpting by the mountain ice sheets. Vegetation of the Hill lands consisted of vigorous bunch grasses such as the well-known *ichu,* or *Stipa ichu,* and at least one dozen other species of the common grass genera *Stipa, Bromus, Festuca,* and *Calamagrostis* (Zimmerer 1989; appendix B.2). Dozens of herbaceous plants, mostly ground-hugging rosettes that

Table 8. The Potato and Ulluco Landraces of Paucartambo[1]

A. Potato Landraces			
Common Name	*Synonym*	*Common Name*	*Synonym*
1. *Alqay warmi*	*alqo qompis*	41. *Pitikiña*	
2. *Ambrosia*		42. *Puka mama*	
3. *Boli*	*p'alta*	43. *Pullwan*	
4. *Chakillu*	*waña*	44. *Puma maki*	*saqma*
5. *Chapiña*	*markillu*	45. *Pusi wamanero*	
6. *Charqanwalla*		46. *Qeqorani*	
7. *Charqo*		47. *Qolla*	
8. *Cheqefuru*		48. *Qompis*	
9. *Chimaku*	*misti pichilu*	49. *Qowisuyu*	*haka wayaka*
10. *Chinchero*		50. *Rosas pata*	
11. *Ch'ilkas*	*phuña*	51. *Rump'u*	*sump'us*
12. *Ch'iya k'utu*		52. *Runt'us*	
13. *Choqllos*	*qachum waqachi*	53. *Saku*	
14. *Ch'orillu*	*ch'uruspi*	54. *Salamanca*	
15. *Cicera*		55. *Sapanqari*	
16. *Conocito*	*conejito*	56. *Sayllus*	
17. *Cuzqueña*		57. *Soqo*	*soqo lomo*
18. *Hamach'i*		58. *Suli*	
19. *Hampara*		59. *Sunch'u*	
20. *Imilla*		60. *Suwa manchachi*	
21. *Inkamanta waq'aq*		61. *Suyt'u*	*maktillu*
22. *Ishpunqa*		62. *Talako*	*kharwis*
23. *Kallwa*		63. *Tayani*	
24. *Kanchalli*		64. *Th'ili*	
25. *Khuchi aka*	*yana pitikiña*	65. *T'oqhe*	
26. *Kuchillu p'aki*	*puka tarma*	66. *Toqlulu*	
27. *K'ewillu*	*amakaylla*	67. *Trumbus*	
28. *Kusi*	*puqolla*	68. *Tukiwayna*	
29. *Kharwis*		69. *Unchuna*	
30. *Leqechu*	*leqe chaki*	70. *Virintus*	
31. *Llama ñawi*		71. *Wakan chilena*	
32. *Llama senqa*	*kondor senqa*	72. *Wakan kayllu*	*wakan killas*
33. *Lunachi*		73. *Waka sanqho*	
34. *Miskela*	*makathu*	74. *Wakoth'u*	
35. *Olones*		75. *Waman uma*	*wamanero*
36. *Oqa papa*		76. *Wayata*	
37. *Paqo*		77. *Wayru*	
38. *Paqocha senqa*		78. *Yana pari*	
39. *Patallaqta*	*garmendia*	79. *Yuraq*	*mestiza*
40. *Phuqolla*			

Table 8. The Potato and Ulluco Landraces of Paucartambo (continued)

B. Ulluco Landraces			
Common Name	*Synonym*	*Common Name*	*Synonym*
1. *Moro chukcha*		4. *Qompis*	
2. *Moro qompis*	*moro papa lisas*	5. *Wawa yuki*	*chukcha*
3. *Puka qompis*			

1. Potatoes based on a sample of thirty fields and ulluco based on a sample of twenty fields. Fields were sampled in 1986, 1987, and 1990.

seasonally bore bright flowers, were sprinkled amid the tussock-forming grasses.

Hill fields climbed from above 12,100 feet (3,700 meters) to the greatest heights of farming. In the Quencomayo high country of Sipascancha and Toqra that banks against the glaciate Cordillera Vilcanota the Quechua farmers perched fields upward of 13,450 feet (4,100 meters). Eastward in Paucartambo, this upper limit of Hill farming dropped due to greater humidity and lower growing-season temperatures that capped all cropping below 13,300 feet (4,050 meters). Hard frost—dry frost (*ch'aki qasa*) to the Quechua—fell in Hill areas nightly during the rainless "winter" months between June and August. Lighter frost whitened its fields during the transitional months beginning in April and lasting until October. The lowest elevations to hold Hill parcels varied from site to site, since they adjoined upper edges of the Oxen Area and Early Planting, themselves subject to widely varying levels in the landscape.

The Quechua in Paucartambo packed their diversity of Hill crops into a pair of suites. The first ensemble featured the modest set of frost-resistant bitter potatoes belonging to the species *Solanum juzepczukii* and *Solanum curtilobum.* Due to peculiar plant morphology and chemistry, the bitter potatoes could withstand the periodic frosts common in the highest fields. Farmers referred to them as *papa ruk'i,* literally *bitter potatoes,* in reference to the trademark bitter taste of chemical substances known as glycoalkaloids. The cultivators removed the bitter principle from their harvested tubers through repeated day-night cycles of drying and freezing. Simple freeze-drying resulted in chuño tubers, while leaching produced the white-skinned moraya. Total diversity of the bitter potatoes in Paucartambo did not exceed one dozen landraces, a sparseness due ultimately to the rarity of cross-breeding. This reproductive trait severely constrained the inception of new landraces.

Far more fecund diversity was fostered in the second main suite of Hill crops. It was crowded with ulluco (*Ullucus tuberosum*); oca (*Oxalis tuberosum*); mashua, or *añu* (*Tropaeolum tuberosum*); and the unrivaled source of diversity, the floury potatoes. The floury potatoes derived from no less than four

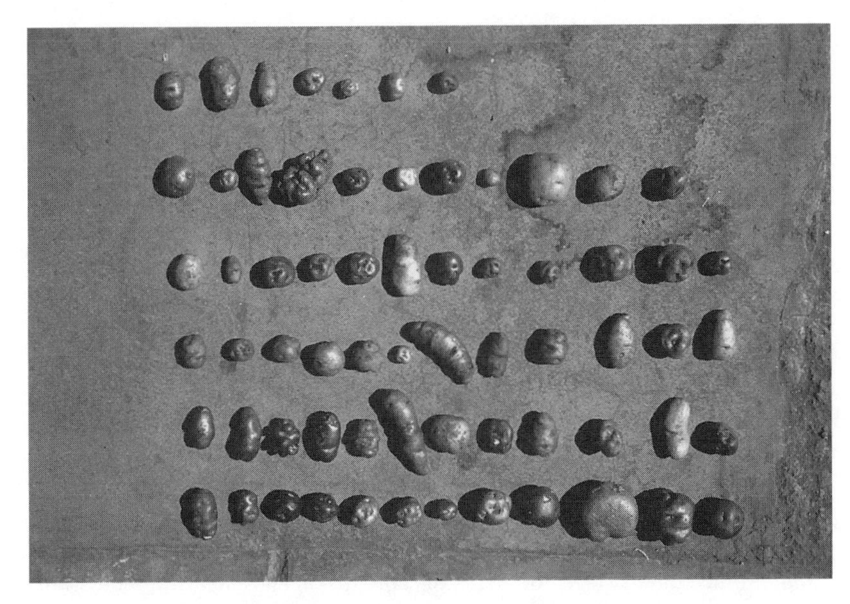

Plate 6. Some of the distinct potato landraces sown in the Paucartambo Andes. The collection contains less than half of the total diversity in the region.

nonbitter species, *S. goniocalyx, S. stenotomum, S.* x *chaucha,* and *S. tuberosum* subsp. *andigena.* Less tolerant of cold, the second suite of crops dressed Hill farming below 13,300 feet (4,050 meters), some fields descending as low as 12,100 feet (3,700 meters). Because the Quechua farmers rotated their floury potatoes among Hill fields, this group of diverse landraces was eventually sited in a variety of elevation-related environments. Their selection over time of the seed from varied sites generated landraces adapted to an ample range of field habitats. Shaping of the general-style adaptation of diversity's main units—rather than being pocketed precisely in narrow niches—posited a degree of flexibility that was often times to prove crucial in de facto scenarios of conservation.

Human facets of Hill farming, such as access to farmland, soil management, and cropping calendars, were well-defined. Hill parcels registered the chief land resource for most Quechua in Paucartambo, including both the better-off and the poor.[16] Some well-off families held as many as two dozen parcels, while even poorer households typically had tenure of at least a few. The Hill parcels at a family's disposal, regardless of the family's socioeconomic status, tended to be scattered. This dispersed patterning of Hill fields evident during the years between 1969 and 1990 was inherited from the earlier systems of sectoral fallow that were once farmed under haciendas and in independent communities. Harvests from a family's Hill fields, both before and after the Land Reform of 1969, primarily fulfilled its needs for subsistence and seed, with commerce ranking third among uses.

Renewing soils in Hill farming led the Quechua in Paucartambo to manage fallow carefully, control the corraling of their livestock, rotate various crops, undertake arduous plowing and hoeing, and, in some cases, fertilize with chemical balanced N-P-K mixtures. Farmers preferred if possible to replenish nutrients in the far-away fields by fallowing them for at least three years after cultivation. Due to frequent fallow that was common through the 1980s, a majority of Hill parcels lay uncultivated at any time. A casual observer might unknowingly have wagered a less farmed landscape than was the actual case due to the frequency of fallow. Temporarily untilled sites did not, however, justify the unfounded association of fallow with the terms *unproductive, idle,* or *abandoned.* Farmers regularly renewed the cropping of fallowed parcels and used them as redeemable rangeland in the meantime.[17]

Concentrated manuring of Hill fields came about through the nighttime use of movable corrals or sheepfolds managed in a technique that farmers called guano-giving, or *wanuchiy.* Herders, cocooned nightly in a makeshift hut, or *chuqlla,* fenced their stocks of cattle, sheep, alpaca, and llamas in the corrals each evening, ensuring a generous deposit of fertile droppings. Crop rotation also shored up the fertility of Hill farming. The standard protocol for rotating Hill crops was to plant potatoes in the first year followed by a second-year crop of either ulluco, oca, mashua, or repeat potatoes. Due to this strong concern of Paucartambo farmers with crop rotation, the sowing of Hill fields could denote as much about soil management as it did about their production criteria and consumption goals.

An elevated bed for planting, locally termed *wachu,* or lazy bed, so enhanced soil fertility, texture, and moisture that the diverse plants in Hill farming were difficult to imagine under another system of cultivation. Plowmen wielding foot-plows sliced and levered sod into chunky wachu beds—perpendicular to contours in order to facilitate run-off—that would nourish the tuber plants and buffet their underground ganglions of tuber-bearing stems known as stolons. At the expense of great exertion and no less expertise, the plowmen then mounded soil on the wachu beds, piling them as high as waist level. They supplemented fertility by grading the alleys between the beds, or *wachu wayq'o,* bed canyons, from which soil was scraped. In a final earthmoving task prior to planting the plowmen pounded soil atop the beds in order to pulverize the clods into a tilthy texture that would nourish the vigorous growth of roots and tubers.

The aims of their soil management were stressed in several of the labor-intensive steps spelled out in tasks typical of Hill farming (Zimmerer 1994b). Farmers followed a sequence of twelve or so careful cultivation steps, bringing Hill farming near to the level of effort indicative of horticulture or gardening. One heart of Hill farming was the seasonal round of a wachu bed:

mounding and pulverizing, seeding and covering, shaping twice, and finally sacking it in order to harvest the buried tubers of either potato landraces or those of the other tuber crops. The Quechua cultivators expressed the inescapable centrality of the wachu beds in their Hill farming via a verb, *wachuy,* that doubled in meaning as *to plow* and *to grow.* In fact, no less than ten terms based on the root word *wachu* filled the everyday language of the Quechua farmers in Paucartambo and in neighboring regions of Cuzco (Beyersdorf 1984).

The calendar of cropping in Hill agriculture highlighted a few innovations that have lent a helpful dose of flexibility to the scheduling of tasks. The advanced plowing of Hill fields in April and May at the tail end of the rainy season, for example, contributed a welcome option. By tilling a full five to six months in advance of the standard plowing, the Quechua farmers could capitalize on labor-time that was available in April and May (Zimmerer 1991c). Earlier plowing also conserved soil moisture by creating a plow layer akin to the moisture-saving style of dry farming, tilled historically by the wheat farmers of the North American Great Plains. By conserving soil moisture, dry farming among the Paucartambo cultivators somewhat relieved the scheduling exigencies of their tilling and planting Hill fields. This innovation added to the flexibility of cropping and thus strengthened the prospects of diverse Hill crops in the labor-scarce peasant economy.

Still another element of flexibility in the Hill calendar was coaxed from its environmental traits and the qualities of diverse plants themselves. Cool temperatures dipping toward or below freezing at night checked the growth of many noxious field weeds, pests, and diseases. As a result of this diurnal climate the Hill farmers could count on a moderate but unaccustomed degree of flexibility in scheduling those tasks otherwise hurried by the pesky biological threats. Such time-consuming tasks as harvest could be loosely scheduled in Hill farming in a way that was not to be found in the other production spaces. Still further elasticity was added to the cropping calendar via the traits of the tuber-bearers themselves. All could resist, evade, and otherwise tolerate minor incidents of the chief climatic hazards, namely hail, frosts, and drought. While differing in detail, the agroecological hardiness of each tuber crop apportioned to Hill farming not only a measure of certainty but also welcome flexibility (Cook 1925a, 1925b; Hawkes 1983).

Hill farming systems in Paucartambo during the years from 1969 to 1990 resembled the ensembles of puna cropping and livestock-raising in the adjacent Urubamba-Vilcanota Valley and in Andean regions beyond southern Peru (Gade 1975; Pulgar Vidal 1946; see also Brush 1976, 1977; Mayer and Fonseca 1979). There were, however, two crop species notably missing from the Hill farming of Paucartambo; the grain-bearing relative of quinoa known as *kañiwa,* or *Chenopodium pallidicaule,* and the distinct *Solanum ajanhuiri* potato with its telltale landrace, *ajawri.* This pair of unique Andean crops continued to be common in the Central Cordillera of southern Cuzco and the

Altiplano basin to the south (Gade 1970, 1975; Johns and Keen 1986); however, neither *Chenopodium pallidicaule* nor *Solanum ajanhuiri* was tilled in the Paucartambo Andes. Curious lack of this crop pair in Paucartambo is shared by the oriental ranges of Puno as well, and suggests a characteristic absence from the eastern Andes in general (Camino et al. 1981).

Valley Agriculture

Dozens of maize variants comprised the quintessential crop of Valley farming in Paucartambo. A sample of seventy-odd fields enumerated twenty-seven maize types, each with a distinct morphology of ear and kernel, and each named particularly by the Quechua farmers (table 9). The diverse maize landraces tilled by Paucartambo growers belonged, in turn, to a roster of eleven "races." Since about one hundred maize races exist in all Latin America—home of the crop's greatest diversity—the Paucartambo cultivators may have fielded as much as one-tenth of its overall variation. In the shadow of maize diversity, but not to be overlooked, was an ample assortment of minor crops headed by quinoa, tarwi, broad beans, amaranth, common or kidney beans, and at least three squash species. The supporting cast of minor crops complemented the role of maize as the cornerstone of Valley farming.

The Quechua in Paucartambo segregated the maize crop into a triad of size-based landrace groups: the slow-maturing big seed, the intermediate-term medium seed, and the fast-ripening small seed (appendix D.4). Planting of each landrace group entrained differences in cropping calendar, growing seasons, and, biologically most significant, maturation period, which varied from a leisurely nine months logged by maize of the big seed class to a rapid cycle of five or six months in the small seed types. While all three groups were restricted to Valley farming, ideally, at least, they covered distinct habitats within it. Big seed fields, for example, were anchored in the warmest climates below 9,200 feet (2,800 meters). Small seed fields, by contrast, scaled upward of 11,650 feet (3,550 meters). Small seed maize could also be sown lower in a special off-season crop known as *miska,* the analogue of the maway planting of potatoes.[18] The loose bracketing of maize landraces at different elevations boasted more specialization than the other diverse crops, although the agroecological associations of maize would prove more malleable than estimated at first glance.

Maize plantings traversed a broad spectrum of Valley habitats. The full gamut of growing environments spanned a diverse series of temperate and subtropical habitats. At the upper limit near 11,650 feet (3,550 meters), short-season fields of maize were laced with a flora that married temperate types, such as native llok'e shrubs of *Kageneckia lanceolata* and eucalyptus trees with the subtropical agave known as *paqpa* (*Agave sp.*) and the *chanki,* or *Opuntia* cactus (Zimmerer 1989; appendix B.2). From its upper limit, Valley

Table 9. The Maize and Quinoa Landraces of Paucartambo[1]

A. *Maize Landraces*

Landraces	*Planting*	*Taxonomic Race*
1. *Chaminko*	ss/ms	Cusco Cristalino Amarillo
2. *Yawar chasqho*	ms	Cusco Cristalino Amarillo/Pisccoruntu
3. *Chullpi*	ms	Chullpi
4. *Ch'uspi*	ms	Pisccoruntu
5. *Ch'uspi-hatun muju*	bs	Huancavelicano/Pisccoruntu
6. *Fallcha*	ss	Cusco Cristalino Amarillo
7. *Hank'a sara*	ms	Pisccoruntu
8. *Huch'uy kullu*	ss	Uchuquilla
9. *Hump'a*	bs	Morocho/Cusco Cristalino Amarillo
10. *K'ellu chaupi muju*	ms	Cusco Cristalino Amarillo/Pisccoruntu
11. *K'ellu hatun muju*	bs	Ancashino/Cusco Cristalino Amarillo
12. *K'ellu huch'uy muju*	ss	Cusco Cristalino Amarillo
13. *Kulli*	ms	Kulli
14. *Paro*	bs	Paro
15. *Perlas*	ss	Confite Puntiagudo
16. *Piricinco*	ss	Uchuquilla
17. *Puka-chaupi muju*	ms	Cusco Cristalino Amarillo/Ancashino
18. *Puka-hatun muju*	bs	Huancavelicano/Cusco Cristalino Amarillo
19. *Puka-huch'uy muju*	ss	Cusco Cristalino Amarillo
20. *Pusaq wachu*	ss	Cusco Cristalino Amarillo
21. *Qoqotoway*	ms	Pisccoruntu
22. *Qosñiy-huch'uy muju*	ss	Cusco Cristalino Amarillo
23. *Qosñiy-chaupi muju*	ms	Cusco Cristalino Amarillo/Cusco Gigante
24. *Sanka*	bs	Morocho
25. *Saqsa*	ss	Cusco Gigante
26. *Waqankilla*	ss	Cusco Cristalino Amarillo/Kulli
27. *Yuraq*	ss/ms	Cusco Cristalino Amarillo/San Jeronimo

B. *Quinoa Landraces*

Common Name	*Some Synonyms*
1. *Kulli*	
2. *Misa*	
3. *Paraqay*	*yuraq quinoa, yuraq paraqay*
4. *Puka*	
5. *Uwina*	*k'ellu*

1. Maize landraces were collected from a sample of sixty-seven fields and quinoa from twenty fields. Fields were sampled in 1986, 1987, and 1990. *Bs* refers to the early planting of big seed landraces, *ms* to the middle planting of medium seed ones, and *ss* to the late planting of small seed types. Identifications based on the classification system devised by Grobman et al. (1961) were made with the assistance of Robert M. Bird, Major M. Goodman, and Ricardo Sevilla Panizo.

farming unraveled down treacherously undercut slopes. Semiarid and spiny subtropical vegetation, such as the *hawanqollay* columnar cactus of *Trichocereus sp.* and the thorny *roqe* shrub, or *Colletia sp.,* dotted the plunging canyon walls until near the humid riverside floodplains. Riverside habitats could be discerned by festooning air plant bromeliads (*Tillandsia*) and by a variety of shrubs including the Bolivian Fuchsia, or *chimpu-chimpu* (*Fuchsia boliviana*), and thickets of the invasive *retama,* or Scotch broom (*Spartium juncaeum*).

Rapid descent through the entire mix of growing sites sandwiched into Valley farming revealed a small overall area that limited the availability of precious fields for maize and the minor crops. Indeed, the access of families to arable Valley farmland suffered severely in comparison to the other production spaces and especially to the expansive Hill areas. Farmers relished small postage-stamp Valley parcels, many pasted infelicitously on the steep inner creases of the Paucartambo and Quencomayo Valleys. One fifth or so of the poorest farm households did not benefit from even a single Valley field in representative communities, such as Mollomarca, Majopata, and Colquepata. In general, the inequality of Valley landholdings was sharper than other production spaces, for while some farm families lacked a single plot others seeded numerous exemplars.

Harvests reaped from each Valley field were partitioned as a matter of custom among the three main uses of subsistence, seed, and commerce. Multipurpose cropping was not surprising given the relative scarcity of Valley sites. The Paucartambo farmers used nearly three-quarters of the maize parcels that were harvested in the mid-1980s to furnish a full spectrum of subsistence, seed, and commerce, evidence of the flexible allocation of harvest. They even dispensed harvest to all three end uses in their most commercial maize fields—the market-dominated plantings of a single landrace variety. Their strong commercial motives in these cases did not preclude the partial allotment of harvest to subsistence and seed as well. Such complexity within fields marked Valley farming in general and, as discussed below, supported its distinct hold on diversity.

The complexity of field management grew out of a paucity of spatial and social cohesion in Valley agriculture. When Líbano alluded to the fencing of his maize-yielding field—the Rumiriyoq Rock Place—in terms of the infrastructure much-needed against his neighbor's cow, he was also delineating the autonomy of an individual household with respect to its maize plots. The family's near-exclusive dominion over these parcels was quite unlike the sectoral fallow commons of Hill farming that were overseen by communities. The small degree of spatial coordination present in Valley farming stemmed from the shared use of small-scale irrigation as farmers shunted springs, streams, and rivulets to their thirsty maize seedlings at crucial moments early in the growing season. This coordination was, however, muted. Farmers were mostly resigned to treating their Valley fields like islands that would need to be shored-up with fences and other protection from the livestock of their neighbors.

Valley cropping permitted much room for variation in field techniques, a versatility that helped establish it as an unmatched arena of field-level innovation. For instance, to replenish the Valley soils beset by heavy maize cropping and blessed sparingly with natural nutrients the farmers created a dizzying variety of fertilization techniques. They designed amendments of livestock manure (*guano de corral*), shorebird manure (*guano de isla* imported from Peru's Pacific islands), and mineral and chemical fertilizers. They also orchestrated field fallow, crop rotation, and intercropping with quinoa as well as with leguminous broad beans and tarwi. Overall, the rich spectrum of fertilization techniques attested to the resourceful and individualized design of field inputs. It showed that the diverse crops could be cultivated where spatial cohesiveness and social coordination were missing; but it also pointed to a few pressure points or potential "clashes of cultivation" bared during the postreform decades.

One sign of resource pressures and the innovative counterefforts of farmers was beheld in the intercropping of Valley fields. Paucartambo farmers often intercropped their maize fields with quinoa, tarwi, and broad beans—and, less commonly, with amaranth, kidney beans, and squash. They sowed the interplantings of quinoa, broad beans, and tarwi in perpendicular "throat" rows, each a *kunka,* or into gaps opened in maize rows by ungerminated seed (*pank'e*). Farmers intercropped for production-related reasons in addition to gaining their favorite foodstuffs. Seeding the leguminous broad beans and tarwi helped to fertilize maize fields, or in the more vivid phrasing of the Quechua cultivators, "nourished" and "sweetened" patches of "bald," or less fertile, soil. The mixtures within fields also shielded the quinoa plants by thwarting bird predation and buffering them against strong winds. Tending to the diverse crops planted within a single Valley field truly was like a complex hybrid of gardening and field cropping.

Special husbandry spotlighted in the case of Valley parcels after the Land Reform of 1969 disclosed another motive for intercropping, that of overcoming land shortages. During earlier decades, the minor crops of quinoa, tarwi, and broad beans were mostly sown in midslope areas at higher elevations than Valley farming (Palacio Pimental 1957a, 1957b). The recent widespread adoption of high-yielding potatoes and barley in the Oxen Area, however, had led cultivators to withdraw the diverse plants from these sites and to relocate them in Valley fields below the middle elevations. By sowing small solo parcels or inserting the minor crops into maize fields, they innovated a locally unique solution to the pressing problem of limited farmland. Farmers in Majopata and their complex of field types illustrated this creative enlargement of diversity in Valley cropping (map 7).

The Quechua farmers of Majopata sowed half a dozen species of diverse crops into small fields on the community's slopes below 11,300 feet (3,500 meters). The diverse crops planted solo included maize, broad beans, quinoa, and tarwi. They also intercropped that group of crops, frequently interplanting them further with mixtures of still other diverse minor plants. As a result, the plots on Majopata's lower slopes abounded with amaranth (*achiwiti* or *kiwicha, Amaranthus caudatus*), arracacha (*rakacha* or *viraka, Arracacia xanthorrhiza*), winter and crookneck squash (*sapayu* and *lakawiti, Cucurbita maxima* and *C. moschata,* respectively), achocha (*achoqcha, Cyclanthera pendata*), passionflower (*granadilla, Passiflora sp.*), common beans (*purutu, Phaseolus vulgaris*), and yacon (*Polymnia sonchifolia*).

The repacking of quinoa and tarwi in the Valley fields of Paucartambo as a result of field shortages helped to resolve a biogeographic enigma of the region. Farmers in many other Andean regions spread the quinoa and tarwi crops across the upper and middle slopes of mountain farming rather than in lower Valley sites. Assuming the ubiquity of these patterns, well-known descriptions of the two crops have highlighted a hardy tolerance of cold climates and a characteristic occurrence in temperate zones (León 1964; Milstead 1928; Sauer 1950, 498; Tapia and Mateo 1992). In the Urubamba Valley west of Paucartambo, for instance, quinoa and tarwi had inhabited field sites as high as 12,500 feet (3,800 meters) (Gade 1975, 154–55). The Quechua farmers of Paucartambo, by contrast, caused atypical distributions of quinoa and tarwi by concentrating them below 11,500 feet (3,500 meters). In so doing they skillfully maneuvered their minor crops into a new landscape-scale pattern, one that was adjusted to the pressures and desires of postreform livelihoods.

A conspicuous variation of cultivation calendars flourished along with the diverse crops in Valley agriculture. Calendars of the maize plantings alone—the big seed, medium seed, small seed categories—were staggered by several months (appendix D.4). The varied planting and harvest dates of Valley farming seemed magnified, moreover, by the large number of Saints' Days that demarcated them to local people. Beginning with the celebrations of Saint Rose of Lima and Saint Mary the Virgin (*Virgen de la Asención*) in mid-August and concluding with the feast of Saint John the Baptist on June 24th, the farmers guided the staggered cropping of their Valley fields according to the Quechua-Catholic religious calendar. Conspicuous variation would not, however, be confused with capacious flexibility. Labor scheduling in Valley farming was rigid due to the regular threats of drought, plagues, and pests; cultivators delayed their field tasks there only at the peril of major yield loss.

Valley farming in Paucartambo's warm canyons conjured a number of images inverting the conventions of Hill cropping on its grassy plateaus. These juxtapositions pitted not only field environments but also the apportioning of harvest; land tenure and sociospatial coordination; and the use of soils, minor crops, and labor allocation. A dramatic dissimilarity also contrasted the "crazy quilt" patchwork of disparate Valley fields with the neatly contiguous patterns of similar parcels coalescing in Hill farming. The Quechua farmers' belief in dichotomous contrasts of landscape symbols appeared no less stark; symbols of "inner-civilized" and "outer-wild" were paired with the production spaces of Valley and Hill farming, respectively. Not least, the cultivators conferred maize, rather than potatoes and the tuber crops, with a truly unequaled significance in their ceremony and ritual.[19]

By intercropping the diverse maize with a variety of minor crops, the Quechua farmers in Paucartambo sowed Valley fields using an array of diversity at least equal to their Hill parcels. Unaware of this regional reality, acclaimed treatises on Andean agriculture have heaped accolades on high-elevation Hill cropping as the unrivaled repository of world-class biological richness (Cook 1925a, 1925b; Horkheimer 1990; León 1964; C. Sauer 1950; Tapia and Mateo 1992). The focus of the treatises on Hill farming may have overlooked competing diversity in the steep recesses of Valley farming due to scholarly and scientific sights fixed on the unfamiliar tuber-bearing crops, the promise of potato germplasm, and exotic tropical peaks. That view was upended, however, in the farming of Paucartambo after the Land Reform of 1969. In that time, the paramount diversity flourishing in both Valley and Hill farming was being pried apart and pressed upon by pressures building on the internal frontier of the region's farming.

Absences of Biodiversity: Routes of Commerce

Oxen Area Agriculture

Quechua cultivators crafted the Oxen Area into a key piece of Paucartambo's agricultural assemblage during the years between 1969 and 1990. Numerous farmers including Faustino and the well-to-do Líbano of Umamarca initiated a flood of new Oxen Area fields. Although the households of Faustino and Líbano differed in terms of their field holdings, they shared an unchecked enthusiasm for expanding fresh fields of the Oxen Area. Both the families seeded improved high-yielding potatoes and certified barley advanced on credit by the Cuzco brewery. To produce these market plantings, they purchased fertilizers and hired as many day-wage recruits from outside the household as was possible (Mayer and Glave 1990, 1992). Accelerated growth of their Oxen Area farming during the postreform decades did not, however, dilute its distinct character.

Biological uniformity, rather than diversity, emerged from the parcels of Oxen Area farming. When choosing seed for Oxen Area potatoes, Faustino, Líbano, and the other Quechua in Paucartambo picked from among a small handful of high-yielding varieties—including mariva, mi Perú, yungay, papa blanca, and *cica,* the last named for the Center for the Investigation of Andean Crops based in Cuzco. Since the market-conscious growers gravitated to a single variety due to yield criteria and prevailing price factors, the majority farmed only one or two of the above. For the same reasons, they had mostly deserted the first generation of improved varieties that were pioneered in the 1960s and the 1970s—*revolución* ("Revolution"), renacimiento, and the first of this type, the mantaro. Within the large spread of new adoptions, less extensive portions of minor crops also sprouted. Although the farmers still scattered fields of quinoa, tarwi, broad beans, and ulluco in the Oxen Area, they were seeding their diverse crops in greater variety in the other production spaces.

A peerless array of farm habitats was swept into the surging Oxen Area, which lapped from below 9,200 feet (2,800 meters) to nearly 12,800 feet (3,900 meters). Within this lofty ascent from subtropical to cold-temperature climates, the Oxen Area fields bunched most densely on the middle slopes above 11,500 feet (3,500 meters). Middle slope fields reposed at a moderate pitch compared to the steep inner canyon below them and thus they could be more easily tilled with the oxen-drawn plow. The slumping there of small slope sections—rotational slipping rather than landslides or mass wasting—created an alternating pattern of deep deposits at slump toes and shallow soils near the points of detachment. Shrubs with broad tolerance—including *chilka* (coyote brush, *Baccharis sp.*), *mutuy* (*Senna birostris*), *cheqche* (barberry, *Berberis lutea*), and *llawlli* (*Bernadesia horrida*)—and scattered bunch grasses like *Stipa ichu* vegetated the unsown patches and balks between fields in the Oxen Area (appendix B.2 in this volume; Zimmerer 1989).

The Oxen Area borrowed its name from the Spanish term *yunta,* meaning an oxen team. True to its name, the area was furrowed using the oxen-drawn traction or scratch plow known as the *arado.* Notwithstanding the tidy logic of yunlla etymology, not all Quechua farmers in Paucartambo put the same term to this production space. Some referred to it as midslope farming, coined literally as *chawpi qhata,* while others opted for big planting, or hatun tarpuy, in reference to its rainfed cropping. Some dubbed it *tayasqa,* or plowed-in-furrows, in contrast to the wachu beds of Hill farming. Another name was *waqa yapuy,* or oxen plowed. Still other Quechua farmers admitted being puzzled about the proper name. Apparent vagueness and outright uncertainty among the people of other Andean regions similarly clouded their naming of the Oxen Area. Titles such as *suni* and *templado* were inconsistently shared and in many cases difficult to define (Brush 1976, 1977; Gade 1975; Pulgar Vidal 1946).

Vagueness in naming did in fact mirror a distinctive lack of clarity in the symbolic meanings that Quechua people in Paucartambo attributed to the Oxen Area. Rather than convey a clear image like the untamed wildness symbolized by the Hill, their sense of the Oxen Area was neither symbolic fish nor figurative fowl. In terms of the landscape itself, however, the farmers apprehended that their Oxen Area fields made up a space of production wholly unlike Hill and Valley farming. Whereas the mental images of the Paucartambo people often ordered their landscape more neatly than the real complexity of its messy character and diffuse boundaries, the perceptions of Oxen Area farming reversed the generality. Its existence was more a fact of matter than imagination.

The murkiness of the symbolic content of the Oxen Area after 1969 did not detract from the sharp historical delineation of that farming space. Its origins dated to the early colonial epoch, when Andean farmers first tilled with the oxen and traction plow newly brought from Spain (Crosby 1986; Gade 1975). Although mandated by law and by colonial demands for tribute, farming of the Oxen Area expanded slowly in Paucartambo. Surveyors of the Colquepata landscape in 1595 pigeonholed all fields into either the Valley or Hill categories without mentioning once the Oxen Area (AAC 1595). In the 1700s, however, depiction of a merely dual division in farming was plainly impossible since accounts commonly described oxen-plowed parcels of wheat and barley (ADC 1784–85b, ADC 1807–8; see also Gade 1975, 1992). By the mid-1950s, the estate peasants of Paucartambo routinely yoked trusty oxen teams in order to cultivate barley and qompis potatoes in the Oxen Area (Palacio Pimentai 1957a, 1957b). Plowing of this internal frontier of production space in the region was subsequently hastened in the period from 1969 to 1990 under the similar impetus to grow barley and high-yielding potatoes of the improved varieties.

Farmers struck changes into the Oxen Area amid a context of little social coordination and a relatively weak cohesion of production space. Earlier, much land use in the Oxen Area had corresponded to sectoral fallow commons; in many cases the sectors were joined to high-elevation ones while at least a few haciendas had run a fully separate series of sectoral fallow commons on the middle slopes (for example, Ccapana in nearby Quispicanchis). Since the 1960s, the Quechua farmers throughout Paucartambo were busy dismantling this coordination by bits and pieces in order to intensify new commercial cropping.[20] By the mid-1980s, they implemented only sparse sections of sectoral fallow in the Oxen Area. Vestiges of its use could be found, however, in the peculiar scattering of the many fields tilled by families of former tenants. Líbano's fields sprinkled evenly across the Oxen Area of Umamarca, for instance, were land tenure's witness to former sectoral fallow.[21]

In the years from 1969 to 1990 many Oxen Area cultivators unmoored their plots of the diverse crops when seas of either high-yielding potatoes or barley fields swelled around them. Finding the asynchronous landraces engulfed by new crops, the Paucartambo farmers were swamped with added costs for crop protection and a definite rise in the risk of failure. If they wished to keep sowing their quinoa, ulluco, and tarwi fields in the Oxen Area, farmers would need to expend more resources. Those families with ample assets could most ably incur the costs and handle the risk. Most others, however, decided the option of continuing to curate the diversity-rich fields amid an unfriendly and weakly coordinated farm territory was infeasible and thus withdrew their diverse crops from the Oxen Area. Many of these families were the ones that innovated new intercropping mixtures of the diverse food plants in Valley parcels.

Weakening coherence of the Oxen Area also undercut the diverse crops by lessening soil fertility. Under the sectoral fallow commons, farmers had ensured field fertility by fallowing and by manuring with their livestock through the technique of guano-giving. After rescinding sectoral fallow, however, the tillers tended to adopt new and more disparate forms of rejuvenating field fertility. They often failed to renew soil nutrients as before. Their diverse crops and especially a ravenous nitrogen-feeder like quinoa did not, moreover, proportion good-size increases of yield in response to fresh amendments of chemical fertilizer. As a consequence of their dilemma, the role of nitrogen-hungry quinoa in the Oxen Area was reduced perceptibly albeit inadvertently. In contrast, chemical fertilizers swiftly lifted the yield of improved potato types to 2.5 tons per acre and higher in the Oxen Area (6 metric tons per hectare; Mayer and Glave 1992, 136).

Due to the twin pressures imposed by a compressed crop calendar and limited flexibility in labor allocation, the Paucartambo farmer faced somewhat inelastic demands for scheduling labor in their Oxen Area plantings. Although considered part of rainfed farming—generalized locally as the big planting—the Oxen Area fields ripened one to three months more quickly than their high-elevation Hill counterparts. The aggravated assaults of pests, diseases, and weeds on crop plants in the warmer climate and more intensely cropped Oxen Area lessened the elasticity of its farming. Unless the Quechua cultivators could marshal the effort for timely tasks such as fungicide and insecticide applications, soil mounding, and weeding, they stood to forfeit unaffordable amounts of yield.

An especially inflexible claim on labor-time was staked by the contracted barley crop. When griñon barley finished maturing in May and June, the dilemma of so-called shattering or dehiscence subjected it to a sizeable loss of seed. Battering slope winds that gusted through the region during the early months of the dry season worsened the threat of shattering. Barley growers thus sought to schedule their harvest as precisely as possible. They even torched their kerosene lanterns to initiate the finicky cut of frail seed heads at night or during early mornings in late May or June when damp air allowed the flailing sickles to shatter less seed. Their urgency, and anxiety, was deepened by the brewery's contracts; fine print stipulated that harvest must be delivered to the Cuzco factory within thirty days after the company's Paucartambo agent had dispensed empty gunny sacks for filling with the new harvest.

Early Planting Agriculture

Off-season crops of potatoes and ulluco were grown in the maway Early Planting. The off-season calendar began with sowing during the rainless winter season in July and August and concluded with harvest by late February. Dates for planting and harvest of the accelerated Early Planting preceded the mainstay big planting by several months. This distinction colored the off-season crop vividly during the dusty months of August and September when its verdant growing patches relieved the otherwise brown hues of dormant and drab-colored vegetation. Emerald fields of the Early Planting were also notable because they extended over the smallest area of the four production spaces while inciting the greatest commerce. For many Quechua cultivators in Paucartambo, the small but exhaustively worked plots were to supply the chief route for expanded commerce after 1969.

Although the Early Planting in Paucartambo boomed but recently, Andean farmers long ago had pioneered farming systems of similar sorts. In an early description of off-season potato cropping Felipe Guamán Poma de Ayala portrayed the maway as a sumptuous foodstuff reserved for the Inca royalty under state regulation (Guamán Poma de Ayala [1613] 1980). Potato eaters of the Inca period, like more recent ones, treasured the tubers newly harvested from off-season plantings, since fresh produce was otherwise sparse during the lean preharvest season. At least as early as the Inca period, the Quechua seeded their maway fields with the fast-maturing potatoes known as papa chawcha, or *Solanum phureja,* that yielded fresh tubers both quickly and reliably. During the colonial period, maway fields hunkered in the perennially humid floodplains of Paucartambo (ADC 1794–96). A quite different base of crops and farm resources, however, ignited the decisions of Paucartambo farmers when centuries later they raced ahead with a revamped form of the Early Planting.

High-yielding, scientifically bred potatoes fueled the pell-mell expansion of the Early Planting in Paucartambo beginning in the 1960s. Agronomic traits contributed crucially to this success. Fast maturation, high yields resulting from fertilizer inputs, and genetic resistance against the dreaded Potato Late Blight, or *Phytophtera infestans,* figured most importantly. The high-yielding potatoes that first reached Paucartambo in the early 1960s—renacimiento, revolución, mantaro—bettered the local landraces and boosted the innovative Early Planting. Implied in the names Rebirth and Revolution, the breeders of early improved varieties were counting on the transformation not only of the Early Planting but of potato agriculture in general. Another series of improved potato varieties was released in the 1980s, including the high-yielding mariva, which soon sprouted in much of the Early Planting. In the Early Planting of ulluco, a crop that lacked improved cultivars, the single landrace known as Cradled Baby, or *wawa yuki,* spirited the off-season crop, since it matured in six to seven months rather than eight or nine like the other ulluco types.

Beginning in the 1960s, the Quechua in Paucartambo innovated various sorts of the Early Planting in a variety of peculiar settings that could sustain cropping during the rainless and cool off-season. Four environments supported the burgeoning Early Planting of potatoes: (1) irrigated sites at elevations below 12,100 feet (3,700 meters), where only light frost could be expected after July, especially in the southern Paucartambo Valley and near Colquepata village: (2) bogs at similar elevations in the Quencomayo upland of Colquepata and its adjoining communities; (3) floodplains lining the main channels of the Paucartambo and Quencomayo Rivers; and (4) fields below 10,850 feet (3,300 meters) in the perhumid cloud forest, known as ceja de la selva, or Eyebrow of the Jungle, which rose along the right bank of the Río Paucartambo north of Espingone (see maps 2 and 5 for details).

The Early Planting of Cradled Baby ulluco, however, could be sited only in the cloud forest habitats of the northern Paucartambo Valley, since it did not tolerate the waterlogged soils common to other settings. The Early Plantings of both potatoes and ulluco, sown in moistened and frost-free settings, were inserted conspicuously into a surrounding fabric of dormant vegetation. Not surprisingly, the perhumid frost-free patches played host to abundant Potato Late Blight and scores of other pests and diseases. It was obvious that the Quechua in Paucartambo could not have tilled their off-season field sites in equal plenitude or with similar profit without the fast ripening and hearty resistance of mariva potatoes and wawa yuki.

Specificity of the Early Planting in terms of agroecological niche was one basis for its highly unequal possession among Paucartambo farmers. Unequal usufruct rights could be traced to previous land tenure in the key settings. Peasants on estates and Indigenous Communities in the region had long coveted the well-irrigated fields, bogs, and floodplains as scarce but reliable sources of watered crop land and especially off-season livestock fodder.[22] The wealthier of the peasant families had already commanded control of the key environments by the late 1960s, when cultivators began converting them to the Early Planting. The socioeconomic schism already at work in the capacity of farmers to craft the Early Plantings opened still wider due to requirements for capital and labor-time inputs as well as the security needed to bear economic risk. Chronic hazards of the Early Planting—such as frost and pest or pathogen epidemics—periodically ruined their fields and could deplete a family's savings.

Most often, however, the grower households invested their profits from bumper harvests into subsequent plantings and thereby catapulted the region-wide rise of the Early Planting (Brush and Taylor 1992; Mayer and Glave 1990, 1992). Growers operated with a go-for-broke outlook in conducting this planting, at least in comparison to the strong rationales for risk aversion ruling the triad of other production spaces. One pivotal investment for an Early Planting farmer was the acquisition of superb seed tubers—disease-free, large, lots of eyes. Since the Early Planting fields rarely returned worthwhile seed due to the afflictions of pathogens and pests, the farmers acquired seed by necessity from extra-household sources. They often procured seed stock from locales outside their community, where the load of shared diseases was less. This acquired seed was the kernel of their Early Planting economy, packaged in an array of related and costly investments in farm chemicals.

Humid habitats and perennial cropping, compounded by the Early Planting calendar, hastened the application of assorted pesticides, insecticides, and fertilizers. Growers repeatedly opened their tins of biocides—including the highly toxic parathion and aldrin—in order to squelch outbreaks of insects such as the coleopterous beetle known as Potato Leaf Bug, locally coined *piki-piki,* swarms of which descended on fresh seedlings. The farmers depended similarly on purchases of fungicides and fertilizers. Notwithstanding the stiffer genetic resistance of improved potatoes to Potato Late Blight, the fungal parasite lived up to its local name as *soqra,* a small deviation from *saqra,* or devil. The regimen of repeated plantings in a single site plundered soil nutrients and especially the available nitrogen, while the year-round humidity of fields diluted the potency of biocidal granules and powders. As a result, Paucartambo growers applied either massive doses of the mineral compounds or costly spray-on concoctions.

By local measures, the Quechua farmers innovating the Early Planting allotted Sisyphean sums of field labor and unmatched outlays of scarce capital. Tillers in the bog sites of the Quencomayo watershed—each denoted as a *wayllar*—sunk an annual average of 149 person-days per hectare into this production space during the mid-1980s (Zimmerer 1994b). That allocation of labor-time greatly surpassed even labor-intensive maize growing; yet Early Planting farmers faced little choice but to augment the labor-time for earth-moving and other inescapable requisites of their peculiar field settings. Waterlogging, for example, was an unending threat. Only by elevating the planting beds to waist height and then grading the low-lying alley surfaces could a farmer alleviate puddling and saturation that would putrefy the potato tubers. Standard tasks—such as weeding, application of pesticides, fungicides, and fertilizers, and monitoring fields—also posed fuller workloads in the Early Planting plots.

Weeds that invaded the humid bastions of the Early Planting typified most tangibly the time-consuming agroecology of its off-season cropping. The disturbance-loving flora ceaselessly defaced all four types of fertile Early Planting environments. The invaders colonized fields and interrupted cultivation most menacingly, however, in the pocketlike bogs of the Quencomayo Valley. Not only did the fast-growing weeds threaten to overtake the bog Early Planting but they also compounded the difficulty of even the most basic farm tasks. For example, an invasive rush evocatively termed *qhari waqachi,* or "That Which Makes the Husband Cry," wove dense mats of low-lying stems and compact roots that impeded plowing mercilessly. Growers complained that the pernicious weed hindered even the sharp blade of their foot-levered hoes, or foot plows. Another mat-forming bog plant, the *q'emilla,* also could quickly dishearten a plowman (Zimmerer 1994b).

Weeds of the Quencomayo bogs presented a flora of species brought unintentionally as seed from the Mediterranean Basin and Europe to the temperate Andes (Crosby 1986; Gade 1975). One prime example of a thriving European introduction in the bogs was philaree, or *Erodium sp.,* which the growers coined as "That Which is Fed to the Guinea Pigs," or *kovimirachi,* dubbed for its main use (Zimmerer 1994b). Other weeds from across the Atlantic settling in the montane bog fields included wild mustard, or *Brassica campestris*; the edible *nabo* or *yuyu* to the Quechua in Paucartambo; and Shepherd's Purse (*michi-michi*), or *Capsella bursa-pastoris.* In addition the field-loving flora was filled with an opportunistic suite of Andean weeds with names that identified them as bog plants such as *waylla llantin,* or Andean plantain (*Plantago australis* ssp. *pflanzii*); *waylla fallcha* (*Gentianella* aff. *dolichopoda*); *waylla totora* (*Juncus sp.*); *waylla pasto* (*Calamagrostis sp.*); and *waylla ichu* (*Trisetum spicatum*) (Zimmerer 1989, 1994b).

If Early Planting growers did not allot labor-time hastily in response to one affliction or another, weeds and the pernicious microorganisms incurred a costly toll in terms of crop loss. Through the 1970s and the 1980s, the Paucartambo farmers relearned the lesson that success in their Early Planting afforded a most meager flexibility. Physical factors of the Early Planting habitats exacerbated still further the rigid exactitude of scheduling tasks and thus of allotting labor-time. Frost or hail late in the cloudless winter season could demolish their new seedlings, while waterlogging threatened the precarious plantings almost continuously. Farmers could employ tasks such as the grading and mounding of additional soil onto planting beds to lessen threats from the physical milieu, but only if the chores were timed with an experienced and unimpeded precision. Proportioned too early, too late, or too little, the tasks failed disastrously.

The Quechua in Paucartambo grouped fields of the Early Planting not as a result of social genesis but rather due to nature's clustering of key niches that promised both enough soil moisture and the adequate absence of frost. Some settings dotted the regional landscape in small postage-stamp patches—such as the narrow floodplain sites in the Paucartambo Valley and the bog fields that many believed to be fed by the eye springs, or *ñawi pukyo,* that bespeckled the Quencomayo Valley. Small-size niches of the Early Planting precluded more than a modicum of spatial cohesion. By contrast, the Early Planting of cloud forests in the northern Paucartambo Valley coalesced moderate to high numbers of adjacent fields. The environmental factors alone did not, however, fully prescribe the areal coalescence of the Early Planting. Farmers labored with innovative techniques to extract their off-season fields from formerly unsown bogs, floodplain, and cloud forest. In doing so they more precisely defined the degree of spatial cohesion.

Shorthand depiction of the Early Planting on the basis of its staggered, off-season calendar revealed a modern-day version of the timeworn modification of the maturation period of crops and the growing season of field habitats. To be sure, resourceful Quechua cultivators had long rooted the off-season cropping of particular food plants in perhumid settings. During the period from 1969 to 1990, however, they renovated the Early Planting through an unprecedented expansion and a definitively commercial cast. Like much agricultural innovation arising at the hands of peasant farmers, new cropping of the Early Planting emerged neither as de novo adoptions nor as mere imitation of previous practices (Denevan 1980; Doolittle 1984; Knapp 1991; Zimmerer 1991d, 1994b). Once founded, the new plantings amply filled a fourth farm territory created by the Quechua in Paucartambo.

Farm Space as Key to Conservation

This foursome of production spaces framed in the Paucartambo Andes—Hill, Valley, Oxen Area, Early Planting—mattered greatly to the course of agricultural change and the de facto conservation of diverse crops. As the basic building blocks wielded by Quechua people in piecing together their farm livelihoods, the production spaces established what farm production was, how it functioned, and the ways it changed. After the Land Reform of 1969, many families managed to brick their farm spaces together in a fashion that still housed the diverse crops. Dozens of species and hundreds of landraces infused Hill and Valley farming; however, their diversity-rich spaces were abutting dynamic new expanses of the Oxen Area and the Early Planting, leading to a reconfigured landscape.

The Quechua in Paucartambo altered their use of the four farm territories in an effort both to gain from the new array of economic activities and to accommodate the cultivation of diverse crops. In this endeavor the farmers cobbled their production spaces and livelihoods reflexively. The double role of land use, in other words, was as both a result of farming outcomes and a conditioner of subsequent changes (Entrikin 1991; Pred 1984; Sack 1980, 1986). In their innovations from 1969 to 1990 the Quechua farmers managed to reinforce the discreteness of the four production spaces. Sprawling commerce after the Land Reform of 1969 did not soften the distinctions between production spaces, but rather it sharpened them. This conveyed a clear message for biological conservation: the fates of production spaces carried with them the fortunes of the diverse crops.

The style of segregating farm space that was at work in Paucartambo also shed new light on land use and the diverse crops. Previous studies of land use among peasant farmers in the central Andes have characterized the contrasts in crops belonging to different tiers, floors, or zones of production. Such contrasts were found in the Uchucmarca community in northern Peru (six tiered crop zones; Brush 1976, 1977), the Río Cañete Valley south of Lima (ten production zones; Mayer and Fonseca 1979), northern Potosí in Bolivia (bizonal cultivation; O. Harris 1982, 1985; Platt 1982), and Pisac in the neighboring Urubamba-Vilcanota Valley (four production zones, one of intensive horticulture; Kervyn 1989).[23] The descriptions of farm space in the other regions, however, dealt little if at all with the evidence of dynamic changes. By contrast, production spaces in the Paucartambo Andes were being cast in the overlapping fashion of an irregular mosaic. They were historically evolving, and they greatly influenced the course of subsequent change.[24]

Production spaces mattered immensely to the biogeography of diverse crops, since Paucartambo farmers used them to outline adequate sites for their planting and exchange. They tended to define the whole of a farm space, or sometimes a subunit, to be well-suited for a certain landrace group. The farmers then rotated fields and supplied seed of these landraces on the basis of such units of land use, thus arbitrating the effective limits of landrace distributions. Scores of floury potato landraces, for instance, occupied the lower Hill, which meant that landraces would at times be sown below 12,500 feet (3,800 meters) and upward of 13,300 feet (4,050 meters). Landrace groups of the maize plantings—small seed, medium seed, big seed—as well as ulluco and quinoa likewise dispersed across whole farm spaces or well-defined subunits. The influence of the farm territories thus assured that the spread of biodiversity's main unit, the landrace, tended to be ample and shared rather than narrow and specialized.

Another role of production spaces was to function for the Quechua cultivators in Paucartambo as a shorthand for estimating their capacities and constraints in everyday decisions about livelihoods. Each space was taken to hold a characteristic series of properties—farmlands, calendars, techniques—key to its farming. A family deciding whether to plant quinoa landraces, for instance, took stock not only of its precious Valley fields but also of its Oxen Area parcels and those belonging to other production spaces as well. A farmer's habit of perceiving each field in terms of a production space helped him to weigh the parcels in shared and commensurable terms rather than in the extreme detail of smaller units.[25] In considering field-level changes in cropping the farmer converted parcels from one production space to another—say from Valley to Oxen Area—rather than introduce smaller modifications that would make fields strangely dissimilar to both types.

The spatial cohesion of similar land use in adjacent parcels demonstrated the power of reflexive-style forces shaping the nature of farm space. An example of the effect of spatial cohesion was apparent when sectoral fallow commons were abandoned in some Paucartambo communities, as well as in Cuzco regions not far from there (Zimmerer 1991e). Dissolution of coordination under sectoral fallow abetted a noteworthy decline of high-elevation farming by inflating its production costs. This effect was already apparent in the Urubamba Valley by the 1960s (Gade 1975). A quite different scenario transpired in the sectoral fallow of some Oxen Area plantings. Many Paucartambo farmers found their capacity to cultivate diverse quinoa, ulluco, and tarwi fields diminished in these production areas. Besieged like small islands by the improved potato varieties and barley that flooded the reformed sectors, the diverse crops became more difficult for farmers to produce. The ordering of farm space in sectoral fallow could thus mediate the outcome of agricultural change and the tilling of diverse crop plants in different ways—in favor of de facto conservation in one case and opposed to it in another.

Farm spaces and their distinctive combinations—configurations in short—differed dramatically among the various places of Paucartambo.[26] Such place-based distinctiveness of farm space suggested that the personality of a place ultimately shaped whether the Quechua farmers replenished the diverse crops or discarded them. As will be discussed in the next chapter, they fared quite differently after the Land Reform of 1969 in the chief places of Paucartambo. By narrating the geographical stories of diversity-rich and diversity-poor farming in the region's distinct places we are able to broach a fuller analysis of diversity's fortunes.

5

Loss and Conservation of the Diverse Crops

Fateful Places in the Paucartambo Andes

The renowned collectors who had ventured to Paucartambo since the 1920s in an agronomic hunt for diverse Andean plants—O. F. Cook, Sergei Bukasov, Jack Hawkes, César Vargas, Carlos Ochoa—would typically depart from the central village to ferret out the most diverse caches of biological quarry. Retracing their footsteps we find that they followed a number of recurring routes through the rugged countryside and its rural redoubts. The Cuzco botanist Vargas, the International Potato Center's Ochoa, and the renowned British solanologist Hawkes all oriented their potato-collecting expeditions toward the southern Paucartambo Valley. There, on the valley walls of Umamarca, Mollomarca, and Huaynapata, and on the adjoining upland of Humana and Carpapampa, the national and international experts pursued their search for the most diverse potatoes in Paucartambo (Hawkes 1941, 1944; Ochoa 1975; Vargas 1948, 1954).

Travel in the region was not easy for the crop-collecting expeditions. Jeeps could safely reach the capital village of Paucartambo, but beyond there the state of provincial roads was perilous at best. Slides of rock and rubble frequently blocked passage. During the rainy season, chauffeurs spent more time freeing the roadways of rubble—or inciting peasants to do so—than they did driving. In March 1986 when I worked with anthropologist César Fonseca each day we passed a human corpse stranded on a gravel bar deposited in the raging torrent of the Paucartambo River. Reporting it to the local police, we were told that sadly there was nothing that could be done due to the floodstage discharge. The crop collectors exploring in Paucartambo did, however, manage to cross rivers, clear roads, and scale mountains. On foot and horseback as well as by jeep, they resolutely tracked their prize genetic game with unerring success.

Ochoa and his predecessors knew that the copius diversity of potatoes did not filter evenly across the Paucartambo Andes, and they would have anticipated a similar effect in the other diverse crops. Their surmise about diversity's checkered fortunes was confirmed (Zimmerer 1988, 1991b, 1991c). Maize diversity, for instance, congregated in the central and northern Paucartambo Valley, while the diversity of ulluco and quinoa peaked in the upper valleys. By contrast, the remaining plots of early chawcha potatoes bunched to the south between Challabamba and Huaqanqa. The conspicuousness of intraregional contrasts did not eclipse the defining roles of farm space: maize and quinoa premiered in Valley farming while diverse floury potatoes and ulluco hugged Hill cropping. The patently uneven geography of diverse crops turned on the basis of place, however, rather than revolving around farm space. Seasoned maize collectors, for instance, those of Grobman's historic expeditions, were not resigned merely to finding Valley farming but rather knew to seek it out in particular locales (Grobman et al. 1961).

Clustering of the diverse crops in key places became accented more sharply after the Land Reform of 1969. Although some striking differences in diversity's distributions were long in place, the conservation via cropping that enticed collectors such as Vargas, Hawkes, and Ochoa to the Paucartambo Andes was being frequently juxtaposed with its inverse—the absence or paucity of diverse plants due to recent farm changes. The result was a new map of place-based crop loss that took shape during the years from 1969 to 1990 (map 8). In the case of diverse maize, for example, a recent downturn of cropping on the steep Quencomayo slopes near Colquepata undercut its biological fortunes. By 1990, only a handful of maize-rich parcels still clung there. The diverse quinoa crop was also undermined severely in Colquepata. Meanwhile, many Quechua peasants in the region's northern reaches curtailed cropping the precocious potatoes, *S. phureja,* and slow-ripening big seed maize. Ulluco cultivation withered in the austral section of the Paucartambo Valley.

Certain places within Paucartambo thus cornered distinction after 1969 for either the cropping and hence conservation of diverse plants or the loss and in some cases extinction. Three central places in these outcomes were the northern Paucartambo Valley between Challabamba and Huaqanqa, the Quencomayo Valley and its upland watershed centered on Colquepata, and the southern Paucartambo Valley surrounding Mollomarca. Diverging fortunes of diverse crops in the triad of mountainous places were cast in the character of each locale. Place-based analysis bridged aspects of the agroecological foundations of farming, its socioeconomic and cultural transitions, and farm space (Blaikie 1985; Entrikin 1991; Pred 1984; Sack 1980, 1986; Zimmerer 1992a).

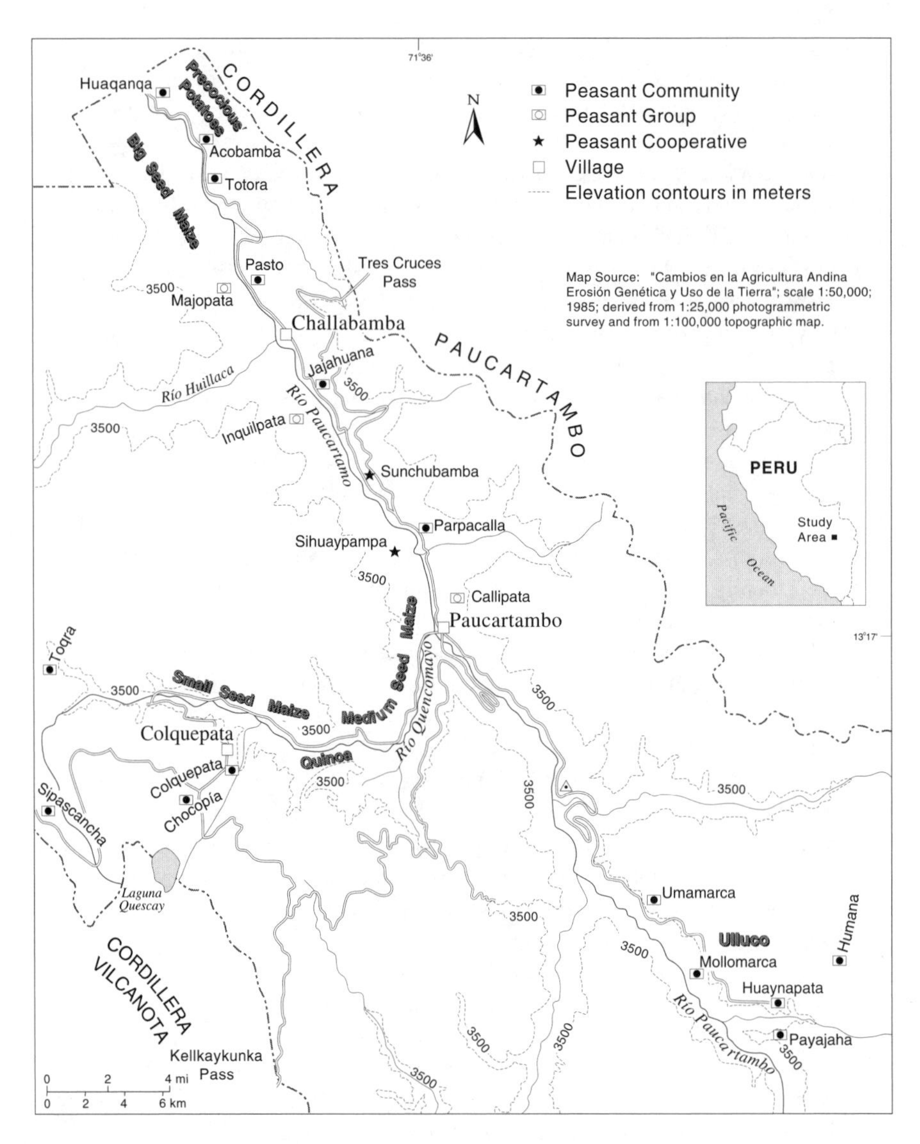

Map 8. Loss of Diverse Crops in the Paucartambo Andes (1969–1990).

Ridding the Odd-Ripeners in the Northern Valley Cloud Forest

The Paucartambo Valley funneled downstream of Challabamba and the roadside communities of Pasto and Pilco into a narrows at the locales of Totora and Acobamba and then widened at the hamlet and Peasant Community of Huaqanqa. The churning Río Paucartambo and the clinging riverside road led to a dynamic frontier of the region's economy. Quickening commerce after the Land Reform of 1969 transformed farming in this periphery of the region, abruptly changing the livelihoods of more than two thousand Quechua farmers who lived there. Farming changes caused the diversity in some crops of the northern Paucartambo Valley to vanish, so that at least a few regionwide extinctions of landraces were tallied. At the same time, however, the evolution of a number of diverse crops profited from conservation via continued farming. In the varied biological fortunes being cast after 1969 the northern valley showed a markedly uneven course of environmental change that was glimpsed more faintly elsewhere.

Pace of economic changes after 1969 quickened due to a propitious pairing of land availability and year-round humidity in the northern valley. Land availability there derived from the tragic and catastrophic history of Indian populations in the vicinity of its settlement nucleus at Challabamba (1981 village population: circa 335). Atop a gravelly base of river terrace and alluvial fan, Challabamba had sprung up as a forced resettlement, or reducción, of local Indians in the 1570s (plate 7). Colonial-era Challabamba and its sierran haciendas serviced the lucrative coca trade of the nearby montaña with Indian workers, way stations, mule teams, alfalfa, and storage facilities from locations conveniently near the heavily traveled Cañac-Huay Pass. Busy muleteers in the northern valley haciendas led Paucartambo Province to claim 1,334 of the 9,000 mules in Cuzco in 1786 (Flores Galindo 1977). After collapse of the coca trade in the early 1800s, the muleteers of Challabamba, including a wave of Scottish immigrants, trafficked mostly in the *aguardiente* rum distilled from sugar cane planted in the Pilcopata and Qosñipata lowlands (Llona 1904; Lyon 1984).[1]

Close ties of highland Challabamba and the northern valley to the montaña had doomed the population of Indian peasants. Catastrophic mortality due to disease during the first century of Spanish rule was worsened by the atrocious exploitation of colonial coca workers. The Quechua Indians of Challabamba were literally wiped out. A Church census in 1690 logged zero natives in Challabamba, the entire population having been "consumed by the coca trade" (Villanueva Urteaga [1693] 1982, 236). Notwithstanding the intermittent immigration of Indians and mestizos, the population of the northern valley stayed the most sparse of Paucartambo. For the few inhabitants access to land was abundant. Members of Huaqanqa, for instance, which numbered fifty-one adults in

Plate 7. Challabamba is gateway to the northern Paucartambo Valley via its lower village road and to the Cordillera Paucartambo and Manú National Park and Biosphere reserve and the Pilcopata-Qosñipata frontier lowlands of the upper road (at cloud line). Since the Land Reform of 1969, some better-off peasants have moved to town, coupling diversity-rich farming in their communities with a residence location for commerce.

1970 and 110 in 1985, still averaged no less than 50 acres per family (20 hectares per family). Although less than one third of their land was likely to be arable, the Huaqanqa people and the northern valley farmers in general enjoyed abundant farmland relative to other Paucartambo cultivators.

Environments cloaked in year-round humidity also foreshadowed the character of agricultural change during the period from 1969 to 1990. Heavy rainfall in the northern valley fell over a dense canopy of evergreen trees laden with vines, moss, and epiphytes. The montane cloud forest, the ceja de la selva, especially dense on the Río Paucartambo's east bank in the areas north of the Uch'uypilco tributary, was controlled by climatic effects of the Cordillera Paucartambo. Tapering northward, it permitted an inflow of moisture-bearing clouds from the Amazon Basin. Over countless eons of geologic time, the heavy rainfall had incised treacherously steep slopes of feral relief into the friable sedimentary strata of the northern valley. Cloud forest trees and shrubs were anchored tenaciously to thin soils and formed thick stands as high as 11,500 feet (3,500 meters), a tree line depressed in elevation because of the perennial cover of clouds and fog.

Forest limits that were imposed by a lowered tree line paled, however, in comparison to the effect of farmers' clearing trees and shrubs for new land use.[2] They felled the cloud forest to open up fresh fields and cattle pastures. Countless clearings in the mosaic of vegetation after 1969 were interspersed with the forest fingers that still wedged into rocky ravines and side canyons—where farmers continued to extract useful forest products—and with the impenetrable thickets of bamboo and fast-growing trees and shrubs that invaded their fallows. Quechua farmers of the northern valley felled the cloud forest mainly to sow Early Planting fields, which intermingled with sizeable swaths of Valley and Oxen Area farming (see appendix D.3 for elevation limits). Most combined the cropping with cattle-raising, especially in the upper-elevation grasslands. Their cattle thrived in the soggy expanse of grassland that crept across the Cordillera Paucartambo, where it bordered the world-acclaimed Manú National Park. In addition to their farming and ranching Quechua of the northern valley also tended sizable fruit and vegetable house gardens.[3]

Early Planting fields diffused swiftly among the Quechua farmers of the northern valley. Harvest season in January and February saw at least a few trucks plying the muddy roads downstream each day, lumbering into Challabamba and sometimes onward to Pasto, Pilco, or Huaqanqa in order to ship cargos of improved potato varieties and Cradled Baby ulluco that were bound for market. The Quechua farmers who launched this Early Planting boom invested capital from their savings, loans from the state-owned Agrarian Bank, and credit obtained from the personal funds managed by villagers and well-off peasants. Not least, the new potatoes, such as mariva and mi Perú, added to momentum of the boom due to a hardy resistance against the ubiquitous Potato Late Blight and a celebrated response to the Green Revolution-style regimen of farm inputs.

By 1990 most farm families in the northern valley were pioneering several parcels of the Early Planting. Many reinforced their new commerce with earnings from short-term migration, which neatly complemented the Early Planting calendar. Migration to work sites away from Huaqanqa and its neighboring communities—Lucuybamba, Acobamba, Totora—involved stays of a few days upward of several weeks. The highlanders searched for wage-paying work mostly in the large-scale agriculture and extractive industries fanning out across the nearby Amazon lowlands. Luring of the laborers for rice, pineapple, and coca growing, logging, and placer gold mining was waged most intensely from late July through September and in late February through March after Carnival; that is, precisely the periods when the Early Planting fields required little effort. Quechua families in the northern valley could thus migrate to the lowlands, either together or by sending a member, without undoing their newly adopted farm commerce.

The vitality of short-term migration set the northern valley apart from other

Paucartambo areas. Sheer proximity of the northern valley to the Amazon lowlands plus the convenience of low-cost transport on the trans-Andean trunk road undoubtedly aided the seasonal migration of northern valley people.[4] Another decisive force was the paucity of hindrance from patronizing villagers in Challabamba. Many Challabamba villagers, in contrast to the larger cohort in Paucartambo, held farmland outright rather than sealing access through patronage with the local peasantry. Cultural tradition also bound the northern valley people to the nearby lowlands. For instance, they regularly sponsored their dance groups, especially the feathered, spear-carrying *chuncho* warriors, in the Saint Elena's Day (*Cruz Velacuy*) celebration in Pilcopata. A marked absence of barley growing also spurred the flow of short-term sierran migrants by ruling out an income source prevalent elsewhere.[5]

Once the Quechua farmers were busy propelling their new commerce in Early Planting fields and short-term migration, they encountered a contracting pool of labor-time available for other work. Labor, both within their families and among their neighbors, was being absorbed in substantial sums by the new activities. A sparsity of labor-time seemed incongruous with the area's rapid population growth, for thousands of Quechua farmers had swelled the northern valley during the postreform decades. In fact, peasant immigration filled its communities in unequaled quantities. Notwithstanding the post-1969 demographic leap, however, the northern valley farmers still possessed larger landholdings on the average than their counterparts (9.3 acres or 3.8 hectares per person; see appendix C.3 for comparisons). The benefits of ample land brought to bear the unwelcome reality of shrinking labor supplies. Scarce labor-time even inflated the daily wage of the northern valley. Its rate typically doubled the concurrent day-wages in other places within Paucartambo.

Environments in the northern valley added to the laboriousness of routine farm tasks. Clearing fields from the dense cloud forest that clad the steep slopes was especially onerous. Thickets of bamboo known locally as *kurkur* tangled with cloud forest shrubs like *chinchilkoma* and *tiri* (species of the genera *Miconia, Tibouchina,* and *Brachyotum*) (appendix B.2). Fast-growing trees like alder (*labran* or *aliso*) grew in close stands. Grasses, too, hindered field clearing, especially a disturbance-adapted pampas grass, or *ñiwa*, and the ineradicable *kikuyu* (*Pennisetum clandestinum*) introduced into the Andes earlier this century from East Africa (Zimmerer 1989). Farmers could rid such vegetation only with swidden techniques, or slash-and-burn, which called for a combination of cutting, burning, and grubbing, using axes and picks in order to remove the forest cover fully.[6]

In "breaking the forest," what was called *rompe* or *chakmeo,* the northern valley peasants were beset by a Faustian dilemma. They could put the thin and nutrient-poor soils of the cloud forest to use only by resting old fields after just a few years in cultivation. By necessity, they tended to repeat the cycle of swidden labors once every ten to twenty years. Dense scrub of cloud forest thus returned to the former field sites at not infrequent intervals. Disturbance-adapted invaders made up the majority of regenerating plants, which included scarce specimens of the slow-growing trees that gave the ceja de la selva flora its renown for biological uniqueness. Flora in the mostly uninhabited reaches of the eastern Cordillera Paucartambo, by contrast, was much richer in the diverse tree species emblematic of the cloud forest, such as ones belonging to the genera *Weinmannia* and *Podocarpus* (Young 1987).

Pests, diseases, predators, and weeds further complicated the burdens of the Early Planting. The infectious Potato Late Blight, or devil, had flourished in the rains and mists of the northern valley—known locally as disease rain, or *onqo para*—long before its infamy was achieved in the deadly Irish Potato Famine of 1848–49. When the area's farmers multiplied the spate of Early Planting fields after 1969, chronic presence of the dreaded Potato Late Blight bedeviled them still more. Cultivators of the Early Planting were pressed to monitor their fields as carefully as though they were finicky plots of garden vegetables. Plenty of other moisture-loving insects and microbes—such as the soil fungus known as *roña*, or *Spongospora subterranea*—likewise menaced parcels. White-tailed deer and Andean spectacled bears also damaged the Early Planting fields and other crops on occasion, although the large mammalian intruders exacted less damage than did the smaller organisms.

Undercutting of the diverse crops due to quickening commerce and shifting labor-time allotments occurred among many peasants in the northern valley. Their predicament was exemplified in a conversation of late May 1987 with a Huaqanqa family. Twenty-five-year-old Cipriano Qoya Qalhuanca; his wife, Tomasa Machacca Tunque; and their young son, Benjamin, were en route home after several weeks laboring in the rice harvest on a large farm near Pilcopata. Resting briefly in Challabamba after paying their fares to the driver of a Cuzco-bound truck, the couple was anxious to plow and manure their Early Planting fields.[7] Cipriano worried if they would be able to buy good-quality mariva or mi Perú seeds, or even those of mantaro, at a reasonable price or whether seed costs had inflated on the eve of planting. He and Tomasa also spoke of changes in their repertoire of diverse crops. Like many neighbors, they had recently relinquished two exemplars: the early chawcha potatoes of *Solanum phureja* and one field of the late landraces of big seed maize.

Of the disappearing pair of landrace groups, precocious potatoes suffered the greater curtailment. Landraces of the precocious potatoes—principally White Chawcha, or *yuraq chawcha,* and Red Chawcha, or *puka chawcha*—had furnished a year-round source of fresh tubers due to rapid ripening. Fast ripening also customized the early potatoes for at least one major ceremonial dish, the timp'u stew (which also contained early maize), served at the annual Carnival celebration in February, a date that was impossibly early for produce of the main rainfed planting. Fast maturation conferred agroecological benefits, moreover, since the swift growth of the early potatoes evaded the damage that crop plagues could wage over a longer season. Furthermore, the *Solanum phureja* species excelled in terms of its resistance to Potato Late Blight.[8] Mindful of its many merits, the northern valley Quechua had grown the early chawcha crop as a staple tuber-bearer rather than relying on the floury potatoes, otherwise common in Paucartambo.

The very agroecological success of the early *S. phureja* potatoes in the northern valley also imperiled them in the years from 1969 to 1990. Two of the crop's most well-suited ecological adaptations became a bane for farmers: short maturation period and the lack of seed dormancy. Since its tubers sprouted shortly after harvest, the precocious chawcha left farmers with little leeway but to bring either three or four crops to harvest during ten or more months each year. The process of obligatory replanting, or relay cropping as it is known, inflated the demands on their labor-time severalfold. A field of the fast-ripening tubers demanded an unstinting series of Augean labor inputs that on an annual basis averaged to 205 person-days per hectare (Zimmerer 1991c). The huge sum of labor-time overshadowed even the second most intensive crop, potatoes of the Early Planting, by more than twenty-five percent.

Penitential demands on labor-time imposed by the early chawcha potatoes simply outstripped the resources managed by most farm families. By the mid-1980s, more than four fifths of Huaqanqa households omitted the fast-ripening potatoes from their thickening work regimes (Zimmerer 1991c). Likewise no more than a few persons in the nearby communities of Acobamba and Totora still seeded the precocious potatoes one decade or so after the land reform. Chawcha growers discovered that the severest bottlenecks of labor-time were lodged in the crucial periods of September–October, mid-January–mid-February, and June–July. In these bimonthly junctures they faced growing demands from the numerous Early Planting fields, short-term migration forays, and still other labor-time pressures from the locally abundant maize crop. Rather than reduce their commercial endeavors, a majority of Quechua planters chose to rid the precocious potatoes from their crop rosters.

A noteworthy chasm separated the northern valley farmers who kept cultivating the early chawcha potatoes from the ones who believed the crop's agroecological dictates had become insurmountable. By the 1980s, well-off peasants numbered disproportionately among those who still tilled the peculiar potatoes (Zimmerer 1991c). In Huaqanqa, for instance, five of the community's remaining six chawcha growers were well-to-do peasants. Their resources could procure the necessary flexibility and labor-time. The well-off farmers sometimes opted for wage payments to recruit workers, even deploying them in other tasks such as clearing patches from the forest scrub while household workers cared for the finicky potatoes. By contrast, Quechua peasants in the middle of the socioeconomic spectrum were beset by the gravest labor shortages. Amid their diversified and expanded commerce, they could not muster sufficient help from either household members, neighbors, or nearby family.

Many of the poorest Quechua also sheared the early potatoes from their crop rosters, although their predicament was compounded by changes in the spatial character of land use. Wielding decidedly fewer means to adopt Early Planting fields and other commerce, the poorest peasants seemed less likely to forego the diverse early potatoes. Unfortunately, the changing configuration of farm space undermined their resolve. The impress of spatial pressures came about due to the pell-mell expansion of Early Planting fields by their more well-off neighbors. Cattle that grazed from March through May in the harvested Early Planting fields threatened to damage any standing parcel of the awkwardly asynchronous precocious potatoes, which typically were relay-cropped through June. Cattle thus became key actors in the clashes and were numerous, too, for the average size of the bovine herds had doubled to more than six head per family by 1987.[9]

The imperative to coordinate livestock-raising and cropping also inclined against early chawcha potatoes in a second way. Field environments covered by the diverse potatoes fell entirely within the compass of Early Planting fields, both cropping systems traversing habitats from the valley slopes below 9,500 feet (2,900 meters) to as high as 10,850 feet (3,300 meters). A standing parcel of the diverse but awkwardly asynchronous chawcha was thus likely to be placed within the vicinity of proliferating Early Plantings. This added vehicle for agropastoral and spatial pressures did not eliminate the early potatoes altogether. Well-built rock walls that fenced a number of parcels helped to protect them partly and at least enough to satisfy owners. The efficacy of their rustic fences, however, spread further the socioeconomic schism in this farming, for they were erected and then repaired with labor-time that could be summoned most ably by the wealthier growers.[10]

A deficit of taste appeal also betrayed the precocious potatoes. Practiced potato-eaters faulted the waxy texture of the cooked chawcha tubers, which they found watery, or *unusqa,* the culinary inverse of flakiness—or so-called mealiness—prized in the floury potatoes. Since floury potatoes failed to produce well in the misty disease-ridden vale, the culinary alternative to early chawcha potatoes was the yield of Early Planting types. Culinary traits of the new improved varieties, such as mariva and mi Perú, did not suffer comparison to the unexceptional chawcha types. While not much whetting their appetites, farmers tolerated the improved varieties as soup potatoes (*bondón papa,* altered from the Spanish *mandar,* or "to peel"). The growers felt able to replace the chawchas with new potatoes without comprising the standards of their daily cuisine. Only a razor's edge, therefore, set apart the norms of a fit livelihood with diverse landraces, a force favoring conservation, and the suitability of new crop substitutes. In the case of early chawcha potatoes the decisiveness of the latter helped make a vulnerable Achilles of even the most ecologically well-heeled crop.

Diverse landraces of the big seed maize crop also disappeared from many

Plate 8. Tomasa and her children after harvest of the family's unique landraces of the big seed category in late June. Many of her neighbors have curtailed their big seed plantings due to work in off-season market cropping and short-term migration.

farms in the northern valley after 1969 (plate 8). Slow-ripening, the big seed landraces were sown between the ritual calendar markers of Saint Mary the Virgin (Virgen de la Asención) on August 15 and Saint Rose of Lima on August 30. In late May or June of the following year the farmers sickled through the giant stalks of big seed plants in order to harvest (appendixes D.4 and E.5). This prolonged season was made possible by environmental features of the northern valley, which offered year-round soil moisture and a subtropical climate on the foot slopes below 9,500 feet (2,900 meters). Only under such unique conditions could farmers ensure the full maturation of big seed landraces. Large ears and canelike stems of the lanky plants yielded handsome harvests and abundant forage for their livestock. Dwarfing the fecund field weeds of fresh swiddens, the tall big seed specimens also could imbibe sunlight and soil nutrients unfettered.

Unfortunately the labor-time needed for big seed befit its name. Work in the lengthy planting exceeded the 121 person-days per hectare typical of the fields belonging to small and medium seed maize, perhaps by one quarter or more—in some cases topping the value of 160 person-days per hectare (Zimmerer 1991c). The prolonged growing season of the slow ripeners extended the duration and upped the intensity of the work routines. Uniquely vital tasks included the guarding of succulent maize plants against the hordes of bird and insect pests, such as parrots and moth larvae. Farmers defended them also from the damaging onslaughts of cattle and pigs, quail and skunks, rabbits and vizcacha, white-tailed deer, Andean spectacled bears, domestic dogs, and crop thiefs. Because peasant tillers of the northern valley preferred to site their big seed fields in fresh swiddens for the sake of special fertility, their labor-time expenditures were often freighted with forest-clearing and other swidden preparation in addition.

Unable to meet these labor demands, farmers began curtailing their plantings of big seed maize so that by the mid-1980s its unique landraces made up less than five percent of the northern valley's crop. The Quechua farmers who ceased production claimed that its prolonged calendar with distended calls for labor-time both early and late in the growing season clashed with the unbending dictates of their expanding commerce. Their family, like many others, was adopting a variety of new farm tasks in the early and late months of the growing season. In August and September they bustled to keep up with new potatoes and ulluco fields of the Early Planting. Similarly, they quickened their already busy work rhythms late in the season during June and July when Early Planting harvest and labor migration vied with tending to the final details of big seed maize. Sensing the labor-time bottlenecks, most farmers preferred to trim the diverse maize rather than cut back their expanding commerce.

Shortages of farmland also gnawed at their means to cultivate the maize landraces of the big seed category. Quechua of the northern valley learned quickly that the subtropical growing season of big seed fields could alternatively produce the coveted Early Planting. Burgeoning numbers of Early Planting fields led many farmers to settle for suboptimal sites when planting their maize. Increasingly they intermingled the three maize plantings—small seed, medium seed, big seed—within the space of Valley farming. This partial mingling of landrace groups within Valley agriculture departed from more static situations such as those in adjacent Quispicanchis Province where an analogous series of maize plantings—referred to there as *llacta sara, wari sara,* and *yunka sara*—were sown in an orderly "altitudinal sequence" (Yamamoto 1985, 89). By contrast, the growers of Paucartambo's northern valley were electing suboptimal field locations in an effort to accommodate new commerce while innovating their maize-growing traditions.

While experimenting with the mingling of production, farmers found that big seed maize was the most expendable of their plantings. Maize fields of the small seed and medium seed categories could, more feasibly, be cinched onto their adjusted livelihoods, since they clashed less with their revamped regimen of resource allocations. The cropping calendars of small seed and medium seed maize, for example, could mesh better with the labor-time demanded in commerce. Sown in October and November and harvested during the following May, the small seed and medium seed landraces stood safely outside the peak periods of Early Planting work. In addition, the northern valley farmers could locate their small seed and medium seed fields as high as 10,150 feet (3,100 meters) rather than to only 9,500 feet (2,900 meters)—the upper elevation of big seed fields. Less flexible and more demanding, big seed maize was at a decided disadvantage in farmers' minds.

While the valuation of farmers helped ensure the cropping of maize in general, their cultural concerns did not secure a role for big seed landraces in particular. Quechua in the northern valley, like their counterparts throughout Paucartambo, consumed the staple cereal in customary dishes and celebrated it in ritual (appendix E.1). Farmers heaped additional esteem on the maize crop because of its cherished secondary uses in beverages such as homemade chicha or aqha beer and finely ground *api* mush, in succulent corn-on-the-cob choqllo, and for its unrivaled repertoire of roles in religious ritual. Yet in prizing maize in general, the Quechua farmers knew that the landraces of the small seed and medium seed categories fulfilled nearly the full breadth of their assorted uses. Although discerning, the customary eating habits of farm families did not protect the big seed maize in particular when their farming was being pinched by resource limitations.

One further impetus for the growing of maize in general, but not necessarily the big seed, stemmed from the cereal's value in the nonmarket exchanges that thrived in the interstices of new commerce in the northern valley. Several arrangements for payment in goods and barter persevered in valuing maize; under one arrangement, farmers paid laborers at the daily rate of one *arroba,* or 25 pounds' worth of unshelled ears (11.5 kilograms). Maize could also be bartered for other goods. Growers sometimes swapped maize for the floury potatoes, freeze-dried chuño, and sheep traded by inhabitants living on the west bank of the Paucartambo River. The maize growers also bartered their product—especially the ceremonial types known as *misa* and *k'uti*—with the long-distance traders who each year drove caravans of llamas and horses packed with wool and dried beef on a six-day 100-mile (167 kilometers) trek from Pitumarca in the southern Cuzco province of Sicuani.[11] Although the varied nonmarket exchanges made maize dear none placed special value on the big seed class.

Without the customs of cultural valuation, market rationales alone would have endangered even the small and medium seed maize, since profits were more rewarding in Early Planting plots (Mayer and Glave 1990, 1992). Northern valley growers knew that the commercially robust Early Planting would return a greater reward than the mixed marketing and consumption of maize. Their decision nonetheless to sidestep the commercial impulse was not an aversion to commerce per se; rather, it exposed the powerful motivator of economic risk-aversion when so many uncertainties marred the local markets for farm inputs and goods. In an ideal economy the farmers would have converted all their fields to the Early Planting and bought the maize types desired. In the northern valley's peasant economy, however, the Quechua farmers gauged their actions against factors such as buyers' price-setting and monopolistic transportation, risks too large to override their cultural affinity for growing and consuming maize.

Diversity of the ulluco crop within the northern valley was compromised little as a consequence of commercial inroads after the Land Reform of 1969. While new commerce of the ulluco Early Planting was wed to a single landrace, the Cradled Baby, or wawa yuki, that emphasis on uniformity for market purposes did not eclipse the seeding of other ulluco types for home consumption. Subsistence growers sowed wawa yuki with medleys of the *Hair,* or *chukcha,* and qompis landraces—the latter identical in name to the well-known potato. Quechua farmers credited the cropping and thus de facto conservation of ulluco to its varied advantages, mostly to do with a hardy tolerance of disease, drought, and frost (appendix E.2). It required only moderate amounts of field labor and soil fertility, being sown successfully with modest effort in second-year fields after potato cropping. By sticking to the role of ulluco in the standard rotation cycle, the Quechua farmers conserved the crop.

Overall Quechua of the northern valley incurred steep biological losses due to the shrinking areas sown with diverse crops. The nearly complete curtailment of chawcha potato cultivation brought the status of the species to the brink of regionwide extinction. Because the early *S. phureja* was restricted to the northern valley and thus distributed in an endemic fashion within the region (see map 6; table 6), its desertion delivered an unwelcome biological outcome. Without accurate records made earlier, an exact estimate of landrace loss in the *S. phureja* crop cannot be attempted, although it can be safely surmised that the species was succumbing to an unmistakable evolutionary decline.[12] Reduced cultivation of big seed maize with its various slow-ripening landraces—exemplars of the races Huancavelicano, Morocho, Ancashino, and Paro—threatened to set in motion a similar scenario of the local but biologically significant loss of endemic landraces. Further from complete cessation than early chawcha potatoes, the noticeable decline in the landrace-rich big seed plantings nonetheless did not show any signs of being slowed.

The Demise of Maize, Quinoa, and the Colquepata Wetlands

Thousands of Quechua peasants residing in the Colquepata area—the ribbed uplands surrounding Colquepata village (1981 village population: circa 425) and the Quencomayo Valley—embarked on a similarly dramatic transformation of agriculture and its biological basis. After the Land Reform of 1969, the

Quechua in Colquepata relinquished good-size areas of three groups of diverse crops: small seed maize, medium seed maize, and quinoa (map 8). Contrasting their shrinking reservoirs of diversity, a pair of new plantings was pooling a growing share of the resources of farmers: the Early Planting of improved potato varieties and the malting barley contracted with the Beer Company of Southern Peru. Dynamics of agricultural change and histories of the diverse crops etched a course in Colquepata that was quite distinct from the northern Paucartambo Valley. Only 15 miles (24 kilometers) apart and both spirited by the upstart Early Planting, the biological outcomes in each place offered salient differences.

Scattered swales that cradled wetland bogs, each a wayllar, were used to launch the Early Planting by 1970 (map 9).[13] The small wayllar bogs in Colquepata held fertile soils derived from material that had washed from the region's sere and extremely erodible landscape of weakly adhered phyllite rock. Of particular importance, the bogs abounded with year-round soil moisture (Zimmerer 1991d, 1994b). The bogs that were located safely below the frost line of the rainless late winter months—approximately 12,100 feet (3,700 meters)—became a leading center of the Early Planting in Paucartambo. By the 1980s, the Quechua farmers from Ccotatoclla community to Sipascancha—including the skilled cultivators Juan Santos Yucra and Fortunata Wallpa Amao (Zimmerer 1994b)—had converted their bog sites into parcels of so-called raised fields. Tirelessly and quite ingeniously, they drained the soggy sites while elevating the planting surfaces.[14] Their efforts fashioned an unusual landscape of wetland agriculture that skirted the middle slopes as low as 10,850 feet (3,300 meters).

Historic leapfrogging of the Early Planting fields across Colquepata bogs escalated in the 1970s and the 1980s. This rapid spread followed a peculiar fusion of environmental deterioration and frequent social clashes that were set in motion during previous decades. Uncultivated wetland bogs, known as virgin or *purun,* had long supplied succulent forage for the herds of llama, sheep, and cattle pastured by Quechua herders. The herders coined special compound names for this wetland pasturage; the three chief grasses were waylla ichu; *fuku-fuku,* or *Dichondra sp.*; and *cebada-cebada,* or *Alopecurus magellanicus* (Zimmerer 1989, 1994b). Numerous herbs and an alga valued for grazing also sprouted in the lush bogs. During the rainless season from May until October, the green foliage of bog vegetation helped sustain hungry herds of the varied livestock. It was the sole exception to a dormant and barren landscape of badlands that elsewhere gave Colquepata a desperate look.

While the Quechua of Colquepata proliferated their holdings of cattle and sheep, unfortunately, the bog pastures were being silently but surely infested by a parasitic liver fluke (*Fasciola hepatico*), termed locally as *alikuya* and *kayotaka.*[15] Teeming liver parasites infected the intestines of mammals including humans. The parasite began taking a damaging toll on the livestock that grazed

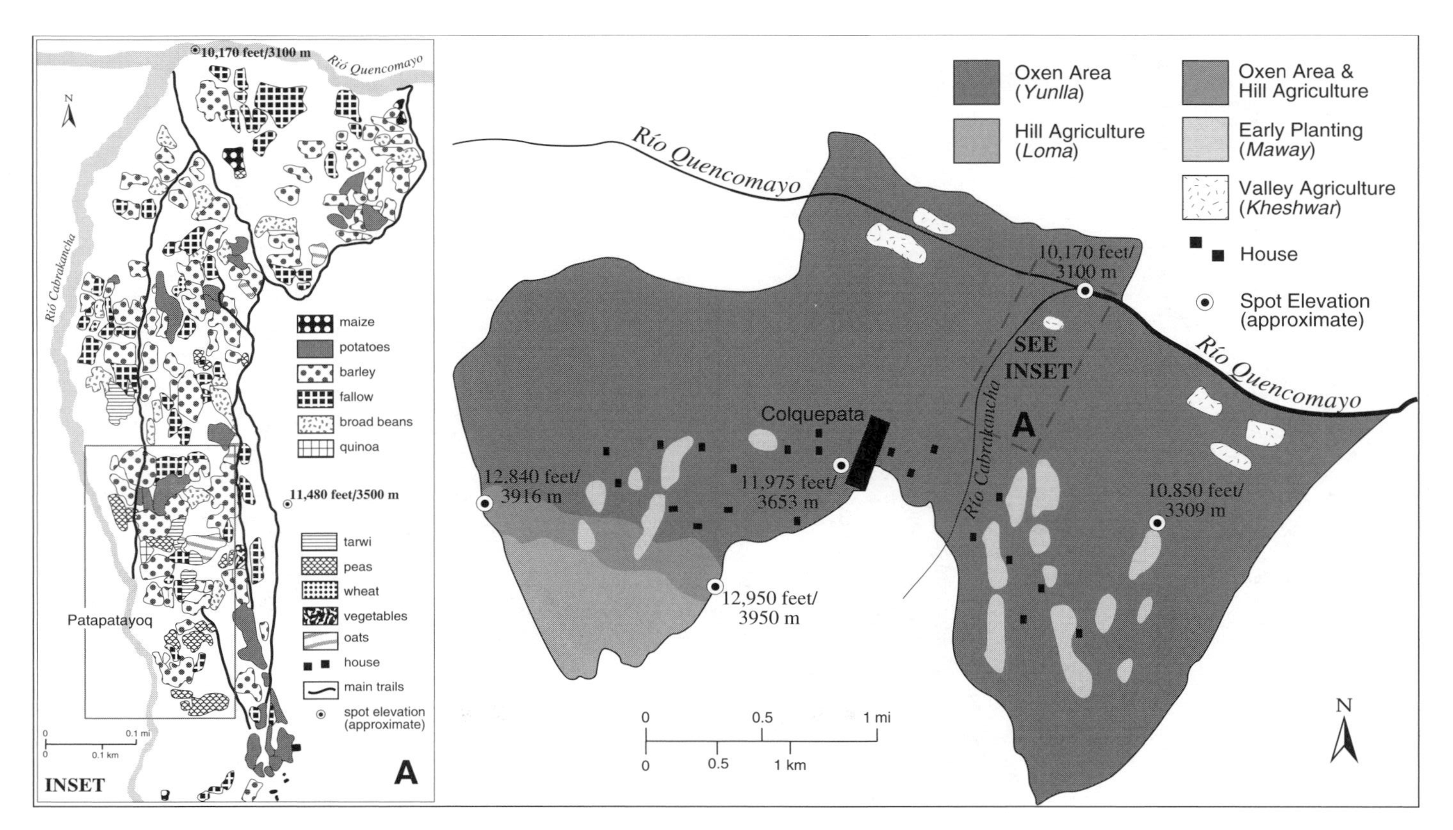

Map 9. Production Spaces of Colquepata Community (Paucartambo) in the Mid-1980s.

the prominent patches of perennial wetland. Due to the worsening debilitation of their livestock through illness and death, the Quechua people ceased putting a premium on bogs for grazing purposes. Indicative of the decline of the livestock economy, communities such as Colquepata and Chocopía were the only ones in the Paucartambo region to cut the size of herds between 1970 and 1987. Families halved their herds, mainstay sheep flocks being reduced to an average of eleven per household and cattle holdings to an average of one. Juan Santos and Fortunata Wallpa recounted their parents' land use in Chocopía by referring to that generation's economy being based on livestock; in their words, the older people lived as "livestock folk," rather than as farmers like themselves.

Coincident with this ecological deterioration and the fall of livestock-raising, Juan, Fortunata, and other Colquepata farmers began in the 1960s to divert growing farm resources into Early Planting plots seeded in the bog sites. The innovative farmers converted their bogs to fields in order to feed the escalating demand of urban markets for a year-round supply of fresh potatoes (Zimmerer 1991d). The Early Planting's remarkable success was also due to the timely liaison of new technologies and production subsidies. Taking the form of farm loans, subsidies were made available in Colquepata as the result of unpublicized links between agricultural credit, local politics, and Peru's national government. The links surfaced when the national Agrarian Bank disbursed small farm loans in Colquepata after the Land Reform of 1969 under guarantees and encouragement from local agents of the nationwide System for Social Mobilization (SINAMOS), an agency of the Velasco military government designed to foster pro-state beliefs and consolidate the countryside's sentiment against hacienda estates.

Paucartambo offices of the antiestate SINAMOS were headquartered in Colquepata village. As a political institution, the presence of SINAMOS there was neither selfless, effective, nor even long-lived. In the wake of the Land Reform of 1969, however, the agency's officials did steer a small albeit crucial flow of funds into coffers of the Quechua peasants who were converting bogs into fields of the Early Planting. The headquartering of SINAMOS in Colquepata helped to initiate a peculiar process of agricultural change. Its decision to locate in this district capital could be attributed to the fact that the local agents found a social atmosphere at least mildly congenial to their mission. The town of Colquepata and the Quechua who lived there and in the nearby countryside expressed a notorious legacy of bitter antipathy toward the haciendas. By contrast, the other Paucartambo villages were more dominated by ex-estate owners, townspeople, and mestizos who vehemently opposed the measures of the Land Reform of 1969.

A traditional ferocity of antiestate beliefs in Colquepata was traceable even to the colonial period when its Quechua peasants carved a presence that was overwhelmingly dominant in numbers and seemingly undeterred in struggles over livelihood-giving territory and resources. Unlike the villages of Paucartambo and Challabamba where many mestizos and Spaniards resided according to a Church census in the 1680s, that of Colquepata had consisted solely of four hundred "natives" (Villanueva Urteaga 1982). Centuries later in the 1980s, the non-Quechua inhabitants of Challabamba and Paucartambo still referred without hesitation to Colquepata as a peasant town or an Indian town. A substantial fraction of Colquepata's populace consisted of village-dwelling Quechua who also held farm land and belonged legitimately to Peasant Communities. Bog-cultivating Juan Santos and Fortunata Wallpa, for instance, lived on Ayacucho Street in Colquepata village. Their presence imparted a strongly indigenous flavor to life in Colquepata village that was distinct in the region.[16]

The Quechua of Colquepata had earned a reputation as violent and subversive when necessary in their conflicts with the small but oppressive clutch of mestizo villagers and neighboring hacienda owners. The educated son of a mestizo family from Colquepata had divulged the fear and hatred that permeated village life in an essay written at the University of Cuzco in 1959: "The Indian is not simply suicidal: he hates the mestizo to eternal torment. In mass rebellion he is ferocious, cruel and bloodthirsty" (in Allen 1988, 28). Their reputation of rebelliousness was rooted in the recurrent conflicts between the thousands of Quechua peasants who had resided outside estate jurisdiction and a medium-size corps of nearby estate owners and landless villagers who intermittently tried to usurp farmland and labor resources from them. One dozen or so conflicts in the 1900s that led to reports archived by regional administrators suggested a landscape of nearly perpetual conflict (Zimmerer 1991d).[17]

The ecological infestation of bogs by liver flukes further strained the tensions in Colquepata over land. To make matters worse, much farmland near the deteriorating bogs was already badly worn. A conjuncture propitious for the Early Planting thus existed in the late 1960s: the ecological deterioration of bog pastures and surrounding farmland combined with the unabated antiestate attitudes of the Quechua in Colquepata to lure the ideologically sensitive and geographically wary SINAMOS officials from General Velasco's government to locate in Colquepata rather than Challabamba or Paucartambo. SINAMOS agents subsequently sponsored a spurt of low-interest loans through the mid-1970s that capitalized the conversion of bog wetlands into Early Planting parcels. Via the local implementation of national government's farm policy, the living legacy of peasant protest was thus translated into a wave of landscape modification.

Conversion of the Colquepata bogs into Early Planting fields benefited from the weaker degree of social domination by villagers. Less encumbered by social debts and economic obligations than their counterparts near Paucartambo, the Colquepata farmers were more able to commercialize farm production following 1969. Their entrepreneurial spirit and economic success drew several truckers—including owners from Cuzco unaffiliated with the Paucartambo trucking mafia—to traverse the potholed High Road between Mica and Sipascancha and negotiate the purchase of bulked crops from farmers along the route. It was along this road that new commercialization was starkly transforming the economy of Sonqo community by the mid-1970s (Allen 1988). The Quechua in Colquepata also sold products at a regular Sunday marketplace in town. Not least, low freight costs to Cuzco favored the farmers, who shipping from Colquepata paid one-half of the chronically inflated per unit price that was charged from Paucartambo to Cuzco.

Local knowledge and innovative techniques fixed the Early Planting in Colquepata bogs as a unique system of wetland or hydraulic agriculture.[18] While SINAMOS agents preached ideology and guaranteed loans, they did not deliver much technical advice of merit. To cultivate a soggy bog, the Quechua farmers like Juan Santos and Fortunata Wallpa engineered an ingenious system of soil drainage and field elevation. Ditches connected in a distinctive herringbone shape drained a series of elevated wachu beds. The combination of dug canals and mounded beds made up a so-called raised field (Zimmerer 1991d, 1994b; see also Denevan 1980; Parsons 1971). The Quechua of Colquepata learned the details of raised field farming through observation, often while working, in communities of the nearby grassland-moor of the Cordillera Vilcanota (such as Pampallaqta in Calca). The farmers then adjusted the borrowed techniques, refining the methods of canal scraping and laying out lateral ditches. While borrowing the basic design for their new raised fields, rather than inventing it, the Colquepata cultivators applied the lessons of trial-and-error experience to polish a new jewel of their farming landscape.

Expansion of Early Planting fields led farm families in Colquepata to reapportion resources and, in the end, to reconsider once more their production and consumption of the diverse crops. To till the wetland Early Planting, the Quechua farmers had to muster labor for no less than one dozen tasks that on the average totaled 149 person-days per hectare (Zimmerer 1991d, 1994b). Families themselves could provide much labor, since a variety of intensive but minor tasks like canal grading, repeated weeding, and careful monitoring did not mandate the work teams that typically were filled by extra-household recruits. In fact, a number of poorer families were able to adopt the Early Planting in bog sites by exploiting mostly their own labor. Many new demands for labor-time were shouldered by women, who weeded and monitored the dicey parcels of the Early Planting. Fortunata once joked that she was married to their bog fields and that her husband, Juan, was the son of this union, his main chore being to keep the wachu beds in good shape.[19]

Meanwhile the Quechua farmers were adjusting other elements of their livelihood strategies. Many diminished and even discarded maize and quinoa, although their Valley farming was located in the Quencomayo gorge and thus distant from the battery of Early Planting parcels that covered the middle slopes. Farm spaces mapped in the Peasant Community of Colquepata, representative of the interior communities, showed the definitive separation of the two production spaces (map 9). A small area of Valley cropping was set deep in the Quencomayo gorge, while field-size islands of the Early Planting were scattered in the distance at intermediate elevations. Located between the Valley and Early Planting regimes, the Oxen Area sprawled expansively both on the walls of the gorge and across the surrounding upland. This image of Valley fields buffered by the Oxen Area from insular Early Planting fields was deceptive, however, for those juxtaposed shapes concealed a string of strong and influential ties.

It was the allocation of labor-time by farm families that especially knotted the expansive Early Planting and the decline of landrace-rich Valley farming. The role of labor-time as a binder of production spaces was woven through the "generalist" mode of land use whereby the Colquepata families held fields in all or at least most production spaces.[20] Many families who owned both Early Planting and Valley parcels were being pinched for labor in September, October, and November at the springlike onset of the rainy season. By the 1970s, that three-month stretch of the work calendar saw the Quencomayo farmers slogging mightily in their wetland fields, mounding the planting beds and grading the all-important canals. The same crucial months, however, were the ones when the maize and quinoa fields of Valley production demanded tillage and planting. Being squeezed more tightly for labor-time at this early stage of the growing season, many Quechua families decided to discontinue the maize and quinoa crops.

Valley planting of maize and quinoa shrunk to a small fraction of its former area except in a few prime sites on river terraces at the base of the Quencomayo gorge. A survey in 1987 of four Peasant Communities with farmland in the gorge recorded the following sparseness of maize and quinoa parcels: Colquepata (five), Huaranca (two), Roquechiri (six), and Paucona (five). The paltry smattering of maize production in the Quencomayo bottoms contradicted a common assumption that maize growing defines the landscapes of Andean agriculture below 11,500 feet (3,500 meters) (Troll 1968). Three decades earlier in that same gorge and its side canyons like the Cabrakancha, however, a predominance of maize and quinoa parcels stuck closer to the Andean ideal. That ideal faded beginning in the 1960s when the Quechua of Colquepata set to deflowering the Valley landscape by pushing the diversity-poor Oxen Area downhill over wide areas (map 9).

By the 1980s, mainly artifacts and place naming told of the former abundance of maize and quinoa in the Quencomayo Valley. Vestigial Inca terraces of dressed stone, earlier grown with maize, lined the base of its southern slopes. Place names and the verbal accounts of elderly inhabitants also recounted the recent history of the Valley landscape. The Quechua *sara,* for maize, bonded the names of once central growing sites such as Sarapampa (Maize Flat) and Sarachimpa (Maize Barter), where the crop had flourished in scores and maybe hundreds of fields. Quinoa farming was not recollected as vividly as maize, although its decline during the postreform decades could be gauged via the fate of maize, with which it was habitually intercropped.[21]

The extent of the quinoa crop in the Quencomayo Valley shrunk during the years from 1969 to 1990, although it was already much reduced due to the earlier dynamics of wheat and barley expansion. The Quechua of Colquepata decried a welter of disadvantages that had recently weakened their resolve to carry forth with the cultivation of quinoa (appendix E.3). Declining soil fertility in fields counted as one critical factor. Regardless of whether they had once intercropped quinoa with maize or sown it solo, Quechua farmers found that the deteriorating quality of field soils was pressuring them anew to desert the nutrient-hungry quinoa. Changes in the use of a modest but key resource, that of livestock manure, was behind the worsening degradation of Quencomayo soils after the Land Reform of 1969.

Prior to 1969, the Quechua of Colquepata had renewed their maize and quinoa fields with a rich mixture of manuring techniques: the herding and stake-and-tethering of livestock in newly harvested fields, the transport and application of manure scooped from permanent corrals, and so-called guano-giving of livestock penned in field areas prior to planting. Farmers had relied especially on manure of the cattle that customarily grazed on maize stubble. Their stubble-grazing cattle not only gained a high quality feed but also could be watched or tethered more easily in Valley parcels than in distant agricultural areas. But the average size of the Colquepata cattle herds was reduced to half during the period from 1969 to 1990, when Quechua families adjusted their herds to reflect the loss of bog pastures. By 1990, the average of one head per family was one-fourth or less the size of cattle herds elsewhere in Paucartambo, adding to the forces on farmers to forego quinoa cropping. Close connections between livestock-raising and cropping—captured in the term *agropastoral*—were once again of clear import to the diverse crops.

The aggressive adoption of malting barley factored no less forcefully in the deterioration of soils and the crippling of quinoa cropping. One diversity-depleting scenario was triggered by changes in field rotation and fallow stemming from the barley boom. Many barley growers chose to sow it in the second year of their rotation sequence. Revising the cropping cycle, they seeded barley in lieu of tarwi and broad beans, both legumes whose roots had pumped nitrogen into worn soils. Replacement of the fortifying legume crops by barley thus diminished soil fertility still further and hastened the decline of quinoa. By the 1980s, the farmers of Colquepata noted the spread of *puka kora* ("red weed" or *Polygonum sp.*) and *mullaca* (*Muehlenbeckia volcanica*), both indicative of poor soils (Zimmerer 1989). Also by that decade the area's cultivators tilled less quinoa than any other farmers in Paucartambo. Their scarcity of quinoa fields registered also at the extraregional scale, since most other Cuzco regions were estimated to cultivate larger areas (Fano and Benavides 1992).

The success of barley in the Quencomayo Valley transformed the crop landscape of the gorge and its side canyons into a nearly unbroken expanse of the Oxen Area. Second only to the southern Paucartambo Valley in barley acreage, the Quencomayo watershed was covered with countless barley fields between the river channel at 10,150 feet (3,100 meters) to upward of 12,500 feet (3,800 meters). The Paucartambo-based agent of the Beer Company of Southern Peru referred to his contracted growers in Colquepata as "good agriculturalists aggressive both in the field and out of it."[22] By adopting a common rotation of their barley with improved variety potatoes, the Colquepata Quechua engorged the Oxen Area and solidified it as the predominant production space in the farm landscape (map 9). Viewing a picturesque panorama of fields and pastures in 1987, the experienced Juan Santos observed dryly: "The barley and 'new potatoes' have eaten up the kheshwar (Valley) and nibbled at the loma (Hill), but we can't eat them."[23]

Despite fleeting reservations about their status as one of Cuzco's chief barley-growing districts, the Quechua of Colquepata admitted that barley fit neatly with the farm changes undertaken after 1969 (appendix E.4). One chief advantage was that it could be grown with less labor-time than the other crops. Of equal import, barley withstood impoverished soils; when planted after the fertilized potato crop, barley yielded well without added inputs. Cultivators could thus generate the beer company's materia prima in degraded soils that were inadequate for the landrace crops, an impetus for barley's widespread adoption in various other Andean regions of Peru and Ecuador as well (Mayer 1985; Weismantel 1988). Thirdly, if cultivators chose, the griñon barley could be made to flourish on a diet of mineral and chemical fertilizers. Barley growers could, therefore, counter declining soil fertility with purchased amendments, exerting a bias in favor of barley and high-yielding potatoes even if the farmers merely intended to fertilize their fields.

Quechua farmers muted the sociospatial resistance that might have impeded the abandonment of diverse crops in this radical reconfiguration of farm space. By creating a landscape of highly fragmented and overlapping farm spaces, they unintentionally aided the course of landrace-eroding changes. The small Valley patches sown along the river channel and the Cabrakancha Canyon were vulnerable since they adjoined the Oxen Area. At higher elevations on the middle slopes, the Oxen Area was interdigitated with fingerlike extensions of the Early Planting. This transfiguration of farm space collapsed former patterns of spatial cohesion in Colquepata farming, including a system of sectoral fallow commons that earlier in the century had spread over the community (Zimmerer 1991e). Spatial transfiguration in the years between 1969 and 1990 thus offered little impediment to the commercial Early Planting and barley growing. Few farmers, even among the more resource-rich families, elected to keep the maize and quinoa crops.

Galloping commerce did enhance the income of the Quechua in Colquepata, even if not as much as they hoped; however, it also subjected them to a hydra head of economic risk, soil erosion, and, not least, dietary simplification. Economic risk rose because of uncertainty in both the cropping and the marketing of their new ventures. Periodic frost, disease, or pest outbreaks could damage even the most expertly scheduled and well-tended Early Planting. Off-season prices in Cuzco could veer unpredictably, especially in the 1980s when fledgling off-season production in other Cuzco provinces began to undermine the seller's market that customarily awaited the bog-field harvests. Barley prices, too, fluctuated sharply notwithstanding the sales contracted by the Beer Company of Southern Peru. An equally unsettling risk for the peasant growers was that company officials could adjust the price of contracted barley abruptly and extra-officially via the dreaded grain-grading that took place at its Cuzco factory.

Soil erosion from the already worn slopes of Colquepata was becoming worse due to the curtailment of field terracing. Fields leveled into small platforms, or so-called lynchet terraces, had long aided irrigation and eased the management of soils on the treacherously steep slopes of the Quencomayo Valley and its side canyons. The Quechua dubbed their rustic yet effective terraces as step-step, or *pata pata,* even naming a terraced section of the steep Cabrakancha Canyon as "place of the pata pata," that is, Patapatayoq (map 9). To construct a step-step terrace, they would level a small-size field and preserve a waist-height wall of sod—or less commonly erect one of undressed rock—at the field's lower hem. The farmers typically stacked a number of the sloping field terraces that then ascended steplike in the canyons.

After several years of sowing in one of their sod-terraced fields, the farmers broke down its walls and those of adjoining parcels in order to weld a single large surface without terraces for planting. Their act of wall-breaking enriched the new field with a flush of nutrients released from beneath the shrubby cover, or *charamusqa,* which was found on and around the former between-field areas. The Quechua in Colquepata anointed their newly joined field with a special name, calling them either pass-through slope (*pasaq qhata*), big skirt (*falderón*), or plowed single field (*huqllay yarapun*). Although field enlargement helped to rejuvenate soil fertility on a short-term basis, the sites without terraces inevitably began to lose nutrients after only a few years and, at a slower rate, the soil itself was being lost.

Step-step terracing lost ground during the period from 1969 to 1990 due to a failing supply of the labor-time that was needed in construction and maintenance. One indicator that farm families felt a scarcity of labor was the noticeable lack of new terraces built with time-consuming rock walls. Becoming more common were the terrace types fashioned by merely digging into slopes and leaving unfinished sod to function as riser walls. The overall quantity of terraces of both types was dwindling on the steep slopes of Patapatayoq and other canyons due to the farmers' withdrawal of labor-time. By the mid-1980s, terraceless fields of pass-through slopes smoothed over areas larger than the stepped and leveled ones, implying more wall destruction than terrace building.[24] A striking predominance of unprotected fields scratched from some of the area's most pitched terrain clearly aggravated soil erosion, and in some sites it jeopardized the medium-term viability of farming.

A dietary shift could also be sensed in the shrinking cultivation of maize and quinoa landraces. The two diverse crops and the dishes prepared from them became scarcer in the diets and cuisine of Colquepata people. To compensate for their loss, the Quechua farmers consumed ever greater quantities of inexpensive macaroni and spaghetti noodles purchased at village stores. While sometimes able to barter or buy the disappearing crops, they could not count on the exchange mechanisms to supply as much as was customary under self-provisioning. In short, the landrace-abandoning farmers simplified their diets and cuisine by substituting a single item—thrifty wheat noodles—for the diverse landraces. Although fairly loathed, the noodles sold for so little they were in effect obtained for less than the cost of producing the dear staple crops.

Of their dietary losses, the Quechua in Colquepata grieved most over the lack of homemade maize beer. Most families had brewed the endearing aqha two dozen or more times each year from a Yellow Maize landrace, or k'ellu sara, harvested from the small seed planting. This stalwart of maize farming was a causality of farm change, however, and aqha or chicha brewing customs changed by the dint of perceived necessity. By the mid-1980s, most Quechua in Colquepata fermented their frothy maize brew on less than five occasions each year. They bemoaned the lack of raw material for their favorite beverage wistfully but were without the will or the way to restore it.[25] Most found themselves producing the raw material for the beer-making industry without too frequently purchasing the bottled commodity in its final form.

The biological consequences of landrace loss in the Quencomayo Valley were moderate when judged in terms of individual landraces. Each landrace formerly sown in the drastically reduced medium and small seed plantings of maize occurred also into the southern Paucartambo Valley (see map 6; table 6). Because seed exchange distributed these maize types well beyond the Quencomayo gorge, a case of disappearance from that area thus imparted an unnoticeable loss of landrace diversity at the regional scale (Zimmerer 1991c, 1992a). A similar lack of endemic distributions in the quinoa crop nullified any extirpating impact of its local curtailment. Despite the buffering effect of widespread distributions in both crops, the local abandonment of landraces might have led unique genes or multiple-gene complexes to vanish, a corollary that could not be assessed without more careful study (Quiros et al. 1990; Zimmerer and Douches 1991).[26]

Hill farming, in contrast to Valley farming, oversaw the de facto conservation of a handful of diverse crops that continued to be seeded notwithstanding significant changes. After the Land Reform of 1969, the Quechua in Colquepata amplified a technique in which they intercropped the minor tuber-bearers of ulluco, oca, and mashua and the predominant floury potatoes. Their successful innovation of tuber intercropping within fields of the grassland-moor was prompted by worsening shortages of amenable farmland. Land shortages at upper elevations in the communities like Colquepata, Chocopía, and Ccotatoclla derived from farmers extending upslope the Oxen Area and a new variant of the Early Planting known as the middle early planting, or *chawpi maway* or *machu maway,* since it was planted outside the wayllar bogs.

Farmers were busy levering the improved potato varieties of the Oxen Area and the middle early planting upslope as high as 12,800 feet (3,900 meters) in parts of the Colquepata landscape such as the ravine of Pantini Wayq'o in Chocopía. Yet, due to their innovative within-field mixing of diverse tuber crops, many Quechua families could still enjoy small albeit substantial supplies of at least some savory foodstuffs. Their resourceful intercropping of the tuber crops—the *haynachu* technique as they called it—established an example of de facto conservation that also proved viable in the quite different setting of the southern Paucartambo Valley.

A Case of Conservation: Innovative Intercropping in the Southern Valley

South and upstream in the Paucartambo Valley, a few thousand Quechua farmers planted more of the diverse crops than were persisting in the loss-prone valleys of the Quencomayo and northern Paucartambo. The cultivators of the southern valley and particularly those of Mollomarca (1985 community population: circa 350) were undeterred in sowing sizeable areas of potato, maize, and quinoa landraces after the Land Reform of 1969. Only their ulluco crop diminished to the detriment of diversity. More obvious was the vigorous planting of diverse potato landraces and maize and quinoa parcels. Communities such as Mollomarca, Umamarca, Humana, and Carpapampa continued to attract a notable clientele of national and international crop collectors through the 1980s. Despite the loss of a few landraces, such as the absence of Fist, or saqma, that alarmed CIP's Carlos Ochoa, the potato hunters still returned regularly to these southern valley redoubts.

Abiding cultivation and de facto conservation of diverse crops did not infer a retardation of farm commerce in the southern Paucartambo Valley. Quite the contrary, its farmers contracted barley and vigorously adopted potato commerce of both the Early Planting and the Oxen Area; however, they also sowed intermixtures of maize, quinoa, and minor crops in their Valley fields, thus differing from their Colquepata counterparts who deflowered Valley diversity. Key to this difference was that the southern valley cultivators managed to intersperse diversity-rich Valley fields with new Oxen Area and Early Planting plots. They adopted the newcomers without pushing landrace-rich cropping below crucial thresholds. Their success hung on widespread access to relatively substantial resources that enabled them to couple farm commerce and diversity-based livelihood customs.

Farming change in the southern Paucartambo Valley was exemplified by the landscape that took shape in Mollomarca during the 1986–87 season (map 10). The community's irregular mosaic of production spaces was formed amid the salient transitions of farming between 1969 and 1990. One shift involved the upslope extension of barley and of high-yielding potatoes. It was apparent in the mid-1980s that the Mollomarca people and other southern valley Quechua had lifted the upper limit of their Oxen Area plantings as high as 13,100 feet (4,000 meters), more than 500 feet (150 meters) above where this limit was notched at the time of the Land Reform of 1969 (Zimmerer 1991c). The farmers were able to stretch the Oxen Area into the abbreviated seasons and adverse climates at higher elevations with aid of the newest potato releases that delivered frost-tolerance and fast-maturation as well as high yields. The most recent trendsetter in this regard was the mariva potato that first arrived in Paucartambo in 1980. By hoisting their mariva-filled parcels of the Oxen Area ever higher, the southern valley Quechua were causing the lower fringe of Hill farming to retreat noticeably.

The contraction of Hill farming was an increasingly common but poorly understood dilemma for the diversity of thousands of potato landraces sown throughout the Peruvian Andes. The general nature of the threat has received a shower of regional studies (Brush 1986, 1987; Mayer 1979, 1985; Ochoa 1975). On the one hand, in the Andes of northern Peru high-yielding potatoes spreading upslope into the highest areas of farming directly replaced a wealth of diverse potatoes by early in the 1970s, resulting in catastrophic genetic erosion (Ochoa 1975). In Peru's central highlands, on the other hand, the diverse potatoes appeared to be conserved during the same period via a process of upslope contraction, or upward adaptation, without the loss of landraces (*adaptación hacia arriba*; Mayer 1985). But claims for the dissimilar scenarios controlling the evolutionary fate of diverse potatoes never assessed the historical or cross-temporal fates of the landraces themselves, a lacunae that could be redressed in the case of the Paucartambo Andes.

By comparing the comprehensive lists of potato landraces collected in the southern Paucartambo Valley by renowned crop scientists during the 1920s, 1930s, and 1940s with my extensive field samples taken in 1986, 1987, and

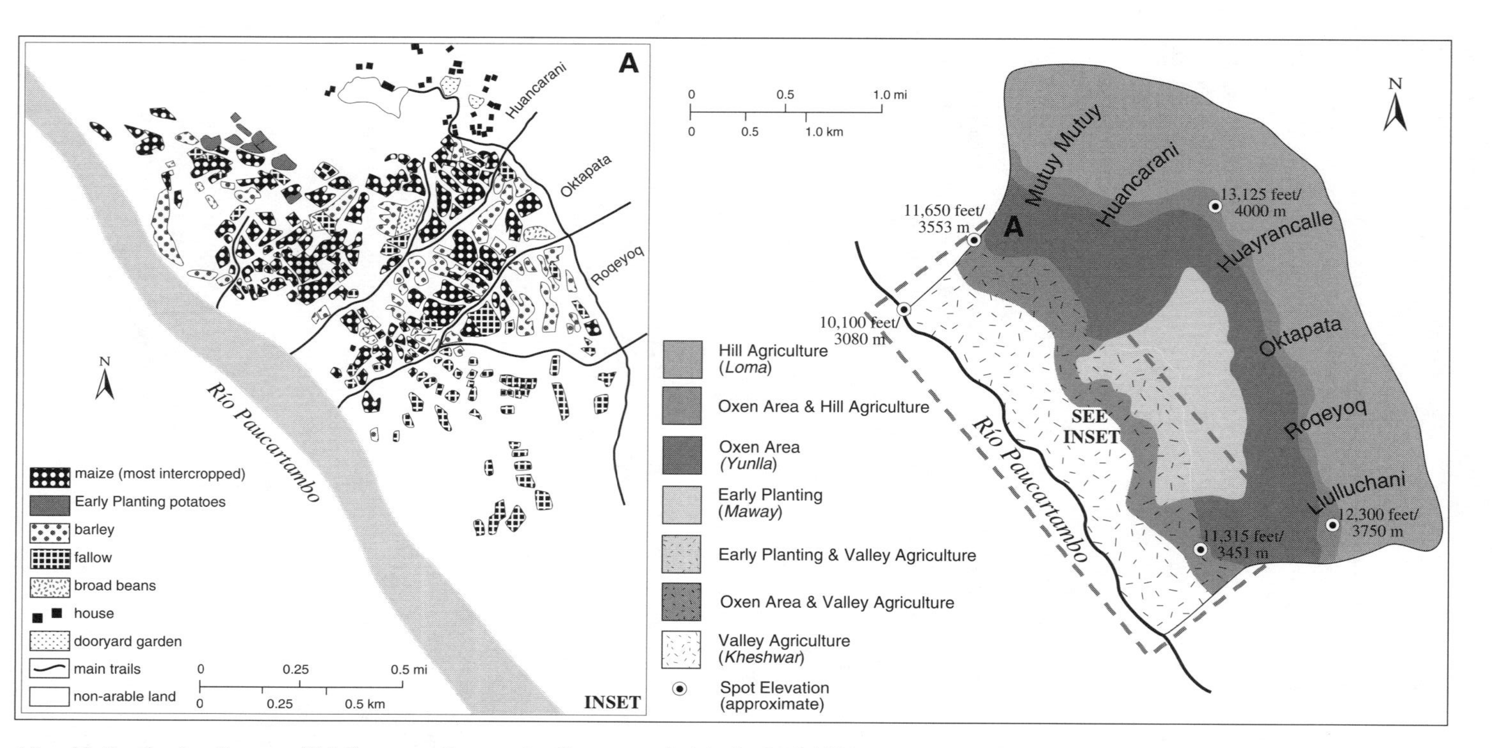

Map 10. Production Spaces of Mollomarca Community (Paucartambo) in the Mid-1980s.

1990, an estimate could be made of the impact on floury potato fortunes when Hill farming underwent noticeable contraction at its lower fringe in the Andean landscape (Bukasov in Villasante Ortiz 1975, vol. 1:43–51; Hawkes 1941, 1944; Vargas 1948, 1954; Zimmerer 1991c). A truly remarkable record of conservation was found to characterize the area's diverse floury potatoes. In fact, the majority have persevered under cultivation since the 1940s. Of 203 landraces and landrace subtypes collected during the 1930s and 1940s, a total of 183 appeared again in the 1986–90 samples.[27] In short, no less than an estimated ninety percent of floury potatoes continued to be cropped.

The estimate of persistence at the ninety percent level may well have underestimated, moreover, the actual extent of living diversity. While the earlier collections aspired to find the greatest number of potato types, the 1986–90 sample was derived from a randomized technique designed to sample fields in a systematic method rather than in a diversity-maximizing fashion. As a consequence, the disparate methods biased the cross-historical comparison toward an underestimate of the floury potato diversity extant in the more recent period. Underestimation likely resulted also from the effect of prevalent synonymy in the naming of floury potato landraces, which was adjusted for in the samples of the 1980s and 1990 but not in the earlier collections (see table 8 for a detailed list of synonyms). In any case Quechua farmers of the southern valley definitely carried forward their cropping of most floury potatoes even while they displaced them to higher elevations during the postreform decades.

Cropping and thus de facto conservation benefited from a variety of basic ecological features in the floury potatoes and in Hill agriculture. The diverse potatoes contributed flexibility to the scheduling of major field tasks because they resisted diseases. Their capacity to fend off rotting diseases such as the ruinous soft spot of the bacteria *Erwinia* was especially helpful in making the timing of harvest more flexible. Welcome pliancy in work routines owed also to the cool climate of Hill agriculture, which tended to shun disease and pest outbreaks and thus permit innovations in the cropping calendar. Unearthing the tubers bit by bit over one or more months rather than in a hasty harvest was one modest but helpful option. The flexibility to be used in alloting labor-time at harvest deviated greatly from lower-elevation farming where chronic infestations of insect larvae—many known as potato worms, or *papa quru*—mandated no real measure of malleability. Hill farming's agroecology, by contrast, helped cultivators to shoehorn the diverse floury potatoes into livelihoods that were increasingly pressed for labor-time.

Innovation of potato farming techniques in the 1970s and afterward also alleviated the exigencies of labor-time scheduling. In advance of seeding their paragons of crop diversity the Quechua farmers of the southern valley sometimes used the chakitaklla foot-plow to dibble finger-deep holes at the cost of little labor—a technique known as *tikpa* or *chuki*—rather than mounding the full-fledged beds. By planting the potatoes in tikpa holes in October, the farmers could then defer their labor-intensive mounding until they were able to muster more labor, typically in November and December. The point of chuki planting was thus to postpone the peak labor-time demands of their elevated wachu beds. Although the innovative technique did not truncate their overall need for workers, it did nonetheless add a much-needed element of flexibility into the cropping calendar and thus cushioned the conservation of floury potatoes. Except for versatility like this, cropping marched to the seasons without much variation.

A second innovation in tillage also shortcut labor-time allotments. In the innovative oxen-plowing technique aptly described as *suka* or *th'aya* (suka being a Quechua version of the Spanish word *surco,* or row), the tillers could speedily furrow fields for their potato planting. Although a proven labor-saver, the planting beds furrowed in this way were often too small to bury the long underground stems, or stolons, of the floury potatoes. In general, therefore, farmers were inclined to apply this row tillage technique to floury potatoes only under the most dire shortage of labor-time.[28] Overall, an impressive suite of innovations that were time-saving and flexibility-enhancing thus aided the floury potato farmers, unlike the rigid and unrelenting calendar of the luckless growers of early chawcha potatoes in the northern Paucartambo Valley.

The common manner of working with seed lightened the load of floury potato farming still further. By treating the full complement of intercropped seeds as a single mixture, rather than as a series of individual landraces, the southern valley farmers were resorting to another labor shortcut.[29] The technique of mixing the seeds of diverse landraces enabled a single collection to be rapidly chosen, piled, and later planted together. Their treatment of the entire set of floury potato seeds as a single mixture definitely saved time in postharvest processing. If each of the twenty or more landraces per field were handled separately, by contrast, the work routines would have been much more complex and time-consuming. That burdening of work routines would have especially troubled farmers during the post-1969 decades.

So widespread was the custom of handling floury potatoes as bedazzling mixtures of multiple landraces that the Quechua farmers often assigned the term *mixed potatoes,* or *papa charqho,* to the landrace group.[30] Quechua farmers especially liked to refer to mixed potatoes when contrasting this cultivation with the high-yielding potatoes, for the latter were customarily sown as a single variety in each field. The popular style of working with the floury potatoes in mixtures did not seem to precipitate the loss of landraces. In fact, the overall diversity of floury potatoes was fairly unchanged in the southern valley while mixing landraces was commonplace. Measly attention granted to single landraces may, however, have jeopardized the future of rare ones. For instance, the saqma landrace whose absence in Humana agriculture had alarmed Carlos Ochoa may have been lost as the result of waning attention to an already rare landrace.

The popularity of floury potatoes in local cuisine helped to fortify this production. Indeed the diverse potatoes enjoyed an unstinting appeal and sociocultural esteem among the Quechua people of the southern valley. Frequent exchange of floury potatoes in nonmarket economic relations reinforced still further this cultural value. Many field owners, exemplified by the well-off Líbano and Natividad in Umamarca, preferred to compensate their farmhands with in-kind payments of their colorful floury potato mixtures. During the harvest of 1987 two arrobas, or twenty-three kilograms, were exchanged for one day's labor. The recruited worker was given a gunny sack half filled with diverse potatoes after his or her labors. In Nova's *Seeds of Tomorrow* Don Pedro, who was interviewed in the central Peruvian highlands, was seen devoting a portion of his floury potato harvest to this purpose.

A persistent custom of bartering floury potatoes was used to help cement friendship and other interpersonal bonds such as social kinship between a field producer and other persons. It also symbolized a kind of complementarity expected to unite the potato growers and the Quechua cultivating other crops, most particularly the specialists in maize-farming. Farmers in the southern valley, like those in other Andean regions, believed that the growers of diverse floury potatoes inhabit a cultural geographic realm that complements the maize growers, and vice versa (Salomon 1985, 1986b; Weismantel 1988, 9). The cultural geographical bonds of crop and people were reaffirmed through the regular bartering of floury potatoes for maize: one chimpu's worth of floury potatoes for the same quantity of maize ears. Finally, barter for floury potatoes often happened to be the most expeditious means of obtaining seeds for future planting or sporadic consumption.[31]

Only a few floury potatoes circulated in southern valley commerce, a contrast to the many that were exchanged via barter. The Quechua cultivators there controlled how a threshold kept the riches of diversity in multipurpose planting while proportioning scant variation—typically a single landrace per field—to commerce. The twenty or so farmers of Mollomarca community who devoted whole fields to landrace commerce sowed just one type or, at the most, a few: typically maktillu, qompis, pitikiña, olones, or *boli.* Nor did the retail vendors of floury potatoes sell the same range of diverse landraces that were customarily exchanged in barter and payment-in-kind arrangements. The vendors charged that while buyers sought floury potatoes they did not value the varied mixtures any more than single landraces. The reluctance of Quechua farmers in the southern valley to commercialize the full scope of their diverse potatoes was thus a rational reaction to market conditions. They were not assured a compensating rise in reward from sales if they cultivated many varieties for market.[32]

Maize cropping also was kept central in the farming endeavors of southern valley people. Cultivators sowed a large number of maize fields belonging to the medium seed and small seed categories. Their fields, higher and drier than the northern valley, did not permit the big seed planting. Maize fields stitched a main fabric into the crazy quilt of Valley, Oxen Area, and Early Planting parcels that covered Mollomarca's inner gorge between the river channel at 9,850 feet (3,000 meters) and an upper limit near 11,650 feet (3,550 meters) (map 10). Three out of four families in the community seeded a maize field each year, which attested to their generally favorable access to farmland. In terms of biological diversity the upstream farmers grew fourteen of the twenty-two medium seed and small seed landraces.[33] The place of the southern valley as a center of maize diversity during the period from 1969 to 1990 was made still more noteworthy by the crop's loss from the Quencomayo Valley of Colquepata.

Quinoa as well as tarwi and fava beans staked secondary but nonetheless vital claims to ongoing cultivation. Innovative intercropping within the Valley fields of maize succeeded in renewing these minor crops. Quechua farmers typically intercropped them either in repeating kunka throats, into openings within maize rows, or as border plants (plate 9). Their common use of intercropping bespoke the firm commitment of the farmers to the minor but heartfelt contribution of quinoa to a fit livelihood. In fact, they curtailed the quinoa crop only when the maize-quinoa pair itself could no longer be feasibly cultivated. Their wide-ranging motives for conserving quinoa ranged from an aesthetic appreciation of its bright, bushy seed heads to its cherished culinary qualities. Intercropped quinoa could also divide quite visibly a field being shared by two farmers. Many growers also found the intercropped quinoa to be useful in ordering the multiperson work team that would labor in a maize field. Quinoa throats split a maize field into a number of sections, each known as a *tabla* or *suyu,* which helped clarify the work quotas assigned to each person (see also Cobo [1653] 1979, 212; Urton 1984).

The widespread adoption of barley mostly bypassed Valley farming in favor of the more fertile and less steep fields reposed on the middle slopes. Commercial transformation of the choice fields above 11,650 feet (3,550 meters) was wrought through the unmatched spread of griñon malting barley being grown under contract with the Beer Company of Southern Peru. After 1969, the southern valley upstream of the provincial capital as far as Mollomarca supported the chief bridgehead for barley growing in the Paucartambo Andes. The existence of countless threshing floors and plentiful horses inherited from the earlier wheat-growing emphasis of hacienda estates had proportioned the southern valley with a nicely suited infrastructure. A further advantage for the barley boom there was the social leverage of Paucartambo villagers. Many villagers benefited from the dense ganglia of patron-client ties that they tended upriver by using their insider stocks of barley-growing contracts.

Plate 9. An intercropped Valley field at 11,500 feet (3,500 meters) near the hamlet of Mollomarca. Early maize of the small seed category is intercropped with rows of quinoa, known as throats, *and a border of tarwi and broad beans (lower right) and flowering roses (center).*

The booming barley crop of southern valley communities such as Mollomarca, Huaynapata, Umamarca, and Queskay cemented a marked cohesion of land use within the outstretched Oxen Area. By sowing their barley in the cohesive units of sectoral fallow, the southern valley farmers subjected the middle slopes to the timeworn practice of agropastoral management that otherwise was typical of high-elevation Hill farming. The cultivators of Mollomarca offered a well-suited example: by the mid-1980s, they planted a pair of the community's six sectors almost entirely with barley (map 10). In the 1985–86 season all farmers with fields in the Roqeyoq and Llulluch'ani sectors planted barley. Together, the pair of barley-cropped sectors created pie-shaped portions of the landscape that widened from the hamlet at roughly 11,800 feet (3,600 meters) uphill to the rolling plateau at 12,800 feet (3,900 meters). Since a sector was cultivated for three years—that is, only three sectors were cropped in any year—no less than two thirds of their Oxen Area agriculture was devoted to barley.

In covering a majority of the highly productive middle slopes with barley the mollomarquinos broadcast the materia prima of brewing over an unprecedented expanse. Its new spread was documented in the transformation of sectoral fallow; since at least 1957, the farmers had rotated crops in the following sequence of sectoral fallow: potato (first year); barley and wheat (second year); ulluco and broad beans (third year) (Palacio Pimental 1957a). In the 1985–86 season, however, the mollomarquinos chose to implement a wholesale shift to a new cycle; potato (first year in the Oktapata sector); barley (second year in the Roqeyoq sector); barley (third year in the Llulluch ani sector) (Zimmerer 1991e). The transformed sequence departed radically from the earlier pattern by substituting beer-bound griñon barley for ulluco and broad beans in the third-year sector. Decline of the ulluco crop was not easily stemmed. It could not, for instance, be simply deferred into a fourth year of cropping due to the accruing deficit of soil fertility.

Ulluco's displacement led it to become the chief casualty among the diverse crops of southern valley farming. The curtailment did not unfold without friction in the Peasant Communities. In Mollomarca members brought intracommunity tensions and divergent views on farm change fully to the surface when they decided to displace ulluco in 1986. Informal discussions, community meetings, and the ultimate decision to double the area of barley cropping from one to two sectors provoked some sharp disagreement. A few mollomarquinos railed against the impending demise of the ulluco and broad bean sector, foreseeing the loss of staple foodstuffs; however, the community's general enthusiasm and the particular interests of powerful members who commanded much barley commerce managed to overwhelm the dissenting faction convincingly. Amid all this, a majority of members agreed that sectoral fallow itself made sound farm sense and that the community's decision, whatever it was, must conform to their social coordination of farming.

Although resigned to a curtailment of ulluco planting, the southern valley Quechua were resolved to retain some pockets of ulluco where possible. Many mollomarquinos and their barley-adopting neighbors were able to relegate a smattering of ulluco fields to elevations higher than barley's upper limit near 12,500 feet (3,800 meters). They kept planting ulluco in its accustomed sector, in other words, but allotted the crop only to the uppermost area. The biological impact of reduced ulluco acreage was mitigated by its regional-scale distribution. Southern valley landraces—qompis; Moor, or *moro,* qompis; Red, or *puka,* qompis—reached widely through other areas, including the Quencomayo watershed and Colquepata where they lost comparatively less ground to the upheaval of farming systems between 1969 and 1990 (see map 6; table 6). The loss of ulluco from Mollomarca and nearby communities did not, therefore, exact a weighty evolutionary or biological toll, at least in terms of the landraces. Like the localized loss of other diverse crops, however, the retrenchment of ulluco may have caused genes or gene complexes to perish.

One minor but noteworthy effect of the explosion in barley growing touched on the fate of soil resources and noncrop vegetation. When the southern valley farmers converted their parcels to barley, they tended to enlarge field areas at the expense of former borders or so-called balks. While plants in the field balks had long betoken the strong impacts of human activities, the hardy flora between fields nonetheless protected the plots from livestock, deterred run-off and stabilized soils, and, in general, hosted a whole array of economic plants (Zimmerer 1989).[34] Market-optimizing barley growers, however, tended to plow up balk areas along with their cover of resilient shrubs—such as llawlli, or *Bernadesia horrida*; mutuy, or *Senna birostris*; and *tayanka,* or *Baccharis sp.* (appendix B.2). The depletion of vegetative cover on field borders led to greater run-off and worsening erosion, hinting that the new pattern of less-balkanized plots would likely incur environmental costs for years to come after 1990.

Changes in Valley farming nevertheless recorded a somber success amid the pell-mell expansion of barley. Southern valley cultivators managed to relocate a number of minor food plants like quinoa, tarwi, and broad beans to their Valley fields. They found sowing the diverse crops within maize fields preferable to planting them amid the incongruous parcels of barley and improved variety potatoes that engulfed the Oxen Area. In order to relocate their diverse plants farmers required adequate Valley fields and labor-time that was sufficient for cropping them. They found that the new intercropping compounded the central tasks of planting, mounding, and harvest by dispersing the plants within fields and adding disparate cultivation calendars. Their innovation was both cleverly inventive and labor-intensive. It reinforced the recurrent irony that wealthy peasants, but not their poorer neighbors, shouldered a major share of the de facto conservation of diverse plants after the Land Reform of 1969.

Women farmers shouldered a major share of the intensified work regimes in diversity-rich Valley fields. Farmers in Mollomarca and other southern valley communities preferred to site their intercropped maize parcels in close proximity to the hamlet since the women divided their labor among many activities (map 10). Women visited the complex plantings often to keep up with the steady stream of tasks. In addition to cultivation chores they paid regular visits to repair fences, monitor crop progress, uncover livestock damage and theft, and determine whether irrigation or fertilizer additions were necessary. Although women farmers sometimes swapped labor through ayni exchange in the principal tasks, overall the new fields amounted to additional work. Flexibility in the familial division of labor that underwrote the gendered transitions in Valley production thus steered greater work with the diverse crops toward women farmers.

Perceptions of Quechua Peasants

The loss and de facto conservation of diverse crops in Paucartambo were cultural acts in the sense that the Quechua farmers shared many attitudes, beliefs, and values about this cropping and its fortunes. Biological losses after

1969 brought their cultural habits into especially high relief. Farmers freely voiced perceptions about the loss of landraces, its causes, and its consequences, which they beheld in the process of making their watershed decisions about whether to curtail the diverse crops. Their beliefs made reference to both material circumstances and supernatural forces. My impressions gained from conversation with two diversity-deserting Quechua farmers from Huaqanqa on March 23, 1987, illustrated their styles of reasoning about genetic erosion:

> It seems that landrace loss in the chawcha [early *S. phureja* potatoes] and hatun muhu crops [big seed maize] is caused by labor shortfalls in farm production here. In speaking again yesterday and this morning with Tomasa [Machacca Tunque of Huaqanqa] and her neighbor Trini, they detailed the laboriousness of economic activities, both older ones and more recent additions. Something in the production system had to be streamlined according to their accounts. Surprisingly, they replied with flat "no's" when I rephrased my question to ask directly "Do labor shortages cause the abandonment of papa chawcha and hatun muhu?" "We farmers wouldn't stop production due to labor shortages," her husband Cipriano [Qoya Qalhuanca] replied, "Land shortages are what is the cause of abandonment . . . also seed was lost due to poor management and pest attacks." I then asked, "But couldn't the seed be replaced?" Once lost, they responded, it was fruitless to restart production.[35]

Their outspoken emphasis on farmland shortages did not seem parsimonious with the proof that labor-time shortages were placing a major pressure on peasant families. Nor could the pair's perceptions be readily reconciled with their own logic stressing the ultimate role of seed shortfalls and an apparent reluctance to restart the diverse crops after loss. Most other Quechua farmers in Paucartambo would, however, agree fully with the sentiments of Tomasa and Cipriano. They too threaded the topic of landrace loss and related changes in agriculture to the issue of farmland access with a steely resolution. Even the Huaqanqa farmers, who as a group enjoyed an unmatched surfeit of planting spaces, gravitated to the same point. Contrary to what the Quechua in Parcartambo alleged, however, a farmland shortage was not exclusively at work in the dilemma of their diverse crops.

Perceptions alleging farmland access to be the sole mover of landrace loss took hold in cultural beliefs and political attitudes, rather than through a cold calculation of local resources. The deeply held beliefs of Paucartambo farmers about the supernatural existence of a powerful Earth Mother founded the primacy of land in their outlook on landrace loss. According to their hybrid Andean Christianity, the *pachamama* Earth Mother tended to the well-being of nature and thus farming by regulating climate and soil fertility. Believing Quechua farmers in the region invoked her image in everyday ritual as well as in special ceremonies, many of them expressing beliefs about agriculture. They imagined this Earth Mother to reside in the land and they paid her frequent tribute in a ritual propitiation: "Earth mother, sacred land, that you too may drink" (*pachamama, santa tierra, qampas tomayki*).

The Quechua cultivators of Paucartambo were also voicing a profane and explicitly political point in their land-centered accounts of biological loss. Centuries-old worries about rights over arable land, concerns more political than pious, surfaced in the farmers' inevitable emphasis of farmland. The peasants of the region had long struggled to secure farmland amid the intermittent despoliations of hacienda owners, vengeful neighboring communities, and, most commonly, chronic disputes among peasants of the same community, a suite of social conflicts not uncommon in Andean peasant life (Larson 1988; Mallon 1983; Mörner 1985; Salomon 1986a; Spalding 1974, 1984; Stern 1982, 1987; Thurner 1993). A chief means of securing farmland against arrogation was to cultivate it fully. By the same token, the family who claimed a deficit of labor-time risked the interpretation that they were unable to deploy their lands. Universally loath to invite an impression that they were lacking labor-time, the Paucartambo people preferred instead to announce a shortage of farmland and thereby reaffirm their existing land claims.

Most Quechua peasants in Paucartambo had engaged personally in conflicts over the control of farmland. Making matters worse, the uncertainty underlying farmland conflicts was reinforced, rather than reduced, by the confused contradictions of the Land Reform of 1969. Peru's new laws unleashed myriad uncertainties about the ownership and control of farmland, such as the initial and ill-fated government decree intending to abolish family holdings and establish cooperative property ownership. The anemic commitment of Paucartambo officials—evidenced in the 1980s by the scandalous yet official dispossession of community land by the Ministry of Agriculture and former hacienda owners on the grounds of disuse (the case of Illichua in Parpacalla community)—heightened the political fury over land tenure. No one could wonder that the Quechua in Paucartambo fingered only farmland access in their discussions of landrace loss and farm change.

At a personal level, the Quechua farmers expressed their perceptions of landrace loss by drawing on a wide-ranging vocabulary of reciprocity—the belief in mutual rights and obligations binding a pair of parties (Allen 1988; Brush 1977; Guillet 1981b; Isbell 1978; Mayer 1977; Mitchell 1991; Salomon 1985; Orlove and Custred 1980). Not just one but various pairs of reciprocal pacts set the cultural referents for the Paucartambo farmers as they pondered landrace loss. Strongly held feelings of reciprocity particularly infused their outlook on four relations: farmer-crop reciprocity, farmer-diety reciprocity, farmer-farmer reciprocity, and crop-crop reciprocity. The reciprocity represented between the related pairs was rarely equitable or ethically symmetrical in actual fact; however, the ideological assertion of mutual aid in these relations anchored each of the local discourses on landrace loss.

A firm belief in farmer-crop reciprocity led the Quechua in Paucartambo to furnish resources such as apt land, proper cultivation, and the performance of ritual rites for the plants and seeds. The farmers addressed their offerings not only to the omnipotent Earth Mother but also to special crop deities such as the Maize Mother, or *sara mama.* Quinoa Mother was another special deity demanding their care, respect, and propitiation. The Quechua farmers intended the ritual acts and ceremony to secure fair returns at harvest by sealing their pacts of anticipated reciprocity with the crops. In return for the proper nurturing, in other words, the crop plants were expected to yield handsomely. Plants failing to uphold their obligations could be blamed. So too could farmers who bore feelings of responsibility for having failed the crops in their care.

If a Quechua cultivator dropped a diverse crop from her farm roster, she could claim that plants themselves had failed to fulfill their duties. Diversity-surrendering farmers frequently reflected on their actions as forced because "there was no seed left" or "the seed was infected and no longer yielded well." To be sure, the Quechua farmers did occasionally lose seed due to damage from pests and pathogens; they could, however, replenish the seed of diverse crops in most instances without much difficulty through barter, work for payment-in-kind, or even gift-giving. Nevertheless, the unenvied cultivators of Paucartambo who had dropped the diverse crops typically shunned the question of why they had bypassed the easy means of restoring them. Instead, a remorseful sense of finality tended to drape over their accounts, often trailing off after a matter-of-fact mention of missing seed. This fateful resignation intoned in their rendition of farmer-crop reciprocity was reinforced by ideas about still other reciprocal relations.

The reciprocity with deities raised a broader interplay of cultural and religious beliefs than the other pacts. Some Huaqanqa inhabitants like Tomasa and Cipriano asserted that the ire of their Earth Mother at the neglect of votary offerings and ritual duties was behind the severe September droughts that allegedly led to loss of their big seed maize. Their reference to climate in believing an irate Earth Mother to cause landrace loss was by no means coincidental. Farmers felt that her disposition toward them was frequently expressed as climate. They understood the crop-damaging extremes in climate, for instance, to be punishments meted out by an angered Earth Mother, sometimes prompted by their personal negligence of ritual obligations. A Quechua farmer's feeling of her own noncompliance then would lead her to fear a climatic retribution from the Earth Mother.

Violations of reciprocity between farmers were also indicated in the local perspectives on landrace loss. Small-scale but nonetheless injurious crop theft by fellow peasants persistently plagued the farmers. Peasants owning fields in Paucartambo like those in other Andean regions habitually posted guard during the harvest season (Gade 1975). The Paucartambo farmers believed that the thievery of crops, and the landrace-bearers among them, was escalating during the period from 1969 to 1990. Some farmers felt that the threats of crop theft forced them to desert their diverse landraces. Since the landrace-rich spaces of Hill cropping in particular lay farthest from residential hamlets, such fields were often most vulnerable. Crop theft also exerted a sort of symbolic tension on the landrace-growing farmers since it reinforced a dread of collapsing social norms that were felt to weaken other traditional customs as well. Customary cultivation of the diverse plants, some Quechua cultivators worried, was doomed by the chronic violations of neighborly reciprocity that ought to adhere between farmers.

Beliefs of the Quechua farmers about reciprocity between the crops themselves also impacted on the course of landrace loss. The farmers felt that the spirits of crop plants held mutual rights and obligations with respect to one another. One animated example of crop-crop reciprocity featured the special roles granted to the fast-ripening crops referred to in general as chawcha, or precocious, and exemplified by but not restricted to the early chawcha potatoes seeded in dwindling amounts in the northern Paucartambo Valley. Quechua cultivators believed that the early potatoes, in addition to the quick-maturing mashua or añu and the early maize landrace known as *perlas* (Pearl), would beckon, cajole, and admonish the other plants to mature quickly and thus ensure full ripening. For this reason, too, the Quechua planters nicknamed their fast-maturing crop a "field boss," or *mandoncha,* an explicit reference to its exhortatory function.

A threshold of diversity-related behavior was governed by these beliefs in reciprocity norms among the Quechua peasants in Paucartambo.[36] The farmers generally adhered to their heartfelt sentiments of reciprocity and thus were reluctant and unnerved to break the reciprocity pacts of all types. Indeed, the perceptions of all four pacts did much to reinforce their reluctance to orphan the diverse crops. Once a farmer decided to sacrifice her cropping, however, some of the reciprocity beliefs subsequently reversed their effect by dissuading the readoption of diverse crops. Quechua farmers voiced this changeover in their overt resignation to landrace loss and in their reservations about resuming cultivation. They believed that subsequent planting, especially of the precocious leader crops, would yield pitiable results and was thus ill-fated. Likely to be part penitence and perhaps post hoc rationalization, this reluctance to readopt weakened the prospect that landrace-abandoning farmers would someday restore their diversity.

6

Diversity's Sum: Geography, Ecology–Economy, and Culture

The Place of Diverse Rationales

The sum of diversity in Paucartambo crops actually exceeded the total of its parts taken thus far. Three main parts have been tallied in previous chapters: the crucial marshaling of farm resources and farming capacities by the Quechua peasants, their persistent albeit historically changing customs of a fit livelihood and kawsay-style subsistence, and innovative production strategies in their mountain landscape. Taken together, the three factors did chronicle much of the de facto conservation and landrace loss that was riven along a variety of social, spatial, and place-based partitions. The three components were akin to enabling "conditions of existence," but they did not determine the magnitude of diversity rejuvenated each year. Instead, the estimated sums of diversity—in terms of landraces seventy-nine potatoes, twenty-seven maize types, and dozens in the minor crops—came about through a crowd of other concerns at work in the lives of Quechua farm people. Additional expertise and experience that figured into diversity's sum were for the most part a function of other forces: ecology-economics, culture, and geography.

Ecological rationales offered one general impetus to farm the particular diversity of landraces that replenished Paucartambo agriculture. Here a pair of agroecological aims guided the choice of distinct variants by Quechua cultivators: they wanted landrace types suited to particular growing sites and they wished a hedge against crop failure in hazard-prone mountain farming. These concerns of farmers not surprisingly emphasized the spatial-environmental criteria of habitats like elevation-related climate and soils. They also steered their goals for diversity based on such temporal qualities as growing season. Because the environmental rationales upheld productive activities, their motives could be denoted broadly as economic. Although extensive, the rationales of the Quechua farmers with respect to the environmental aspects of production were not acting exclusively or at the expense of other forces in their lives.

More specifically cultural motives endowed a second force in defining the scope of diversity desired by the Quechua in Paucartambo. In their familiar custom of kawsay they laid claim to the expectations of a fit livelihood and an appropriate cuisine based on a bevy of the best landraces. Their eating habits, similar to many Andean peasants, specified foodstuffs with a variety of cooking and consumption qualities (Allen 1988; Gade 1975; Isbell 1978; Mejía Xesspe [1931] 1978; Weismantel 1988). Religious customs of the Paucartambo people also sanctioned special crop types in ritual and ceremony. In addition, a cultural or moral aesthetics of farming played a vital role in shaping the sums of crop diversity. A locally shared sense of the good and the beautiful in farm nature guided the Quechua cultivators in deciding the sums of diversity. This desire of cultivators to handle crops the right way in accord with a cultural or moral aesthetics was broadly similar to widespread sentiments among farmers and others about the proper character of farm nature (Berry 1977; Ehrenfeld 1987, 1993; Jackson 1984; Leopold 1949; Sauer 1956; Tuan 1993).

A geographical rationale created in their sense of place aided the Quechua of Paucartambo in reckoning their quantity of diversity in yet a third way. Their geographical rationale was less specific than the ecological-economic and cultural motivations, but was more unifying. It translated the idea of a fit or customary livelihood, the kawsay resource ethic, into a sense of place (Zimmerer, n.d.). Inspired by their resource ethic, the Paucartambo farmers voiced customary expectations not just of cuisine but also more generally of cultural self-identity, social justice, and environmental health. Cultivators used diversity to share their ideas of being able to belong and to farm in the place they lived. The geographical rationale encompassed by their place-based ideal of fit livelihood shaped the sums imparted to biodiversity. Geographical factors worked the other way as well, influencing the nature of places. By farming certain diverse crops and expressing particular styles of production and consumption, the Quechua people in Paucartambo were actively creating their sense of place.

Changes in the geographical nature of places and the diverse crops was illustrated in an anecdote told by Natividad about the loss of the saqma, or Fist, variety from her Hill fields in Umamarca:

> I had cultivated the saqma variety, although I didn't single it out for particular attention. It was part of the floury potato mixture, like one child in the family. There were lots of children in that family. Anyway, I know that saqma disappeared from my field. You see, the Hill is turning into a different place. There's not as much cultivation of floury potatoes and there's more stealing from the fields that are left. I think the Hill of Umamarca is more dangerous now.[1]

Natividad was soldering a sentiment about diversity-defining work with the floury potatoes to her familiar feelings of Umamarca community as a place.

Both her work and her sense of place were undergoing salient changes. Natividad had thinned the sum of diversity by letting saqma slip from the mixture of floury potatoes. Although she noted its absence, Natividad knew that her style of working the floury potatoes could permit such losses to occur. Concurrently, the nature of saqma's place, that of Hill farming in Umamarca, was itself being altered. While they were mostly outside of the bloody path of warfare between Sendero Luminoso guerillas and government antiterrorist police, the Quechua in Paucartambo frequently felt their landscape was insecure. So, while Natividad and her neighbors acted as architects of both diversity's fortunes and geographical places, in each instance they followed ways and means that were only partly of their own design.

Scripts that were both commonplace and complex guided Paucartambo's Quechua farmers in summing the diversity of their crops. Like the "routinized behaviors" of other peasant cultivators, their work with the diverse crops could not be reduced to a matter solely of environmental rationales (Alcorn 1984; Guillet 1992). Many otherwise masterly treatises on Andean agriculture have erred on this point. Redcliffe Salaman, the plant breeder and eminent historian of the potato, and Cuzco potato authority César Vargas subscribed to an oversimplifying assumption that ecological rationales alone determined the sums of biodiversity (Salaman 1985, 10; Vargas 1948, 9, 1954). A fuller and more fair approach to the Quechua farmers' scripts for handling diversity needs to address not only *what* their diversity-related techniques were but also questions such as *who* practiced them and *why* and *where* they were practiced (Brush et al. 1981; Johannessen et al. 1970; Zimmerer 1991b). Less complete rendering of the wealth of farm nature would overlook much richness.

Nature too was a protagonist in the deciding of diversity's sums in the Paucartambo Andes. Illustrative of nature's own role, a pairing of diverse crops was found to shape the Quechua farmers' management of diversity: potato and ulluco, on the one hand, and maize and quinoa, on the other. In the case of potatoes and ulluco the farmers defined a large share of diversity on the basis of a cultural or moral aesthetic. By contrast, they harkened more toward utilitarian rationales owing to ecological adaptation and culinary preferences when deciding the diversity of their maize and quinoa. The contrast in diversity-defining rationales coincided with differences of reproductive biology between the tuber crops and the grain-bearing pair. The nature of nature itself, in other words, entered into the determination of diversity's sum.

Voice of a Cultivator: Willful Words

In mid-June, 1987, Natividad pointed out how to choose floury potato seed while we sat beside a sprawling heap of multihued and many-shaped tubers dug yesterday. Natividad swiftly swept her seed tubers into the long folds of her sturdy outer-skirt, telling me that the chosen ones had "good eyes like flowers," or, as she said, "*chawfra ñawi tikahina.*"[2] "Good eyes," I asked, "what do you mean?" Shuttling a lapful of tubers to the growing mound of future seed, she replied, "Well, these round, disease-free eyes are pretty enough for seed, they'll give good yields." Líbano, Natividad's husband, held a quite different standpoint on seed selection. Once when this topic was raised Líbano skirted it like a momentary distraction, "Eufemia does that, I have to work and worry now that they're happy in this field so they yield well." Líbano's response could have launched any primer about the deciding of diversity in Paucartambo: the Quechua farmers cultivated verbal expressions as much as diverse plants, and in doing so the women and men toiled at different tasks.

Gendered distinctions in the seasonal round of farm work distinguished a variety of spaces in the Paucartambo countryside in the years between 1969 and 1990. Women farmers partitioned the yield to seed and to various other purposes while still working in a field at harvest or near the family's storeroom afterward. Their labors with the future seed then passed through a series of processing and storage steps in work sites that were sanctioned as women's spaces. After harvest, the women farmers labored with the family's seed in open-air drying and separating areas—momentarily quite visible when colorful maize cured on the picturesque drying floor, or *tendal*—and then in the nooks and crannies of storerooms and kitchens. While they bustled through the selection and storage of new seed, the men in their families were transporting produce for sale, tilling new fields, and in some cases migrating in search of short-term work.

Language for handling the floury potatoes and other diverse crops was salted with metaphors that gave life to the myriad pacts of personal reciprocity between a care-giving and willful cultivator and a crop plant that could be capricious. Farmers' metaphors ranged from the explicitly contractual to the merely teasing. Líbano expressed the responsibility of keeping "happy" his field of floury potatoes, for instance, which he felt would be redeemed by respectable yields in harvest. Natividad also tended verbally to the reciprocity pacts with her crop. In the example of choosing potatoes with "good eyes like flowers," her florid metaphor transposed human and plant images, turning the tuber more humanlike. Language thus helped make an endearing image and, as importantly, one more subject to the reasoned entreaties of reciprocity.

Natividad, Líbano, and the other Quechua in Paucartambo used the wide-ranging and at times willful language of mutuality to express their routines that figured into diversity's sum. Such mutuality could be heard in the vernacular of landrace naming. Farmers tagged a unique name to each landrace—all seventy-nine of the potato types, all 180-odd subtypes, and so on in each of the diverse crops. Pervasive naming of every tuber and grain clashed with the lack of a name for the term *crop* in general, a rubric that was missing from their Quechua vocabulary. The complete christening of landraces by farmers, however, helped communicate the know-how for working their mountains of diverse crops. Their penchant for naming also aided in remaking the insentient crops into the image of animate beings. Animation through naming enhanced the prospect that reciprocity would be respected; an unnamed object, by contrast, could not be expected to comply with the mutually pertaining duties of reciprocity.

Naming kindled a lively interest in the character of landraces. Careful deliberation, unwavering curiosity, and jocular competitiveness inevitably enriched the simple acts of landrace identification and naming. Dozens of Quechua farmers in Paucartambo who discussed the naming of landraces during the mid-1980s brandished an avid interest in the topic (Zimmerer 1991b). Not one single person failed to affix a name to each potato or ulluco tuber, maize ear, and quinoa stalk being talked about. If unsure about an identification, they preferred to substitute the name of a similar landrace rather than leave a tuber or ear unnamed. The farmers debated the attributes and names of landraces with similar enthusiasm. Natividad, for instance, had once chided Líbano about mistaking the potato *rosas pata* (Step of Roses) for saqma if he did not grasp how the latter's surface resembled the protruding claws of a mountain lion—its synonym was puma maki, or Puma Foot—rather than the roselike rosas pata, with its pattern of petal-like protrusions.

A language of animation enlivened their studious evaluation of diverse crops for use as seed. In naming hybrid seeds and tubers many deemed to be unsuited as seed, the Quechua cultivators voiced descriptors of racial admixture that were borrowed from the analogies demarcated historically by the ethnic and racial mixing of various peoples in the Andean world (Mörner 1967, 1985; Stern 1982). They appliqued the mixed types with terms such as Mixed Breed (*cholo*), Mulatto (*sambo*), and Moor or Spotted (*moro*), as well as Dog (*alqo*). For example, a hybridized ear of the maize landrace *qosñiy,* distinguished by the atypical speckling of its kernels, was qualified as Moor or *moro qosñiy,* literally Moor Smoke, or more figuratively, Spotted Smoke. In similar fashion the farmers referred to the hybrid of qompis potato with the name *alqo qompis,* or Dog *qompis.*[3] Their habit of labeling hybrids, even when tempted to leave the aberrant types undesignated, attested to the imperative that the objects of all cropping be named.

Language that animated the hybrid variant and enlivened the world of its cultural context did not, however, translate into identical meanings in the actual tasks of seed selection. When the Quechua in Paucartambo encountered the hybrids of maize such as Moorlike qosñiy it was for good reason that they did not seed them. Farmers knew that the planting of maize hybrids would have threatened to dilute the identities of their customary landraces, a risk that was to be strictly avoided. By contrast, the potatoes growers in the region did not fear the effects of further hybridization because their crop was sown as tubers, keeping outcrossing to near nil. They customarily planted the hybrids of potato landraces such as Dog qompis as seed, even mixing it with the unaltered qompis landrace. Names aired in their animated language thus carried meanings that were tempered by the nature of the reproductive biology behind diversity.

Distinctions drawn between crops and noncrops exemplified another role of verbally imposed animation. While the Quechua in Paucartambo deployed images of humans, plants, and animals in naming their diverse landraces, they reserved the terms Fox (*atoq*) and Wild (*kita*) for dubbing the wild crop relatives that grew uncultivated. The image of the wily Andean fox designated the wild potatoes (Fox Potatoes), in addition to the wild ulluco (Fox Ulluco). Close relatives in the seed-bearing crops were named Wild Quinoa and Wild Tarwi (Zimmerer 1989). The presence of these relatives around and sometimes in fields triggered the flow of unique diversity into the diverse plants through genetic introgression; however, the Quechua in Paucartambo focused more on difference than similarity. Their choice of the terms *fox* and *wild* for the wild relatives served to label a linguistic and symbolic territory that was separate from the domesticates and outside the realm of reciprocity. Farmers were in effect drawing lines between the suitable forms of a crop and those of its undomesticated and undesirable relative.

Selection of seed broadcast vividly the willful effort of a Quechua farmer to cement the reciprocal bonds with her crops. Her discourses on seed disease and anatomy illustrated the effort through an imaginative corpus of analogical and metaphorical references to the human body (appendixes F.1 and F.2). The idiom of corporality represented a drawing of her crops more closely than allowed by animation alone. Embodying the crops in a world of analogies that emphasized human design, the Quechua cultivator was able to exhibit her conspicuous domestication of a highly visible part of nature. Her domestic feeling for the anthropomorphic plants thus not only brought farm nature into her home but also made it like a member of the family. By referring to human conditions, she was ably versed in a variety of symbols that could communicate the wide assortment of technical skills needed for seed selection.

Cultivators described the diseases that plagued their seed with a throng of allusions to the human body and its ailments.[4] They identified the various skin infections on tubers as *verruga* and *qak'una,* terms applied also to human skin maladies. Commonly, farmers described as "blind" the tubers with eyes so disfugured by disease that they were deemed to be unworthy of planting. Other crop diseases were encoded in terms of the cultural and ethnic groups familiar to Paucartambo people. The farmers deployed the term *qolla* for tuber seed that was wrinkled and darkened by plasmolysis and viral infection. They maintained that qolla tubers were worn-out, similar to the way in which the dark skin of the qolla people in southern Cuzco, Puno, and northern Bolivia was thought weathered by the harsh Altiplano.

Beliefs in crop disease held by the Paucartambo people adopted the tenets of a broader Quechua etiology (Larmie 1993). The farmers believed the causes and origin of disease to be strongly influenced by the provenance of seeds. They reasoned that the offspring of a tuber or grain seed would resemble the parent whether hale or frail. Likewise, the condition of seed from a certain geographical place was taken to be symptomatic of the place itself. Places that provided healthy and virile seed in the past were more likely to do so in the future. So instrumental was their belief that the Quechua farmers easily recollected whether certain locales had shipped salubrious or diseased seed even twenty or thirty years before. When the farmers decided about subsequent seed procurement on the basis of their mental maps of seed health and disease, they implemented geographic criteria that helped to pattern the all-important distribution of landraces.

The idiom used to describe seed anatomy, similar to disease, embodied a number of analogies to human bodies. Cultivators referred to particular features of tubers and grains exemplified by eyes, ears, nose, face, skin, head, heart, veins, and other body parts (appendix F.1). They surveyed the status of nearly all the anthropomorphically labeled features in the process of selecting seed from the best tubers and grain. The farmers even broke seed on occasion to check its interior quality. On the one hand, a person who found her potato flecked with "black streaks in the heart"—rendered as *sonqompi yana venayoq*—was pleased by its show of vigor. On the other hand, she quickly discarded the tuber with an empty heart, or *t'oqo sonqo.* Other descriptions of anatomical states mixed with an aesthetic appreciation of the diverse crops. Choice seed typically showed a beautiful face (*munay uya*) or flowerlike eyes (*t'ika ñawi*). Voicing her affection and terms of endearment also made more tender the reciprocity bonds between a cultivator and her crop.

Eyes of the tuber-bearers such as potatoes and ulluco eclipsed all other anatomical features as the focal points of seed selection. The word *eye,* or *ñawi,* could even connote "initial" or "beginning" in a general sense. A Quechua farmer scrutinized the size, shape, and disease-status of a tuber's eyes for signs of its suitability, readily tossing aside a tuber with suspect eyes even if its other attributes appeared flawless. She coined precise terms for liplike eyes and eyes about to sprout that pinpointed the exact traits that left a tuber unsuited for seed (appendix F.2). When assuming the tuber eyes to be comprehensive indicators of seed quality the farmer and her neighbors echoed their beliefs about the capacity to divine future happenings on the basis of human eyes. Their divinatory focus on eyes was common also among earlier Quechua peoples: "The Indians gazed at the eyes and the flickering of the upper and lower eyelids, where the Incas and all their subjects drew good or ill auguries according to which eyelid twitched" (Garcilaso de la Vega [1609] 1987, 221).

The Quechua farmers in Paucartambo also animated the crops plants themselves, adding to their already lively vocabulary of analogies and metaphors. Farmers admonished new seedlings to be growing with force (*kallpachawan winan*) and, in the case of tuber-bearers, to sprout branching and healthy "thick bone" stems. Subtle concepts of reciprocity also guided the language used with their plants. The farmers regularly asserted that they would *make plants grow* (*wiñachiy*), rather than stating more simply the *plants would grow.* In their reciprocity pacts with crops they exercised an unending aesthetic critique of cultivated plants. While trekking through the heavily farmed countryside, many opined ad lib on parcels as pretty and like a garden or, for that matter, ugly and like sticks. The farmers were often announcing pride and pleasure in their own successes when they delivered such aesthetic judgments.

Pacts of reciprocity were also believed by farmers to govern the crops themselves. The perceptions of rights and obligations among groups of the diverse plants invoked animated images and metaphor similar to other reciprocity pacts. The farmers' ideas about their sowing of seed mixtures abounded with examples of the between-plant dimension of reciprocity beliefs. The Quechua in Paucartambo believed that between-plant reciprocity was at work in three broad categories of intercropping mixtures: within-landrace combinations, multiple landrace intercropping, and multiple crop mixtures. In each arrangement the farmers' concepts about between-plant reciprocity gave the impetus and guided the implementation of intercropping and thus mediated the sums of diversity.[5]

Within-landrace mixtures offered the least conventional category of intercropping, although the one most common in Paucartambo. The Quechua farmers often decided on a variety of tubers belonging to a single landrace when choosing seed for a single planting hole. Typically, a farmer would apportion two seeds to some holes and three to others. The farmer selected tubers so that larger ones were planted in pairs, a category of two seed tubers, or *iskay muhu.* She could shrewdly choose slightly smaller tubers for planting in groups of three, the ones termed three seed tubers or *kimsa muhu.* Farmers animated the tuber seeds by imagining them to interact, mostly by competing with one another. Their belief in the dynamic competition between seeds was a main reason that farmers rarely sowed with only one tuber, which alone were thought to languish for the lack of motivation. By skillfully distinguishing two seed and three seed tubers, the farmer could better stretch her limited supplies. Careful selection would ensure that seeds could cover much ground or, in her words, walk far (*hatunman purin*).

The farmers' assortment of intercropped landraces was also styled on between-plant reciprocity. The Paucartambo cultivators planted landrace mixtures in all their major crops, as well as several minor ones. They believed that the landraces complemented one another's capacity to withstand hazards such as frost, drought, and hail. If a farmer planted two dozen or more floury potatoes in a high-elevation Hill field, for instance, she figured that some might perish, while others and probably most would persevere. Her belief in reciprocity between landraces did not, however, prescribe a particular mix of specific complementary kinds. Instead, between-landrace complementarity was thought to result from the general nature of differences in the mixture rather than the contrasts of certain sorts. Farmers held that it was the overall degree of distinctions among intercropped landraces, rather than the combination of specific types, that reduced the risk of crop failure in a field.[6]

Their third concept of interplant reciprocity governed the aggregations of diverse crops. Here, the Quechua cultivators in Paucartambo mostly admixed the tuber-bearing mashua, known locally as añu, into fields of potato, ulluco, and oca. They sowed small quantities of the nasturtium relative by either spacing its seed tubers every few paces within rows in the *haynachu* technique or by interspersing a single mashua furrow among several rows of the main crop. Cultivators rationalized their intercropping based on the belief that the quick sprouting and early emergence of mashua beckoned persuasively to the other crops that they follow suit. The commanding personality of the intercropped mashua was reflected in its frequent designation as the prodding field boss, or mandoncha, of intercropped fields, a term alive with animated qualities.

The Quechua in Paucartambo discussed their farming with language drawn from an immense agricultural vocabulary and a rich array of metaphorical and other figurative expressions. The cultivators' artful use of farm language created the expressive verbal medium for knowing diverse crops. Their large lexicon of metaphorical references granted an impressive range of referents for the cornucopia of agricultural objects and processes that established a world center of crop diversity in the Paucartambo Andes. The use of anthropomorphic references in particular nurtured the diverse crops by casting such objects in the most familiar of terms. The farmers could refer to the familiar terrain of the body to express nuanced observations of their diverse crops. Less intimate or less plentiful linguistic devices would not have so fully accommodated the specialized farm knowledge of the Quechua people.

The language of Paucartambo farmers was chiefly Quechua with a sprinkling of Spanish terms. While farmers there conversed in Spanish when necessary, their Quechua mother tongue mattered most to the diversity of crop plants. It offered one of the richest vocabularies of metaphors in any language. Many metaphors communicated the expertise and experience of farming the diverse crops. Since a familiarity with metaphors opens the avenues for complex knowledge, the metaphorical richness of the Quechua spoken in Paucartambo no doubt enabled farmers to know and farm better their wealth of diverse crops. It must not be forgotten also how the Quechua language provides unique and highly specialized terms about farming—such as the disease symptoms identified as liplike eyes and eyes about to sprout. So ample is this vocabulary that one out of every three or four Quechua words related to agriculture lacks a concise equivalent in Spanish and English (Beyersdorf 1984; Cusihuaman 1976; González Holguín [1608] 1952; Lira 1945).

The Quechua farmers of Paucartambo deployed the full extent of their linguistic prowess in voicing the willful words to inform and implement the main rationales behind diversity's sum. Their willful words formed a common language applied with equal ease to each of the diverse crops, however, the Paucartambo people did not pursue the same objectives in their farming of the two major crop complexes. One set of rationales gave the impetus for potato and ulluco diversity—mainly a cultural and moral aesthetic—while ecological-economic and culinary-religious motives guided the farming of maize and quinoa diversity. The farmers not surprisingly used language differently in working with each crop complex. While their linguistic practices aided immeasurably in the general managing and maintaining of diversity, contrasts in crop identification and classification resulted in widely disparate sums.

Cultural and Moral Aesthetics

The Quechua in Paucartambo defined the full scope of potato and ulluco diversity through their sense of cultural and moral aesthetics. This basis for the diversity of tuber-bearers rested first on a few decidedly utilitarian objectives. The farmers expressed their concern for utility in a handful of portmanteau linguistic categories. They construed a total of four such metacategories in the potato crop and two in ulluco, labeling each class according to its utility: hence the titles boiling potatoes, soup potatoes, freeze-drying potatoes, money potatoes, big planting ulluco, and early planting ulluco (table 10). By grouping myriad landraces into the aptly named use-categories, they framed the linguistic architectonics of diversity. Within each use-category, the cultivators expressed their diversity-defining moral aesthetics.

Farmers would begin framing the linguistic and mental architecture for potato and ulluco diversity by classifying scores of individual landraces into use-categories. They placed most landraces in a single use-category, of which the largest were the boiling potatoes and the ulluco big planting. In surveying tuber traits such as shape, the wateriness of pulp, and eye features they could quickly distinguish the use-category of a landrace. Farmers knew in a moment that mealy or flaky pulp, rather than a watery texture, indicated a boiling or floury potato. Because their criteria such as mealy versus watery happened to match several traits that are key to the taxonomic divisions set out in biological science, the use-categories of Quechua cultivators actually corresponded to scientific units. In the potatoes, for instance, each use-category accorded with a specific species or series of them.

Farmers chose names for each use-category that verified its main utility in consumption (boiling potatoes, soup potatoes, freeze-drying potatoes), commercialization (money potatoes), and production (big planting ulluco, early planting ulluco). Their naming of use-categories admitted many synonyms: by resorting to the term *boiling potatoes,* equivalent to *floury potatoes,* for instance, they could add a pinch of extra precision to contrast more sharply the

Table 10. The Primary Usefulness of Potato and Ulluco Diversity

Use-Category	*Taxonomic Affiliation*[1]	*Production Space*	*Disposition of Yield*
Boiling potatoes (*wayk'u papa*)	Sta, Sa Sg, Sc	lower Hill	mainly subsistence and seed
Peeling potatoes (*bondón papa*)	Stt	Oxen Area	subsistence, seed, and commerce
Freeze-drying potatoes (*chuño papa*)	Sj, Sc	upper Hill	mainly subsistence and seed
Money potatoes (*qolqe papa*)	Stt	Early Planting and Oxen Area	much commerce
Big planting ulluco (*hatun tarpuy*)	Ut (all landraces)	upper Oxen Area and lower Hill	subsistence and seed with moderate commerce
Early planting ulluco (*maway*)	*wawa yuki* (Cradled Baby)	Early Planting	subsistence and seed

1. Abbreviations stand for the first letters of the genus and species names given in table 1 and appendix B.1. Sta abbreviates Solanum tuberosum subspecies andigena and Stt is an abbreviation for the improved, high-yielding potatoes in Solanum tuberosum subspecies tuberosum.

other consumption categories like soup potatoes. Another expression for boiling potatoes was *wataychana papa,* or literally potatoes for all year, which highlighted a distinction with the potatoes slated for prompt marketing. It was the wide sharing of use-category names, rather than the fixing of a single name, that mattered to the farmers in Paucartambo. Cultivators did not find the multiple designations a source of confusion provided their neighbors shared the synonyms and thus kept the language of cropping readily communicable.

The farmers designed each use-category to fit into a certain farm space: boiling potatoes in Hill farming, money potatoes in the Oxen Area and Early Planting, and so on (table 10). The match-ups meant that each use-category was managed with a certain farm space in mind. Knowing that boiling potatoes belonged in the lower section of Hill farming, the farmers thus anticipated that fields would range between roughly 12,500 feet (3,800 meters) and 13,300 feet (4,050 meters). The match-ups also inferred a characteristic mode of disposing yield among the purposes of seed, consumption, and sale. Cultivators could confidently infer the disposition of yield based on use-category, since each unit of farm space was managed, at least ideally, with a customary allocation of harvest in mind. In the same way the relation of use-category to farm space augured other facets of production for each landrace group, such as the labor schedule and soil management administered by farmers.

The linguistic architecture of potato and ulluco classification created by Quechua cultivators gaped wide open between the grand scale of use-category and the finer scale of distinction at which each individual landrace was identified. Absence of classificatory divisions between the ranks of use-category and landrace—coarse classification and fine classification, respectively—was rife with implications about the nature of diversity in the two tuber-bearing crops. Imagine the coarse structure of classification to arch spaciously, each use-category like a soaring cathedral. The immensity of a use-category towered well above its finely classified constituents, the individual landraces. The uninterrupted opening between the coarse and fine levels of classification was especially apparent in farmers' designation of the boiling potatoes, which amassed at least seventy-nine landraces into a single use-category without an intervening linguistic distinction.

One implication of the gaping atrium in this classification was that the Quechua cultivators treated all their diverse landraces as equal members of the boiling potato complex. Their failure to classify subgroups of potatoes was noticeable consonant with their lack of farming methods aimed at an intermediate level in the hierarchy of taxonomic ranks. By lumping all their finely identified landraces into the great boiling potato group, their classification helped to define diversity's relation to growing habitats broadly rather than narrowly (table 10). Mental constructs thus stood behind the distribution of nearly all floury potato landraces across the wide range of elevation-related environments covered by Hill agriculture. In sum, farmers were guided mostly by the gaping spaciousness of the use-category designation rather than the detailed level of landraces.

The classification architecture of the Quechua in Paucartambo, their so-called folk taxonomy, was similar to the linguistic practices of diversity-rich farmers in other Andean regions. A surprisingly limited assortment of use-category cathedrals appeared to characterize the linguistic landscape of potato classification. In the neighboring Urubamba Valley, for instance, the Quechua cultivators classified potatoes at the use-category rank "in a surprisingly limited number of ways" (Gade 1975, 392). On the Altiplano of northern Bolivia, Aymara farmers grouped their diverse potatoes according to no more than a pair of use-categories (La Barre 1947). The farmers of central Peru utilized a total of three use-categories to classify their diverse potatoes (Brush et al. 1981).[7] Simply because a potato landrace could be classified in terms of limited use-categories did not imply, however, that diversity was allotted to them in a helter-skelter style.

The Quechua farmers in Paucartambo sheltered the great mass of varieties in the boiling potato category. Containing seventy-one or more landraces, greater than ninety percent of the regional total, the boiling potato group towered over the diversity of other use-categories. Estimates for the minor use-categories ranged from ten varieties in the soup potatoes (a few landraces and several improved varieties) and eight varieties in the money potatoes (all improved varieties) to five types of freeze-drying potatoes (all landraces). Linguistic spaces spanned by the utility-based categories were of course no more than outer shells for potato diversity. It was the work of farmers with the varied landraces within each group that defined diversity's full scope and much of its substance. If farmers were imagined to construe the use-category classification like a handful of cavernous cathedrals, then they drew the diversity-related details of their sheltered masses once inside.

The field, or chacra, was the focus for the Quechua in Paucartambo in their deciding the extent of landrace diversity in each use-category. Farmers' guidance of landrace dynamics in the boiling potato fields engineered the lion's share of potato diversity. Individual boiling potato parcels grown in the 1986–87 season averaged twenty-one landraces and, within that number, more than thirty-five sublandraces (from samples of approximately two hundred plants per field) (Zimmerer 1991a, 1991b). Since many families held at least one field of boiling potatoes, the effort behind the overall wealth of landraces was socially broad-based. While the quantitative estimates of diversity answered to the question of what diversity flourished, they addressed neither how nor why the full sum took its shape. The latter questions shifted attention to the techniques and the rationales used by the Quechua farmers to determine diversity's sum.

Farmers decided the number of landrace numbers in a boiling potato field after the harvest when they chose seed for sowing the next year. Women cultivators carefully carded disease-free tubers of appropriate size in the initial step of seed selection. Size rather than the sort of landrace was thus the main desideratum at first. They rid the minute-size potatoes together with bruised and heavily diseased tubers for use in livestock feed, and in the case of poorer families, for freeze-drying into chuño that could be eaten later. They also seized the extra-large wañlla, or gift tubers, for the purposes of gift-giving as well as proud display and special eating. Remaining tubers comprised the size-ranked second class and third class groups that were stored for seed and consumption. It was from this pair of intermediate-size and medium-quality groups that the cultivators subsequently corralled the landraces to be seeded in their boiling potato fields.

A definitive crossroads was met when the Quechua cultivator culled the exact set of tubers for replanting from her shrewdly chosen stocks of second class and third class potatoes. In the field at harvest or in a corner of her family's windowless storeroom, a farmer sequestered seed from her taqe baskets or straw-lined piles that kept the mixtures of seed-and-consumption tubers (plate 10). Working quickly, she did not exert a precise control over the specific land-

*Plate 10. Santusa (far right) aids her neighbors one misty afternoon in sorting ulluco, or Potato Lisas (*papa lisas*), as they call it, between the second class and third class for seed and consumption. The en masse raking of future ulluco seed resembles the technique used for the diverse potatoes.*

races that were skimmed toward the mound of new seed tubers. The farmer, like her neighbors, mostly composed the ensemble of landrace types through a raking of future seed en masse. She was, in other words, choosing her landraces in mixtures rather than selecting them specifically as single variants. She selected only a few landraces individually, or in some cases, none at all.

Their techniques of seed selection jarred with the prevailing and widely held interpretation of how Andean people gird diversity in their world-renowned potato crop. Conventional wisdom broadcast in Nova's *Seeds of Tomorrow* has triumphed how Andean and other peasant farmers pinpoint the exact nature of their potato diversity via the careful selection of individual landraces (Altieri and Merrick 1987, 1988; Brush et al. 1981; Brush and Guillet 1985; Salaman 1985; Vargas 1948; Webster 1973). It was believed that nothing less than precise landrace-by-landrace selection was needed to fit each unit to a matching series of specific environments both between and within fields. Farmers were thought to handpick the sums of diversity, therefore, by finding an individual landrace for each niche in the varied spectrum of environments. That presumed mode of tailoring diversity to minute microenvironments, and for that matter the rationale that purportedly guided it, did not, however, adhere among the

Paucartambo farmers. For them, the answer to whether to replant most kinds of boiling potatoes was not distilled into a calculus of fine-tuned adaptation.

On the contrary, farmers in Paucartambo governed the selection of boiling potatoes with general desiderata. Their corralling of seed tubers en masse, rather than collating them with cybernetic precision, guaranteed that the abundance and even the presence of a particular landrace would depend first on its availability in the seed-and-consumption mixture. The experienced Natividad voiced it this way: "The good seed tubers are the ones that appear in the harvest."[8] As a consequence, one or two higher-yielding landraces—such as qompis, pitikiña, and Narrow One, or *suyt'u*—invariably comprised the majority of seed in a field of the boiling potatoes (Zimmerer 1991a, 1991b). In fact, a single landrace made up one half or more of the plants in all thirty fields sampled and studied between 1986 and 1990. Farmers recognized the above-average yield properties of the most common landraces, and they appreciated them, but they did not find it necessary to steer selection in their direction, leaving the agroecological outcome of harvest to accomplish the same goal.

Similarly, the farmers selected adeptly for general criteria to lessen the constant threat of crop loss. They realized that admixtures of variants with distinct tolerances of hazards such as drought, disease, and frost lowered the risk of loss, since some types would succeed in yielding while others failed. Their risk aversion concern did not, however, motivate them to specify the landraces that would optimize its reduction. The Quechua farmers in Paucartambo did not, for instance, construe an image of optimal landrace mixtures, each type chosen for its special risk-averting tolerance. Their boiling potato fields in this case would have consisted of one or a few drought-resistant types added to those specialists with excellent tolerances of disease, frost, and other hazards. Diversity's sum in this hypothetical yet unsubstantiated scenario would equal the total of risk-reduction specialists, but the Paucartambo farmers did not choose the whole slew of landraces for specific risk-reduction rationales. Their diverse boiling potato fields in any case tended to contain far more landraces (twenty-one types) than would be enumerated solely by the optimization of risk-aversion.

Handpicking of individual landraces was not entirely absent from the farming of boiling potatoes, notwithstanding the predominant mode of culling seed en masse. Farmers were especially keen to finger a few of the most choice sorts from their mixed seed-and-consumption stores, because these landraces promised an unmatched culinary aspect, such as extra flakiness, sweetness, or fast cooking (appendix F.3). For example, many so enjoyed the crumbly flakiness and delectable nutty flavor of kuchillu p'aki (Broken Knife), which rasped while being eaten, and the brightly colored llama senqa (Llama Nose), that they intentionally handpicked a few members of each landrace when composing their seed mixtures. Farmers' delights in consumption thus piqued a special interest in certifying that the most heralded landraces would be on hand. Handpicking also assured that their comprehensive landrace-specific knowledge and refined appraisal of exceptional boiling potatoes would enter into the deciding of diversity's sum.

The Quechua farmers in Paucartambo also added to diversity's sum in another minor albeit vital act of direct selection, the addition of new landraces. New landraces of floury potatoes derived from the occasional sowing of above-ground seed that was set as a result of the sexual flower-to-flower hybridization of existing landraces, or in some cases, the sexual union of a crop plant with its wild relative. The incorporation of tubers grown from hybridized botanical seed directed a crucial flow of additional diversity to the Andean potatoes (Cárdenas 1966; Hawkes 1990; Hawkes and Hjerting 1989; Jackson et al. 1980; Zimmerer and Douches 1991). Since seed tubers otherwise arose asexually and thus were made genetically identical to the parent plants, it was this seed of botanical genesis that was the main conduit for new landraces.

At least a few farmers searched their fields for the offspring of botanical seed. These extra-observant farmers often found the new tubers in between-row alleys, or wachu wayq'o, where above-ground seeds had fallen, germinated, and grown tubers of sufficient size that they could be replanted. Combing for the new landraces that derived from botanical seed was not a widespread custom among Paucartambo cultivators. Natividad, for instance, expressed no familiarity or interest in it. Other Quechua women, however, expressed their interest and skills in finding if the tubers formed from above-ground seed could add something new to their family's larder. Even the apparent infrequency of incorporating the new variants was enough to enrich the region's diversity of boiling potatoes.

Sophisticated appraisal of crop quality did not, however, inspire the sowing of the prized landraces in large amounts. Proof of an incongruity between the farmers' unquestionable expertise in the matter of boiling potatoes and their limited manipulation of landrace mixtures was found in the composition of fields. Surprisingly perhaps, the most preferred variants of boiling potatoes were actually the rarest (Zimmerer 1991b). The truly prized landraces, such as chimaku, choqllos, ch'orillu, *cheqefuru,* and *leqechu,* regularly ranked last in terms of frequency. From the samples of two hundred or more plants in a field, the exceptional landraces rarely numbered more than ten, or one out of twenty seeds. It was clear therefore that farmers deciding diversity's sum and the proportions of its individual parts drew no more than moderately upon direct selection. In reality, indirect selection and the general criteria of production accounted for much of the biological arithmetic that farmers etched in assembling the diverse plantings of boiling potatoes.

The relative lack of landrace-specific selection among the Quechua farmers of boiling potatoes in Paucartambo ran counter to the assertion that diversity-related farm techniques were geared to individual landraces. It did not appear so surprising, however, on review of another careful study. An ethnobotanical field project undertaken in Puno, south of Paucartambo in the Peruvian sierra, also described the surprising incongruity of landrace rareness and cultural esteem (Jackson et al. 1980). In fields of the Quechua peasants in Puno "three varieties classed as poor were grown at higher frequency than some of those which were considered good or intermediate" (Jackson et al. 1980, 113). Such findings did not suggest an absence of sophisticated management but rather they raised the likelihood that the determination of potato diversity mainly took the guise of complex and sometimes subtle rationales.

A subtle process of selection did in fact steer the diversity and the composition of boiling potato fields in Paucartambo. Each year when the Quechua farmers chose seed, they minded the appearance of diversity as a whole that was evidenced by the variation of colors and shapes in their tuber piles. Following local aesthetic canons, they especially judged the conspicuous contrasts within the striking potpourri of colors and shapes. They then judged whether they had garnered the right diversity for the next planting based on their overall impression of diversity. The yearly selection of landrace mixtures in this manner no doubt reinforced the varied kaleidoscope of colors and shapes among the diverse boiling potatoes, which indeed was often the most memorable impression for a first-time viewer. A similar style of seed management among Aguaruna manioc growers of the upper Amazon Basin in northern Peru was termed "selection-for-perceptual-distinctiveness" (Boster 1985). Like the boiling potatoes chosen in Paucartambo, manioc diversity among the Aguaruna cultivators was represented by a complex collage of colors and shapes.

The culling of landrace mixtures as groups by the Paucartambo farmers, akin to "selection-for-perceptual-distinctiveness," did not infer that their management was a random process. Indeed the farmers shared a remarkably common vision about the amount of landrace diversity that was fit for an individual field. It was their culturally shared vision that inspired the uncanny similarity of diversity levels across the boiling potato fields of the region. A concise statistical measurement was the low standard deviation of 5.7 around the mean of twenty-one landraces per field (Zimmerer 1991b). The fact that nearly all fields contained between fifteen and twenty-seven cultivars gave both an agroecological and indeed statistical verification that the Quechua in Paucartambo shared a common notion of how many landraces properly belonged in a boiling potato parcel. Their shared cultural sense was built on the particulars of work experience, observation, and communication. As a result of being both personal and individual, as well as social and culturally shared, the endeavor of seeding produced a wealth of diversity.

The Quechua farmer in Paucartambo expressed a sentiment of cultural and moral aesthetics in her widely shared notion of apt boiling potato diversity. Her seeding of the field was not, therefore, purely a utilitarian act but rather one entwined with feelings of identity and public purpose. She and her neighbors felt that what was good and beautiful in a chacra of boiling potatoes was part of being a Quechua farmer, both enjoying its benefits and meeting its duties. Their creation of identity, solidarity, and public purpose in the keeping of diversity could be heard in the acts of naming individual landraces, the habits of speech that helped to cement reciprocity ties between the cultivator and her crops.Those farmers still cultivating the landraces expressed the names of their landraces annually, thus keeping their symbols alive and vital.

Names of many boiling potatoes evoked poignant features of the high-elevation grassland. Its landscape, known as the loma in Paucartambo and puna elsewhere in the southern Peruvian sierra, was rich with meaning for the Quechua people. Symbols such as llamas (*llama*), alpacas (*paqocha*), mountain lion (*puma*), flowers (sunch'u), and birds (leqechu, *kondor*) were voiced in the names of potato landraces. Giving expression to such names, a person in Paucartambo was asserting his or her cultural identity, since the symbols expressed were Quechua (Allen 1988; Isbell 1978). He or she was also expressing individual effort and a solidarity of public purpose in the acts of naming loma landmarks. The Quechua in Paucartambo felt themselves embarked on an unending effort to domesticate the loma, hoping to mitigate its inherent character as an uncivilized space of wild and forbidding nature (Allen 1988; Isbell 1978).[9] The moral aesthetic expressed in landrace naming was thus not just an expression of one's sense of place and personhood but also an effort to remake them.

Symbols of social-geographic identity also infused potato nomenclature. When a cultivator referred to the landrace *ch'ilkas,* for instance, she called to mind the peasant inhabitants of nearby Quispicanchis and Canchis Provinces, who regularly traveled to Paucartambo to work and trade.[10] *Cuzqueña* (Cuzco), *hampara* (Amparaes), *chinchero* (Chinchero community near Cuzco), and *patallaqta* (Patallaqta community in nearby Calca) readily conjured other geographic settings. Her voicing of the kind known as *suwa manchachi* referred to "Thief Scarer," and was thought to guard against crop robbery. Other symbols of social-geographic identity referenced earlier epochs of potato growing. The name *garmendia,* for instance, descended from a high-ranking colonial official whose descendents owned prosperous haciendas in nearby Calca (Macera 1968). Symbols of social-geographic identity also gave meaning to more general metaphors, familiar in the cases of cholo (Mixed Breed), sambo (Mulatto), and moro (Moor or Spotted), which marked the hybrid forms of tubers. The term *chola,* which dubbed the imposing market women of cities and villages, was also a pregnant source of metaphor for the crop plants; a vigorous plant was said to "grow like a Cuzco *chola.*"

Public purpose gained its loudest voice in the names of various potatoes that evoked the figures of social and ethnic domination. Characteristic of their jesting, the Quechua fashioned many such names with humor and even satire.

They tagged one small and conspicuously narrow landrace as *misti pichilu,* or Penis of a White Person. Another landrace was entitled One Who Cries for Her Inca, or *inkamanta waq'aq,* which still in the twentieth century conjured the poignant mourning of an imagined past (Flores Galindo 1988). The farmers also joked about That Which Makes the Daughter-in-Law Cry, or *qachum waqachi,* the tuber whose tortuous convolutions could unnerve a nervous newlywed peeling potatoes alongside her proverbially watchful mother-in-law. Names of the boiling potatoes thus managed to wreath acts of rote recollection—what, after all, could be more prone to habit than reciting a common name—with the peasants' jocular expression of resistance against overly forceful and often cruel others. The most local persons as well as others who were quite remote were represented in their dialogues of resistance, a not uncommon feature of peasant worldviews (Scott 1985).

Farmers most often voiced such socially explicit names in out-of-the-way recesses among mountain fields and rural farmsteads. Persons other than the Quechua farmers rarely or never heard the gamut of their landrace naming. Many names, it seems, were almost as buried underground as the tubers themselves and thus hidden from the scrutiny of outsiders including those persons spiked verbally by the pointed labels. The farm landscape of boiling potato landraces and the peasants' naming them, moreover, was far removed from the regular sites of social domination, such as villages.[11] Indeed, if a landrace needed to be referred to in more mixed society or a village, its name could be substituted for with a more suitable synonym. For example, farmers dealing with the boiling potatoes as social capital in relations with Paucartambo villagers tended to rely on the synonym chimaku in place of the barbed and hilariously explicit misti pichilu.[12] Cultivators conveniently had such synonyms ready for those landraces with the most piercing names.

Their rationale of moral aesthetics was not at odds with utilitarian aims even in Paucartambo communities that were said to be ecologically specialized in high-elevation Hill farming. Farmers there tended to work with individual landraces rather than in groups. They selected, stored, and even planted the boiling potatoes separately, reserving a few rows or row segments for each landrace type in a style of sowing called *taka-takasqa.* Their style of direct selection predominated in communities like Humana and Carpapampa where inhabitants could grow little other than potatoes. Directed selection and site-specific planting suggested that sowing was based solely on utility, a function, in other words, of each landrace being chosen for a particular purpose. But discussions with the ecologically specialized farmers and field-level sampling revealed that they too lacked a rationale for each landrace variant (Zimmerer 1991b). In describing their activities with the boiling potatoes they expressed a moral aesthetic that resembled their counterparts residing in lower communities.

Farmers' determination of diversity in the ulluco crop closely resembled the techniques and rationales behind the farming of diverse potatoes. Nomenclature in the ulluco crop revealed the close-knit character of this resemblance. Its quintessence was that the Quechua farmers frequently reserved the name *papa lisas,* literally potato ulluco, when referring to the ulluco crop in general. Their names for specific landraces also disclosed a unifying resemblance to the potatoes. Nearly one half of common ulluco names originated in labels applied to potato landraces. Labels such as maktillu, mantaro, qompis, and suyt'u not only derived from potatoes but from common ones at that. Quechua cultivators thus created a language for the ulluco crop that treated it as a minor incarnation of the grand dame of tuber crops, the potato.[13]

The linguistic practices of farmers that wed ulluco and potato cultivation were consonant with the social organization of work activities in the two crops. The familiar need for farmers to share their apprehension of work routines sealed the similarity of language. A woman aiding her neighbor in selecting ulluco seed, for instance, would not be instructed about the rules of drawing distinctions between tuber sizes in the second class and third class ranks. The Paucartambo farmers were thus pressed to standardize the linguistic and technical practices of seed work and the management of diversity, similar to the social shaping of work techniques and knowledge in fields. To make their work routines sufficiently uniform, the Quechua cultivators not surprisingly patterned ulluco production after the pervasive potato crop. Oca and mashua, the other tuber-bearing crops, likewise seemed to be produced in a manner akin to potatoes, both materially and culturally.

One notable deviation of ulluco farming from potato agriculture was in the former's simplification. Quechua farmers erected barely two use-categories in the ulluco crop, sharply paring their suite of terminology in comparison to the potatoes (table 10). Their "Early Planting" category, moreover, housed only the Cradled Baby, or wawa yuki. The lack of classification structure in the minor crop was wholly incongruous with the fact that the farmers concocted a menu of six ulluco-based dishes that dwarfed the main list of potato plates.[14] Farmers even preferred a few landraces for certain forms of ulluco dining; qompis, noted for a less mucilaginous (*llawsa* or *flemosa*) texture than the others, supplied the best stuff for preparations with little processing like *lisas sakta,* a dish of ulluco tubers flattened and boiled. Rather than delimit use-categories, however, the farmers construed their ulluco roster minus Cradled Baby in terms of a single landrace group. Their choice to handle them in a single use-category was said simply to make sense for a minor crop. It offered a prime lesson, showing how linguistic structures and work routines in the minor crops were less complex albeit less standardized.

Even the one or two sizable contrasts between potato and ulluco farming seemed small, however, when compared to the chasm between the tuber-bearing pair and the maize and quinoa crops. Farmers' practices, rationales, and patterns of diversity evident in the grain-bearing crops revealed whole inversions of some of the most familiar traits of tuber agriculture.

Ecological and Culinary Utilities

Paucartambo maize fully deserved the glowing tribute awarded to Andean maize in general: "A triumph of cultural and environmental adaptation" (Gade 1975, 111). Quinoa in the region boasted a noteworthy but lesser sum of diversity befitting its status as a minor crop. The Quechua farmers tilled the tandem of diverse grain-bearers with an in-depth knowledge versed in familiar concepts of reciprocity; however, their rationales for deciding the sums of maize and quinoa diversity gave priority to a host of utilitarian aims. Differing rationales added to several other contrasts between the maize-quinoa complex and the potato-ulluco pair. Cropping calendars, field habitats, and production space, as well as crop biology and religious belief, were uncannily inverted between the two pairs. These various and inverted features of the maize-quinoa complex guided a distinct working of diversity's sum.

Reproductive facets of maize and quinoa bore sparse resemblance to potatoes and ulluco. On the one hand, the maize and quinoa plants set seed through frequent outcrossing; self-fertilization or "selfing," on the other hand, was common in the tuber crops. Given irrepressible outbreeding, the grain-bearers begged a greater effort by farmers to stem hybrids from swamping the identities of their desired landraces. This biological imperative led the Quechua farmers to create an unfamiliar partitioning of diversity in the maize and quinoa crops: diversity was packed within each landrace rather than between them (Zimmerer 1991b, 1992a). At first glance, the twenty-seven varieties of maize sampled in the Paucartambo Andes paled pitiably compared to the seventy-nine or more boiling potatoes. Each of the maize variants, however, actually housed a greater sum of diversity within it. Nature in the guise of sexual biology thus acted a full-fledged role in the determination of diversity.

Farming of maize in Paucartambo posed a single pair of use-categories: hard-kernel boiling maize, or *mot'e sara,* and soft-kernel parching maize, or *hank'a sara.* The cultivators' twin landmarks of maize eating matched other Andean regions while parting signally from other maize centers, such as the tortilla and pinole cuisine of Mexico, where the crop also exhibits a prodigious diversity (Bird 1970; Grobman et al. 1961; Manglesdorf 1974; C. Sauer 1950). The Quechua farmers in Paucartambo distinguished between a boiling maize and a floury maize on the basis of seed texture. They matched eighteen landraces, or two thirds of the total, with the hard-kernel boiling maize group, while the remaining soft-endosperm types served for parching; but simplicity of their twin category classification belied the role of yet another first-order distinction that the Quechua regularly etched into the maize crop.

The other classification cataloged maize landraces according to ripening time. Per this criterion, the Paucartambo farmers keyed each landrace to one of three plantings popularly known as big seed, middle seed, and small seed, the equivalents of early, middle, and late plantings, respectively. They divided the diversity of maize unequally along the lines of time-to-ripening; the common middle and late plantings housed more cultivars, although the early one was most unique (appendix D.4). The Quechua farmers used their production-based classification to crosscut, so to speak, the division based on use-category. As a consequence, farmers enriched each maize planting—early, middle, and late—with landraces belonging to both the boiling and parching use-categories. Their crosscutting criteria overlay maize classification with a much finer mesh of distinctions than was at work in the cathedral-like structure of potato taxonomy where the ranks parted so widely.

Putting both boiling and parching maize into the three maize plantings was a complex, versatile, and uncommon, perhaps even unique, system of classification. The Quechua of Paucartambo innovated their complex classification in order to maximize the usefulness of relatively sparse Valley fields. Less subdivided classifications, by contrast, appeared in the maize farming undertaken by peasant cultivators in other Andean regions such as Huánuco in the central Peruvian highlands (Bird 1970), the Urubamba Valley adjacent to Paucartambo (Gade 1975), and Sandia to the south in Puno Department (Camino et al. 1981). Like Paucartambo people, farmers in the other Andean regions draped as many as three maturation-based plantings across their mountainous balconies of maize farming; however, they sited only a single use-category in each planting. The Quechua of Paucartambo, by contrast, managed to subdivide each maize planting along the second axis of use-category.

The Paucartambo farmers used this subdivided mesh of six first-order distinctions to specify an ample diversity of landraces. The crosscutting mesh gave their cultural ordering of maize a linguistic architecture that was more akin to closely packed cells rather than the spacious cathedrals of potato classification. Maize farmers put their cellular style of classification into practice by intercropping multiple use-categories within a single parcel (plate 11). In the intercropping of use-categories the Quechua cultivators treated the maize subgroups rather like minor crops, such as quinoa, tarwi, and broad beans, which were often combined within other plantings. Their innovative combinations of maize presented new flexibility to the farmer who frequently held no more than one coveted Valley parcel that could sport maize and its host of intercropped companions.

Eufemia and Faustino—poorer peasants of Umamarca—intercropped maize in a field in the 1986–87 season by sowing a middle planting into its fertile, or

Plate 11. Juan and Fortunata curing new maize. The solid-color Yellow Maize and White Maize are prepared as hominy, the speckled ones (Fly) are parched, and the Dark Purple landrace is brewed. The first two landraces belong to the small seed category while the latter is a medium seed type. All are sown in a single field.

qhechi, portion. Nutrient-rich soils, they knew, would hasten the ripening of a triumvirate of parching landraces. A few weeks after this planting, near All Saints' Day, the couple seeded the fast-maturing late planting into the less fertile, or literally "bald," sections of the same field. By segregating the two plantings within their sharecropped parcel, Eufemia and Faustino could more feasibly undertake the staggered work calendar of each. Other farmers similarly reserved specific landraces to cope with certain differences in soils. A number of people seeded the early popcorn known as Pearl, or perlas, in the poor quality soils and the slower ripening Dark Purple, or *kulli,* into more nutrient-rich patches. By intercropping with two categories, many Paucartambo farmers could compensate at least partly for their limited Valley farmland.

The purposeful sowing by farmers of specific landraces matched their regimes of seed selection and maintenance in the maize crop. A Quechua farmer scouted her landraces carefully in order to meet the crosscutting criteria of use-category (boiling, parching) and planting (early, middle, late). She partitioned such groups after harvest and kept them separate in storage.[15] The women farmers exerted a firm control over the storage and selection of maize. Their authority not only prohibited men from drawing ears out of the consumption-seed stores but also sanctioned them to stay clear of the storage areas. The authority of women over stored maize was solidified by their increased roles in the diversity-rich fields of Valley farming. In their work with maize landraces they demonstrated a degree of cultural precision that was not practiced on the diverse potatoes. It was clear that inverted images of the crops owed to culture as well as nature.

Utility of consumption as much as production inspired a series of strong preferences for special maize landraces (Zimmerer 1991b, 1992a). Luminescent k'ellu, or Yellow, known in some places as *uwina* or pusaq wachu, was widely liked for boiling because of its meaty texture and undiluted flavor (appendix E.1). Sugary *chullpi* and the tasty Milk Paraqay (*leche paraqay*) portended the best materia prima for parching; in fact, chullpi was so sought after that farmers planted it in the middle of fields to deter thieves. The landraces desired for the daily diets of farmers, however, were not necessarily the ones that yielded best. The sweet chullpi, for instance, returned notoriously low yields compared to the tougher White, or *yuraq.* As a result of their assorted preferences, the Paucartambo farmers liked to sow a precise plurality of cultivars in each planting. Chullpi and yuraq, as well as one or two other varieties, were stalwarts in their middle plantings.[16] That the landraces best for consumption goals differed from ones ideal for production meant that the strong likings of farmers were summing toward a diverse total.

A relished spread of one dozen uses beyond boiling and parching also whetted the appetites of the Quechua in Parcartambo, and they found that certain landraces were best suited to each secondary use.[17] To tint the otherwise earthy froth of their maize beer, for instance, the resourceful grower colored with the rich hues of kulli. The speckled Fly, or *ch'uspi,* provided a parched viand that could be easily carried to fields as a snack or finger food. Ch'uspi was, in fact, a favored item in the genre of lightweight snack foods known in Quechua as qaqaw (Zimmerer 1992c; see also Weismantel 1988, 139). The list of secondary purposes enumerated a diversity of uses that was moderate but not extensive. Farmers did not, however, subscribe uniformly in their tastes for the cultivar that was best-suited to each purpose. In fact, the maize growers divulged a wide spectrum of personal tastes for items in the established culinary repertoire (Zimmerer 1991b). They shared their esteem for maize and its preparations in general but not in its particulars. That individuality of personal tastes helped to enhance further the fortunes of maize diversity.

Religious ritual and ceremony sanctified a small number of landraces that added to the culinary and ecological-economic logics of maize diversity. The Quechua in Paucartambo and their local shamans—known as *paqo* and *hampeq*—worshiped with a handful of special types in fertility and healing rituals (appendix F.4). The spiritual leaders anointed two aberrant kinds in particular: the bicolor red and yellow Mass, or *misa,* and the three-eared Store, or *taqe*. Medicine men hunted for the ceremonial types in their own fields and stalked them more widely through the means of generous barter and cash payments. Some publicized their standing offers of compensation for specimens of the ceremonial maize. The Quechua farmers as a whole took enough interest in the misa ears that they regularly paired Red, or puka, with Yellow, or k'ellu, varieties in the same field in hopes of harvesting valuable hybrids.

The lording of spiritual value thus attached extra impetus to cultivate diverse types of maize. It also hinted at how the Quechua in Paucartambo lionized maize nonpareil in their contemporary religion, a modern-day emphasis in iconography akin to the crop symbolism that was once proselyted by the Inca cult (Murra 1960). Although morphological patterns analogous to the ceremonial maize types were recognized in other diverse crops—the bicolored misa pattern in potatoes, quinoa, and ulluco and the branching taqe form in potatoes—none was granted religious symbolism as deeply as maize. The modern-day worship of maize was another close tie between diverse crops and the hybrid Andean Christianity of the Paucartambo people. It even furnished a prime example of the vitality of their religion that was forged upon joint icons such as the Mass, or misa, maize, an explicitly Christian designation that was transferred to an unconcealed pagan rite.

Farmers' expression of moral aesthetics in maize-containing religious ceremonies bore only a faint resemblance to the representations of potato and ulluco farming. Maize ceremony was steeped in far more formality and narrowly circumscribed ritual than the decidedly everyday evocation of moral aesthetics in the other crops. Oddly, the lexicon of maize names paled in its litany of symbolic allusions, since the majority of landraces were straightforward color terms such as Yellow, White, Gray, Red, Dark Purple, and Smoke. In voicing a few familiar symbols through maize names the Quechua farmers styled commonplaces of their Valley habitats, such as Fly. Overall, they created a maize nomenclature largely devoid of the heftier symbols holding sway in their lives. The farmers were deriving utilities from maize that were mainly material and explicitly ceremonial but not mental like the moral aesthetic.

Diversity's sum in the Paucartambo maize crop could be totaled via its utilities, which in the crop came from the crosscutting system of multiple plantings and use-categories. Single maize plots, however, did not average more than a meager 2.7 landraces (Zimmerer 1991b). Farmers' frequent sowing of their two or three preferred varieties in each field met pointed production and consumption criteria. In the middle planting, for example, many growers confined their efforts to the premier parcher chullpi and the high-yielding yuraq. Aggregating the number of landraces in each of the three main plantings and in the two main use-categories led to more formidable totals of diversity. Families cultivating the three main plantings—early, middle, late—might easily cache as many as nine landraces, an impressive sum by any measure.[18]

The farming of diverse maize could not, however, escape the crop's peculiar biology and its titanic capacity for outcrossing or cross-pollination. Quechua farmers realized that the cross-pollination from one maize plant could demote the ears of its neighbor into an undecipherable mix of kernels of different colors and shapes, an outcome that maize biologists refer to as xenia. Notwithstanding their recognition of the threats of xenia, most Paucartambo farmers interspersed completely the seed of two or three landraces. They referred to their style of intersown mixtures as *michisqa* (mixed). By sowing the mixed fields, the farmers seemed to be overlooking the laws of nature and endangering the distinctness of their diverse varieties; however, this appearance of tossing caution to the wind belied a few techniques that kept the maize kinds confined within the bounds of sound varietal management.

Their main means of limiting the hybridization of landraces was to combine a few sets of certain cultivars regularly in customary field mixtures. The farmers found, for instance, that late plantings of Yellow (k'ellu), Red (puka), and Smoke (qosñiy) showed only a small degree of cross-pollination, each desired landrace keeping true to its type. Widespread awareness of the peculiar advantage of this particular mixture led the colorful threesome to paint a majority of late plantings. Such compositions probably constrained cross-pollination owing to biological factors that restricted genetic compatibility and hence cross-pollination (Bird 1970; Grobman et al. 1961). Know-how among the Quechua in Paucartambo did not justify their landrace mixing in the scientific terms of genetic incompatibility, although they adeptly observed the paucity of unwanted hybridization. They also praised the beauty of the three-tone palette, especially when the shucked ears lay exposed curing on the open-air drying floor, or tendal.

The farmers' images of ideal types and landrace identities served to crystallize their aversion to maize hybridization. Cultivators routinely discarded the cross-fertilized ears recognized to be variants and thus stabilized the selection pressure for an ideal type. They sized up an ear on the basis of its overall resemblance to a variety's ideal form. Criteria included its row number and the colors of its kernels and cob. Cross-fertilized ears were depicted with the qualifiers Mestizo (cholo); Moor or Spotted (moro); Defender (*amachayoq*); or the untranslated *saliasqa*. Techniques that hewed maize seed to the ideal form of a variety ruled out a selection style that was based on selection-for-perceptual-distinctiveness, which predominated in the farming of diverse potatoes. The very premise of selection-for-perceptual-distinctiveness was inimical to the basic biology of the diverse maize crop, since it would quickly procreate an unwieldy number of hybrids.

Their skillful techniques and knowledgeable customs that restricted hybridization were enabling the Quechua cultivators to persevere in the habit of interspersing diverse maize varieties in a single plot. Such interspersion in the maize field resembled the complete intersowing likewise apparent in the boiling potato parcels. In both cases, the farmers were planting seed that was not separated by landrace type. In the maize crop, like potatoes, the prevalence of intersown mixtures grew during the years from 1969 to 1990 at the expense of the "row-by-row" segregation of varieties. Row-by-row segregation, called *melga-melgantin* in the case of maize, entailed the careful and separate management of each landrace. Yet the maize crop, once again similar to the potatoes, did not show signs that its diversity declined due to the slow demise of row-by-row techniques.

Absence of a noticeable contrast of landrace diversity between mixed and row-by-row plantings lent evidence that the interspersed sowings were not undoing the crop's diversity. In the northern Paucartambo Valley, where many farmers segregated each landrace into a block of rows according to the row-by-row, or melga-melgantin method, the diversity of their fields did not differ significantly from those of farmers who intersowed landraces in the completely mixed, or michisqa, plantings (Zimmerer 1991b). Quechua maize farmers were succeeding in making innovations for the sake of convenience and time-saving without incurring the apparent loss of landraces. Still unknown, however, were the gene-level and medium-term consequences of their increasingly common techniques.

Quinoa cropping and its panoply of utilitarian aims resembled maize in much the overarching way that ulluco farming came close to mimicking potatoes. Quinoa attested to an even closer connection to maize, however, since it was mostly sown within the very fields of the latter. The Quechua farmers who tended quinoa thus found good reason for dovetailing the details of its tillage with maize in order to simplify the work routines of laborers toiling in their intercropped parcels. As a consequence, several farm tasks designed for quinoa bore the closest resemblance to maize, beginning with planting and evident again at furrow mounding and the cutting of harvest. A sharing of landrace nomenclature also united the two grain crops. Many quinoa names derived directly from those of maize: for example, paraqay, misa, k'ellu, choqllos, uwina, puka, yuraq, kulli. The practices and language of quinoa cropping reaffirmed that in the eyes of Quechua cultivators in Paucartambo their maize was the matriarch to which the other grain-bearers were referenced.

In choosing their quinoa landraces the Quechua farmers mimicked their efforts with maize by plainly giving priority to utilitarian criteria. They mostly produced the quinoa crop for its supply of two landraces, Yellow (uwina or k'ellu) and White (yuraq), which were favored in a hearty quinoa soup known as *kinwa uchu.* A heaping handful of grains from the two quinoa landraces gave ample flavor and substance to the thick nourishing soup. Soup-making was specially suited to the quinoa crop, since it diluted a bitter saponin principle that required the seeds to be leached in a series of rinses. When the farmers concocted quinoa into the floury viand and field snack known as *p'esqe,* however, they preferred the smoother tasting landrace of Red, or puka, quinoa. Utilitarian aims thus ordered the main rules of quinoa farming, although unlike maize the routines were not cast in an elaborate crosscutting taxonomy of multiple use-categories and plantings.

The bedding of seedlings in a special nursery—known in the vernacular as a lake, or *qocha*—and subsequent transplanting to intercropped fields were a unique feature of the quinoa crop. Mostly farmers in the northern Paucartambo Valley applied this technique of transplanting, a measure of decided advantage there since the quick growth of tall-cane maize easily dwarfed the slower-growing quinoa. Quinoa growers found they could counter their crop's predicament by tending nurseries and then transplanting young seedlings, giving them a head start. The resources subsumed in their activities were kept in check by the limited planting of quinoa, rarely more than fifty seedlings. The modest efforts of farmers with a field of quinoa unveiled how they could sometimes marshall the resources for a diverse minor crop due to its small scope whereas the greater effort with a major crop might prove untenable.

The Cusps of Recent Cultural Change

The full suite of diversity-determining farm practices drew on a tremendous wealth of cultural expressions: a common language of reciprocity concepts that was couched in Quechua metaphors, the kawsay ethic of proper livelihood and cuisine, beliefs about production techniques, informal moral aesthetics, and Andean Christian religious ritual. During the years from 1969 to 1990, the farmers in Paucartambo altered a number of their complex cultural practices that were related to the defining of diversity. Although the region's diverse crops did not disappear beneath the "lone and level sands" of biological uniformity, the changes in farmers' culture were of consequence to biodiversity's role in their lives. Cultural changes were not shared equally, moreover, since certain groups in Paucartambo like better-off peasants and Protestant converts were at the forefront.

More well-off Quechua farmers were the first to adopt the modern goods of national culture. They embraced manufactured clothes, shoes, household wares, rice and bottled beer, and a cooking oven or two. When electricity reached a few farmhouses on village outskirts in the late 1980s—after installation in villages as early as the 1930s—a few even purchased television sets. The wealthier Paucartambo farmers conversed with greater ease in Spanish which facilitated their commercial affairs; however, they also wielded the cultural ways of language, livelihood ethic, production beliefs, aesthetics, and religion that guided them in sowing diversity and defining its prodigious levels. In their role as a conspicuous class of diversity's cultivators they were reinventing the culture of the diverse crops amid a post-1969 cascade of environmental, social, and economic transitions.

Their reinventions were making the culture of diverse crops into a traditional-style luxury. Well-to-do Quechua managed to recast and invigorate the role of crops rather than rejecting it in the context of their growing medley of multifaceted dealings in villages. The diverse crops were becoming items that helped to create a comfortable middle-ground cuisine useful to the prospering farmer who sought to gain favorable treatment from powerful people in provincial society. While the better-off family occupied a rented room or bought a village house, its men especially adopted mestizo ways. They would dress in machine-made trousers and don brimmed hats. Although still recognized by villagers to be peasants, they were filtering part way through the entrenched divide between a rural populace of Quechua peasants and a virtual caste of non-Quechua villagers. While residing at home in their rural hamlets, the better-off peasants remade their diverse crops in a newly expanded role that aided them in recruiting labor and profiting from commerce.

The reinvention of diverse crop culture was a scenario quite different than the unexamined assumption that recent change triggers cataclysms of cultural erosion that besiege diversity's growers and lead to the sweeping onslaught of genetic erosion. To be sure, the well-to-do Quechua peasants in Paucartambo wished to adopt some food habits of their village acquaintances. Families like that of Natividad and Líbano enjoyed plates of rice and village-baked bread, but their new eating customs did not convert all their food preferences. They did not, in other words, covet all the aspects of village life. For the wealthy peasants, the character of village life, including culinary habits, did not foretell an image of their own future in the region. Among their poorer neighbors who felt more pressed by the choice between the unrewarding hardships of farming and the opportunities of urban life, even the attitudes of those who chose to migrate did not dismiss diverse crops as might be assumed.

Many Quechua emigrants from Paucartambo still kept ties to the diverse crops. A large number of Paucartambo out-migrants relocated permanently to the San Jerónimo neighborhood of greater Cuzco (1981 population: circa 200,000). Their new urban lifestyles as members of the urban poor continued to attach many strings to their rural backgrounds including food habits. Umamarca-born Segundino Condori Castillo, a taxi driver in San Jerónimo, regretted that his brother Faustino back home no longer sent an occasional sack of floury potatoes on one of the trucks that departed from Paucartambo. Segundino noted that while he could readily buy them in Cuzco's large *Mercado Central,* or Central Market, the floury potatoes on sale were less tasty than his brother's. The vendors mainly sold suyt'u (Narrow One), pitikiña, and qompis; there was not one who offered mixtures of his favorites cheqefuru and llama senqa.[19]

Nor was there an absence of altered roles for diverse crops in the countryside. Farmers were remaking one of their central cultural images of diverse potatoes and maize. By the mid-1980s, most growers conceived of potato and maize landraces in terms of both intersown mixtures and single variants. Although some people had long seeded interspersed combinations of landrace into their fields, the intersowing technique was spreading at the expense of fields with segregated landraces. Paucartambo farmers even used the term *mixed potatoes,* or *papa charqho,* in place of floury potatoes. A farmer might say, for instance, that her mixed potatoes were planted in a particular field or that she was preparing to harvest her mixed potatoes. Yet the farmers' prevailing concept of landrace mixtures did not eclipse their equally rich familiarity with the landraces as individual types. They still held firmly to ideas of each landrace and its properties.

History recorded in the diverse crops showed how the Quechua people had often brought new cultural habits into their farming culture. Vernacular crop names rooted in the colonial era such as One Who Cries for Her Inca and the racial and cultural admixtures originating at that time, such as Mestizo, Sambo (mulatto), and Moor, energized a moral aesthetics that still rolled with momentum in the years between 1969 and 1990. Similarly, but in a curious inversion, the Quechua in Paucartambo voiced their familiar language of reciprocity and ritual ceremony with respect to the improved potato varieties that were widely adopted after 1969.[20] They could be heard kindling their interest in the new objects of farming with knowledge fueled by customary concepts. Such change highlighted how crop elements of the fiercest Quechua identity in Paucartambo were not necessarily Andean natives, much the way barley dishes had long symbolized a core of Quichua Indian identity in Ecuador (Weismantel 1988).

More dramatically, Paucartambo and its Quechua people were not untouched by the conflagration of Evangelical Protestantism that has swept over the central Andes of Peru, Ecuador, and Bolivia during recent decades (Isbell 1978; Mitchell 1991; Weismantel 1988). After 1969, foreign and national missionaries as well as return migrants spread Protestantism in a number of small pockets in Paucartambo. Bastions of Protestant converts resided in the communities of Humana and Callacancha in the southern Paucartambo Valley and Acobamba in the northern valley. The new Protestants among the Quechua in Paucartambo recanted their beliefs in the nature gods such as Earth Mother and the mountain deities, or *apus*. They also disavowed the special crop protectors Maize Mother and Quinoa Mother.

Paucartambo's Protestants also refused to fund the religious offices, or cargos, of Saints' Days. Such cargos anchored the Quechua-Catholic religious calendar and affirmed the civil-religious hierarchy of communities. Many cargo posts in the villages and larger communities were quite costly, although the local churches had once granted farmland to office holders in order to defray expenses. The Protestants in Paucartambo believed, however, in an ethic of personal advancement that was at odds with such ceremonial expenditures. The converts steadfastly declined to be an officer, or carguyoq, disavowing the Catholic celebration and its support of drinking chicha and other alcoholic beverages, dancing, and listening to raucous band music. Their different and often antagonistic beliefs split communities and sometimes even families.

The Protestant converts in Paucartambo did not, however, demonize the cuisine of feast day celebrations. They proved as likely as their Catholic counterparts to grow the diverse crops. The few Protestants living in Umamarca, originally from nearby Callacancha, sowed floury potatoes and diverse maize, quinoa, and ulluco, like their neighbors Natividad and Faustino. They also carried out a number of conventions with the diverse crops that were closer to Quechua-Catholic customs than Protestant orthodoxy. Special food, or *comida especial,* the tradition of serving a crop in special dishes to workers, friends, and neighbors during its planting and harvest, carried a meaning that was undiminished among the Protestant converts. At least a few also made ritual blessings, or *ch'allasqa,* for the crops, in at least one case substituting bottled soda for the standard cane alcohol, or *trago.*

Protestants and Catholics alike were attuned to a Quechua view of the cropping calendar and celestial sphere that oriented their farm decisions. Both groups commonly timed the chief tasks of the agricultural cycle, such as harvest and planting, according to the Saints' Days (appendixes D.4 and E.5). They also looked to the celestial sphere for guidance in the timing of their most important farm tasks. Lunar phases and the locations of constellations, both the familiar star-to-star and the Quechua dark cloud type, aided in knowing when to plant and select seed.[21] Conventional wisdom held that the waning moon, in their view a growing "dead moon," or *wañu killa,* was best for planting potatoes; however, farmers sometimes sowed crops in either a waxing or a waning moon, what they called half moon, or *chawpi killapi.* They would readily admit that the busy farm calendars of the years from 1969 to 1990 did not always allow precise coordination with the lunar synodic cycle.

Cultural change in the countryside was due also to the greater schooling of children. After 1969, nearly all peasant children attended primary school in the rudimentary one-room schoolhouses of their communities. No more than a handful, however, received schooling at the secondary level, since parents could not afford their further education. Mainly boys benefited from the rare opportunity of a secondary education, and only those of the wealthiest Quechua in Paucartambo were sent to Cuzco for training in high school. Most children, therefore, spent ample time on the farm and gained their work skills through experience and observation. Their expertise with diverse crops often began when parents assigned the young children special rows to tend in a few of the family's fields. It was not uncommon for children by the age of five or six to be set aside rows in floury potato fields.

A child's ample time in learning the skills of tending diverse crops came mostly with his or her mother. A Quechua woman in Paucartambo shepherded her children across the farm landscape since infancy when they were tightly swaddled and carried in a sling; a child's learning of farm life thus started early. Although much knowledge about the diverse crops was later derived through labors in the field, customs of kitchen work and consumption also supplied a share of instruction. Children learned the names of landraces with curiosity and sometimes gusto. The names of particular landraces could be made especially intelligible to a child. Santusa described suyt'u, or Narrow One, to one of her grandchildren as Snout, in reference to one of the farm animals. Special meals, like the much-awaited roasting of potatoes in special sod ovens, offered harvest-season treats when the Quechua children in Paucartambo partook of pleasures from their families' diverse crops.

Only the children of some families, however, were regularly joining in diverse crop repasts and approaching the expected norms of a fit livelihood. Particulars of the success of the more well-to-do families in Paucartambo and of the losses incurred by their poorer neighbors must be considered in light of both the development forces and environmental transitions that enveloped the fortunes of biodiversity. To look at general insights and implications means venturing once again to where the specifics of the Paucartambo people and region can be discussed with respect to the broad issues of diversity's role in small-scale peasant farming, past, present, and future.

7

The Vicissitudes of Biodiversity's Fortune

A Mixed Lesson: Less Certainty, Greater Flexibility

> *I have great faith in a seed. Convince me that you have a seed there and I am prepared to work wonders.*
>
> (Thoreau 1993, 7)

Diverse crops and the livelihoods of Quechua peasants in the Paucartambo Andes offer a story of mixed fortunes. The rich variety of crop species and landraces make the region a world-class center of biodiversity, but the generations of Paucartambo peasants tilling them have long struggled to fulfill their expectations of a fit livelihood. Their efforts have relied on the diverse crops as key assets in production and consumption, and they have managed these crops resourcefully by modifying the environmental and social properties of their agroecosystems. Yet the much-heralded biodiversity and peasant farming of Paucartambo were neither as unchanging or fragile nor doomed to extinction as may have been assumed.

The illustrative story of biodiversity and peasant livelihoods in the Paucartambo Andes, however, did not foretell a certain future. While many Quechua peasants husbanding the diverse crops managed enough flexibility to accommodate the rocky paths of their livelihood changes, the diverse crops were definitely not niched safely in a harmonic balance of rural society and nature. Interest and capacity to grow the crops varied among the farmers. Since their success differed according to social and environmental forces, the diverse crops were never preordained a fixed place in Quechua peasant life. Widespread genetic erosion did, in fact, occur, although less than the seventy percent estimated for the Americas in general and the ninety-seven percent said to have occurred worldwide (Fowler and Mooney 1990, 63; Nabhan 1992, 146).

Various changes in production drove genetic erosion more than did the modification of consumption habits. Such loss of landraces in the Paucartambo Andes was registered before 1969. The region's Quechua farmers had already curtailed quinoa and the early chawcha potatoes by mid-century. After the Land Reform of 1969, worsening resource scarcities were pressing the twenty thousand-plus members of the Peasant Communities in the region to forego their diverse crops. Shortages of farmland, labor-time, and capital often dominated their decisions. Recent shortfalls of specific resources similarly have led the peasants of other regions in Peru and Mexico to surrender their diverse crops in decisions of last resort (Bellon 1991; Brush 1986, 1987; Gade 1972b; Zimmerer 1991b, 1992a). Although the outcomes for biodiversity in Paucartambo were not prefigured, that uncertainty could not overshadow the definite outlines of genetic erosion processes and pressures.

The scenarios of genetic erosion underscored how the diversity-deserting farmers felt unable to pursue their preferred mix of production in the years from 1969 to 1990. A family unloosed genetic erosion primarily through the curtailment of one or more of its landrace-rich fields. By contrast, little landrace loss was resulting from the replacement of individual crop types within them. Rarely did the process of genetic erosion take the form of a single improved variety being substituted into a field still sown with landraces. Most commonly, the cultivators who decided to relinquish the diverse crops were opting to forego a field in one or both of the landrace-rich production spaces and adopt new production. They typically withdrew their fields from Hill and Valley farming at the expense of the Oxen Area and Early Planting.

A "scissors effect" or "reproduction squeeze" imparted by the Peruvian political economy put peasant farming at a disadvantage and instilled the dilemmas of resource scarcity and possible genetic erosion among farmers in the Paucartambo Andes. Designed to secure inexpensive food and labor for the urban and industrial sectors, the scissors effect in Peru has deepened rural poverty and underdevelopment since mid-century (Alvarez 1983; Figueroa 1981, 1984; Gonzales de Olarte 1987; de Janvry 1981; de Janvry et al. 1989; Thorp and Bertram 1978; see also Bernstein 1979; Jennings 1988; Netting 1993; Watts 1983).[1] Peruvian economic policies after the contradictory Land Reform of 1969 imposed new variants of the scissors effect, and aggravated the shortages of some resources among Paucartambo farmers. Those national policies helped to create the group of resource-strapped farmers in the middle socioeconomic segment of the region's peasantry. Hard pressed to couple the unprecedented opportunities of new commerce with the merits of their familiar diverse crops, many chose to be rid of the latter.

Agribusiness growth, urban market demand, and international food aid—symptomatic of even the most rural Peruvian hinterland—fueled further the dynamics of biodiversity depleting changes in the countryside of Paucartambo (Alvarez 1983; Figueroa 1981, 1984; Hopkins 1978, 1981). The transnational Beer Company of Southern Peru converted the Paucartambo region into the chief producer of malting barley for its Cuzco factory. Thousands of Quechua peasants regularly contracted with the company under exploitative terms that they nonetheless deemed better prospects than their other commercial venues. Some migrated for short periods to the Pilcopata-Qosñipata lowlands that boomed with plantation agriculture (rice, fruits, coca) and resource extraction (logging, mining). By 1980, most every farm family in Paucartambo wished to produce the off-season crop of improved potato varieties, the burgeoning Early Planting that fed Cuzco's strong demand for staple foodstuffs.

Acute resource scarcity impinged on many Quechua farmers of Paucartambo by exerting a seasonal shortage of labor-time. The farm families throughout the region suffered such bottlenecks due to their new labors in the fields of contracted barley and in the Early Planting. Along with short-term labor migration, the new work routines bunched at the beginning and end of their agricultural calendar (August–September; June–July). Enterprising but resource-strapped farmers in the middle of the socioeconomic spectrum were most unable to renew those diverse crops having cultivation calendars that inconveniently extended into the peak periods of labor demand. They were driven to curtail early chawcha potatoes and big seed maize for this reason. Their sacrifice of the odd-ripening crops was a sign that the new work of many farmers was clashing with the staggered cropping and varied maturation at the heart of much biodiversity.

A decline in soil fertility sometimes triggered the decision of farmers to let go their diverse crops. Numerous cultivators in the Paucartambo Andes discovered that the impoverishment of field soils was inimical to their landraces and that it was not easily remedied. Landraces were found to be less responsive to mineral and chemical fertilizers than were standard improved varieties. Soil deterioration in the years from 1969 to 1990 stemmed from overgrazing, overcultivation, and erosion. Farmers throughout Paucartambo doubled their herds of sheep, cattle, llama, and alpaca on limited rangeland. They also reduced the rotation of legume crops and curtailed terrace building. Widespread deterioration of soils even posed a threat to the wealthier farmers, who otherwise could fund both commerce and the cultivation of diverse crops. In Colquepata, for instance, the declining fertility of Valley soils undermined the diverse maize and quinoa crops decisively.

Impact of farm changes during the years from 1969 to 1990 varied widely among the places in Paucartambo and gave added weight to arguments for the geographical unevenness and place-based specificity of environmental degradation and conservation. Quechua farmers in the northern Paucartambo Valley (Acobamba, Huaqanqa, Challabamba), the Colquepata interior (Colquepata, Chocopía, Cotatoqlla), and the southern valley (Mollomarca, Huaynapata, Umamarca) crafted unique responses to the social and economic reordering of the postreform period. Much confusion about the fate of biodiversity was removed in the place-based analysis. While farmers in each place were facing a number of common circumstances, such as the booms in Early Planting and contracted barley farming, they responded in different ways. Southern valley farmers stood out in their successful coupling of new commerce with skillful innovations that kept the diverse crops. Their success was due mainly to ample resource endowments.

Biogeographical processes redoubled the definitive impact on diversity of the places in Paucartambo. Regular exchange of seed in place-based trade networks shaped the distribution of the diverse crops into either endemic or widespread patterns. The immense diversity of floury potatoes, for instance, clustered in a moderately endemic fashion around high-elevation places inhabited by specialist growers who periodically supplied seed to a network of nearby farmers. The paramount role of place in diversity's distributions certainly added to the critique of adaptationist assertions about the predominance of microenvironmental patterning. It also suggested a cornerstone concept for the fuller biogeography of the diverse crop plants that has long been called for (Altieri and Merrick 1988; Harlan 1951, 1975b; Soulé and Wilcox 1980; E. Wilson 1988). The biogeographical importance of places also indicated that the fortunes of biodiversity could not be thought tied solely to individual families or peasant communities.

An unfavorable prospect for the diverse crops was delivered by the force of economic markets and the other demands on extrahousehold production. Since the onset of Inca governance about 1400, the Paucartambo farmers supplied a combination of goods and labor for extra-regional rulers, local elites, and markets. Their extrahousehold flows of farm goods and labor met the various demands of labor taxes (Inca, viceroyalty of Peru), tribute (viceroyalty of Peru, Republic of Peru, hacienda landlords circa 1600–1969), and marketplaces. However, the particular goods and labor of the Quechua farmers that were entrained in the extrahousehold flows did not encompass much biodiversity. Instead, the rulers and markets specified a narrow range of crops and landraces in order to fulfill certain needs. The diversity-poor nature of extrahousehold flows was not, however, an intrinsically fixed feature of their markets but rather it took shape culturally and historically (Orlove 1977a; Polanyi 1957).

Still many Quechua cultivators in the Paucartambo region managed to adjust their farming, seed the diverse crops, and reap the benefits of their harvests notwithstanding the vast uncertainties and imposing obstacles that confronted them. Their successes owed in large part to the flexibility they crafted in their production systems. The farmers redesigned field and storage techniques as necessary to ensure that diversity-rich production for subsistence could be combined with their undertakings in commerce and with other extrahousehold outlets. By modifying the wildly diverse habitats of their montane tropical territories into a handful of managed environments, the farmers simplified production and thus made it more flexible. The consequence of their unending labors was an extremely transformed farm landscape.

Repeatedly since at least the 1500s, the Quechua cultivators had innovated fields in all units of their farming. By 1970, the chief challenge for farm families was to produce their parcels for commerce (Oxen Area, or yunlla; Early Planting, or maway); and, concurrently, their diversity-rich fields for subsistence (Hill, or loma; Valley, or kheshwar). The successful coupling of varied farm production by many Paucartambo families renewed a prodigious diversity of potatoes, ulluco, and other tuber crops in Hill fields and a no lesser array of maize, quinoa, and assorted minor crops in Valley parcels. Innovation was commonplace in both Hill and Valley agriculture, pace claims of subsistence farming being relict or "natural" (Bradby 1975; Grumbine 1992). Due to their skillful innovations, many farmers cultivated the diverse crops notwithstanding the quickening pace of change in the years from 1969 to 1990.

The Quechua cultivators in Paucartambo routinely gained production flexibility by not restricting their farm spaces to a strictly tiered or layered arrangement. The actual pattern of their land use thus jarred with the widely discussed vertical zones or tiers taken to be a trans-Andean design (Brush 1976, 1977; O. Harris 1985; Mitchell 1991; Murra 1972; Orlove 1977b; Platt 1982, 1986). While the Paucartambo farmers took account of the environment, they also standardized their land use due to a pair of powerful forces. First, state and regional rulers since the Inca Empire purposefully shaped the nature of farming by imposing political and economic policies. While many policies legislated extrahousehold farming, several others were aimed directly at the subsistence of commoners in Hill and Valley farming (for example, topo-based land allocation under the Inca, *República de Indios* fiscal policy under the Peruvian viceroyalty, hacienda labor exactions until 1969, agrarian reform laws after 1969). The interests of imperial rulers and local landlords in everyday customs of resource and land use thereby helped to shape a variety of similarities that have remained as vivid features of the Andean farm landscape.

Secondly, a uniformity of land use at the field scale was enforced by the reliance of farmers on extrafamily laborers. Rudimentary technology and the marginal climate and soils of their mountain fields long led the farmers in Paucartambo and other Andean regions to recruit additional workers from outside their households. When the Quechua in Paucartambo paid or otherwise recruited farm hands for time-consuming tasks like plowing, they depended on those persons' knowledge of "common sense" cultivation styles and techniques. The Paucartambo farmers also standardized land use due to the agro-pastoral imperative of coordinating livestock-raising with cropping. Such standardization actually aided flexibility, since otherwise the farmers would have been less able to comply with the demands of their already varied production spaces.

Flexibility was also gained from the ecological character of the diverse crops. Quechua farmers in the Paucartambo Andes chose the diverse components of their crops—the landraces or traditional folk varieties—to suit a whole production space or a major subportion. For instance, every landrace among their scores of floury potatoes was ecologically adapted to a wide range of growing environments spanned by the lower section of Hill farming. Paucartambo farmers constantly used the ecological flexibility of their broad-based or "generalist" crops to aid in averting risk, securing a regular supply of desired foodstuffs, and, not least, staking a claim to their territory. In demonstrating ecological adaptations to wide-ranging and variable conditions, their plant domesticates echoed the discordant harmonies of nature. Although subtle compared to many organisms, this discordance nonetheless deviated from the popular notion of an equilibrium-tending balance and microenvironmental specialization in the farm nature of indigenous peasants (Botkin 1990; Zimmerer 1994a).

The broad-based adaptation of crops in Paucartambo must not be taken to infer that environmental limits and biological forces were absent. Potatoes, Andean maize, ulluco, quinoa, and several other crop complexes evolved principal traits in the distinct landscapes of Paucartambo and other eastern Andean regions. Ancient farmers guided the uphill evolution of crops and diversified them into staggered suites of maturation periods (Bird 1970; Brandolini 1970; Grobman et al. 1961; Hawkes 1978, 1990; Hawkes and Hjerting 1989; Pickersgill and Heiser 1978; and H. Wilson 1978, 1988). Ripening was not, however, a flexible trait except over the course of long-term evolution. When farmers expanded and diversified commerce in the years from 1969 to 1990, they found their new work schedules could clash irreconcilably with the genetic inflexibility of crop ripening. As a consequence, many withdrew labor from their early chawcha potatoes and late big seed maize, both odd-ripeners.

During the decades after the Land Reform of 1969, the flexibility in several minor crops did permit farmers to innovate the techniques of intercropping. Many Paucartambo cultivators pioneered mixed plantings of more than one dozen species—including quinoa, tarwi, broad beans, amaranth, arracacha, winter and crookneck squash, achocha, passionflower, common beans, and yacon. In order to compensate for the conversion of their other farmland to malting barley and improved potato varieties for commerce they overhauled the intercropping of diverse species in their maize-containing Valley fields. Their innovation of Valley fields highlighted the versatility of intercropping techniques.

Features of their self-provisioning could thus be skillfully adjusted, a capacity that complemented their better-known capacity to make commercial innovations (de Janvry et al. 1989; Figueroa 1981, 1984; Larson 1988; Lehman 1982a, 1982b; Mallon 1983; Reinhardt 1988; Sheridan 1988; compare with Richards 1985, 1986).[2]

Cultural customs at once flexible and resilient also enabled the Paucartambo farmers to keep diverse crops even in epochs of tumultuous change. Farmers' ethic of a fit livelihood known as kawsay was a prepotent cultural concept in the saga of diversity's fortunes. Since the 1500s, the Quechua in Paucartambo retained an unabated esteem for the majority of diverse Andean crops. While they added to their customary diets with new "Old World" crops such as wheat, barley, broad beans, peas, and, most recently, rice, the Andean farmers found the majority of diverse crops to be irreplaceable in their cuisine and still vital symbols of their cultural identity.[3] In a few cases, however, the farmers' changing definition of customary cuisine failed diversity. Early chawcha potatoes and big seed maize, two casualties of the period from 1969 to 1990, did not bestow culinary qualities of any particular advantage and were thus held to be expendable.

The tragic colonial history of Paucartambo offered a special testament to the resilient accommodation of diverse crops by its Quechua peasants. Indigenous people of the region suffered catastrophic mortality during the early colonial period. New diseases and forced labor in coca fields of the nearby foothills, or montaña, decimated their population. Thousands of peasant immigrants flooded into Paucartambo from other regions, while Spaniards and colonial creoles staked out a vast cover of manorial estates by 1650. The impress of Spanish colonial power and hacienda rule on the everyday lives of the Quechua in Paucartambo was recorded in their farming. Even today the Quechua peasants there commonly use Spanish loan words like suerte (a unit of the sectoral fallow commons) and loma (the grassland-moor or Hill), rather than customary Quechua expressions such as laymi and puna, respectively, which are characteristic terms elsewhere in Cuzco and the southern Peruvian sierra.

The tragic juncture of the colonial history of Paucartambo notwithstanding, its peasant farmers continued to care for uniquely diverse congeries of crops. Their special concentration of biodiversity owed in certain part to the region's environment. Like other eastern Andean regions, Paucartambo was likely a home of ancient domestication and early agriculture. Over the ensuing millennia the uncultivated flora of wild crop relatives in the region intermittently added unique diversity to the crop gene pool. A rugged and highly varied terrain led the farmers to spread their cropping over sizable environmental gradients. While they did not pocket special intraspecific types of crops in each microenvironment, the Quechua farmers of Paucartambo nonetheless fostered a wide range of food plants in order to make full use of their broad spectrum of field sites and farm territory.

The Quechua people of Paucartambo also reinvented their cultural traditions in keeping the diverse crops. Well-to-do peasant farmers have comprised a surprising cadre of diversity's cultivators. Inverting the tenet that the traditional use of environmental resources is invariably immersed in mass society, the better-off Quechua in Paucartambo have been both socioeconomic modernizers and the keepers of crop diversity during recent decades.[4] The well-off farmers replanted their diverse crops not for the sake of conservation per se but rather to enjoy their agronomic, culinary, cultural, and ritual values. Diverse crops in the Paucartambo Andes thus inserted an environmental dimension in the process of socioeconomic stratification among the peasantry. The same environmental force might be at work in other Andean and Latin American regions where solely nonenvironmental forces have been credited dynamic roles in the socioeconomic differentiation of peasantries (Collins 1988; Deere 1982; Deere and de Janvry 1981; Larson 1988; Lehman 1982a, 1982b; Mallon 1983; Orlove 1977a; Sheridan 1988; Spalding 1984; Stern 1982).[5]

That a cherished environmental custom was not cornered by the peasant masses was rife with other implications as well. The well-to-do Quechua in Paucartambo not only inverted traditions but also reinvented them, using their diverse crops increasingly as traditional luxuries. Reinventing the cultural uses of diverse crops has helped them to strengthen their existing advantages in recruiting labor and solidifying commercial and political alliances with powerful townspeople. Their new tradition, tending toward the sumptuous rather than the strictly staple, was not an imposition of elites other than themselves (compare Hobsbawm and Ranger 1992). Those Paucartambo peasants in the middle of the socioeconomic spectrum, by contrast, wished to benefit from the reinvented symbols and their material rewards but were less able to do so. For many of these farmers, flexible environments and vibrant traditions in favor of the diverse crops could not buffer the resource shortfalls that triggered genetic erosion.

Their loss of the unique landraces through genetic erosion placed many Quechua cultivators in the throes of new and unprecedented environmental degradation. Loss of unique landraces was often irreversible. Even the loss of landraces that may have been shared with other farmers represented sizable setbacks to rural families. Farmers curtailing the diverse crops suffered diminished access to the agroecological, nutritional, culinary, and cultural resources that remained vital in Quechua peasant life. Their loss of world-renowned landraces also brought the case of the Quechua peasants in Paucartambo to the doorstep of scientists, conservationists, environmentalists, and a broad public concerned about the entwined fates of biodiversity and indigenous peasants.

The Future of Sustainable Development and In Situ Conservation

Thus man's wisdom, or his lack of it, alone decides whether even the richest of nature's gifts shall serve as a blessing or a curse. It is but a league that separates the mountains of Gerizim and Ebal.

(Salaman 1985, 602)

The entwined concerns of conserving biodiversity and aiding indigenous and peasant people are a cornerstone of the growing interest in sustainable development. Also known as "conservation-with-development," sustainable development is defined by the influential Bruntland Report of the World Commission on Environment and Development of the United Nations as "development that meets the needs of the present without compromising the ability of future generations to meet their own needs" (WCED 1987). This much-publicized report highlights the protection of biodiversity and the enabling of indigenous and peasant people to pursue sustainable development. These entwined concerns also anchor later landmarks established in the global sustainability movement such as the so-called Earth Summit of the United Nations Committee on Environment and Development (UNCED) in Rio de Janeiro in June 1992 and the Global Biodiversity Strategy formulated by the World Resources Institute (Grubb 1993; WRI 1992; see also Adams 1990; Braatz 1992).

The changing fortunes of diverse crops and Quechua peasants in the Paucartambo Andes can be seen to offer a number of insights for critical thinking about sustainability. First, the findings beg to be squared with general ideas on the environmental relations of indigenous peasants. Sustainability thinking is much influenced by the evidence of harmonious accord between the state of biological nature and the activities of indigenous peasants. Well-researched examples of such environmental harmonies can be found in the studies of tropical forest diversity and its enhancement rather than destruction by contemporary forest people. Similarly, harmonies between nature and indigenous people resonate from a number of cross-historical comparisons of current human-induced catastrophes to some of the less degrading modes of occupation witnessed in the past (Balée 1992; Gómez-Pompa and Klaus 1992; Grossman 1984; Grumbine 1992; Hecht and Cockburn 1990; Nietschmann 1973; Posey 1983, 1992a, 1992b).

Sustainability thinking on the sound and sophisticated use of agricultural biodiversity by indigenous peasants is able to call on a similarly large repertoire of well-documented case studies. Research in the field of geography reaching back for decades has shown the wisely managed role of diverse crops in Indian and peasant life, especially in Latin America and often in Peru (Chang 1977; Clawson 1985; Gade 1969, 1970, 1975; Johannessen 1970; Johannessen et al.

1970; Richards 1985, 1986; C. Sauer 1950, 1952; J. Sauer 1993; Watts 1983; Zimmerer 1991b). These contributions of geographers to the idea of sustainability in diverse crop use is supplemented by a trove of ethnobotanical accounts of wise indigenous farming authored by anthropologists and plant scientists (Alcorn 1984; Balée 1992; Bellon 1991; Boster 1985; Brush 1980; Brush et al. 1981; Geertz 1966; Nabhan 1985, 1989, 1992; Oldfield and Alcorn 1991; Vickers 1983). Proposals for sustainable development involving the diverse crop plants are thus in command of a wealth of case studies they can consult and even champion.

At times, however, the details of the case studies have been lost to a diffuse idea that the resource management of peasant and indigenous people is intrinsically sound. That assumption of a primordial style of harmony between "nature and native" has exerted a strong influence on some wings of the sustainability movement. Mythologized in modern Western thinking, the essentialist assumption is the "myth of the ecologically noble savage" (Parker 1992, 1993; Redford 1990, 1992).[6] A number of convincing counterarguments by commentators like Redford stoutly dispel this myth. The refutations range from contrary evidence to concerns about the troubling political repercussions of making disingenuous claims. These counterarguments merit a brief introduction to aid in clarifying the implications and insights to be garnered from the changing fortunes of diverse crops in Paucartambo.

Disproof of the myth of the ecologically noble savage is amassed almost effortlessly. Redford's research demonstrates how some tribes in tropical rain forests destructively clear their trees and deplete game populations (Redford 1990, 1992). Meanwhile, abundant studies attest to the historical extinction of Pacific island fauna by indigenous people, inspiring Diamond almost one decade ago to rail editorially against "The Environmentalist Myth" (Diamond 1986). Their perspective of a modified and, at times, degraded nature at indigenous hands echoes the decades-old insights of Carl O. Sauer (C. Sauer 1938, 1956, 1958). More recently, Denevan assessing "The Pristine Myth," Butzer commenting on "No Eden in the New World," and Lewis writing on "Green Delusions" also cite ample instances where indigenous peasants, past and present, act none too nobly with respect to their environments and resources (Butzer 1993; Denevan 1992; Lewis 1993).

Redford caps his empirical counterproof by cautioning that unfounded belief in the myth of the ecologically noble savage can appear to deal a trump card to those persons advocating the political rights of indigenous and peasant people (Redford 1990). He points out, however, that the claim of able environmental management where there is none might easily be turned against them. Powerful forces opposed to conservation and to the cause of indigenous peasants could easily gain credibility by attacking such unsupportable claims of the innately sound use of resources. It is important to add that the opposition forces would stand to gain also by craftily manipulating the counterexamples to the myth of the ecologically noble savage of Redford and others. In Latin America both resource extraction corporations (mining and logging companies in particular) and the development arms of governments have found the fodder of well-publicized counterexamples to be convenient for their anticonservation canons.[7] The idea of the ecologically noble savage, a pillar of current thinking both for and against biological conservation and sustainable development, must therefore be addressed in any case study like the present one.

The changing fortunes of diverse crops and peasant livelihood in the Paucartambo Andes offer a critique of the myth that might aid future efforts in sustainable development. The critique warns against assuming exemplary environmental behavior among the Quechua peasants there, rather demonstrating both their environmental successes and failures over time. Their changing fortunes underscore that the varied forces at work in the lives of both plants and people are not adaptationist by definition. Indeed, the well-founded rejection of an adaptationist perspective helps to expose the "myth of the ecologically noble savage" as untenable in the terms of contemporary social science and environmental philosophy (Ellen 1982; Orlove 1980; Zimmerer 1994a, 1995). By contrast, the Quechua peasants in Paucartambo and their diverse crops tell a story of human environmental relations that is more complex than either the myth or the assumption of adaptation.

Complexity is the norm for most "keepers of biodiversity" in Third World countries. Representing the complexity of their changing fortunes and human environmental relations is ethically necessary when we recognize their political situations. Generally speaking, the rights of these people to choose their future course of development ought to be exercised within the context of democratic institutions. Simplifying assumptions in an era when environmental issues are weighted heavily could jeopardize the efforts to establish these rights. Claiming the indigenous peasants to be ecologically noble savages for polemical reasons might indict the subjects of attention for a subsequent failure to conserve. Equally unconscionable would be the manipulation of ecologically noble savage avowals to guide development other than that being democratically expressed. Environmentalists would violate human and civil rights in permitting or perhaps even endorsing impoverishment for the purpose of conservation (Buttel n.d.). Conserving biodiversity, one of sustainable development's greatest goals, could lead to either the sacred pinnacle of Gerizim or the cursed peak of Ebal, as versed in this section's epigram by the potato chronicler Redcliffe Salaman.

Changing fortunes in the diverse crops and livelihoods of the Quechua peasants in Paucartambo also offer a second generation critique of general proposals made earlier to integrate in situ conservation with plans for sustainable development (Altieri and Merrick 1987, 1988; Brush 1986, 1987; Cleveland et al. 1994; Nabhan 1985, 1989; Oldfield and Alcorn 1991; Prescott-Allen and Prescott-Allen 1982; Wilkes 1983, 1991). Since the mid-1980s, first generation proposals have extolled the general merits of conserving the diverse crops and agricultural systems of indigenous peasants by in situ arrangements. Their broad descriptions advocating the need for in situ conservation and sustainable development are echoed in recent landmark pronouncements on biodiversity planning. Agenda 21 of the Earth Summit report of 1992, for instance, issues a first-order priority for the "Conservation and sustainable utilization of plant genetic resources for food and sustainable agriculture" (quoted in Braatz 1992, 25; see also UNCED Agenda 21 in Braatz 1992, 25; Grubb 1993; WRI 1992).

A second generation critique of earlier in situ proposals is intended to revise and refine them. One advance of second generation efforts has highlighted how genetic erosion is often uneven among geographical regions (Brush 1989). Its unevenness raises the possibility that existing in situ conservation can be combined with agricultural development. The present effort of a second generation critique that is derived from the legacy of diverse crops and peasant livelihoods in Paucartambo helps to further this discussion with specific recommendations. The region's crops and its Quechua peasants have differed in a number of ways from widespread assumptions, distinctions that were highlighted in previous chapters. Implications of these distinctions for sustainable development and in situ conservation are far-ranging and possibly crucial. More than in preceding discussions, the emphasis must be on the existing successes of farmers and how to make them more certain and widely beneficial.

Cultivation of diverse crops by the better-off Quechua farmers in Paucartambo indicates a certain core of compatibility between biodiversity in agriculture and some forms of socioeconomic development. Farmers there with adequate resources of land, labor-time, and capital eagerly grow both the bridgeheads of new commerce and their bulwarks of esteemed foodstuffs. Their preferences convincingly refute the assertion that in situ conservation is inherently at odds with all types of development (Frankel 1974; Querol 1993). In situ conservation need not fetter the prospects of socioeconomic development and can, in fact, enhance them. Poorer Paucartambo farmers forced to forego the diverse crops feel robbed of goods that in their minds ought to accompany a decent life. Their predicament signals that a key task for in situ conservation will be to enable farmers to gain adequate resources. Adequacy broadly defined refers to resources sufficient for the once deemed odd couple of income-raising commerce and diversity-rich self-provisioning.

Resilience and reinvention in the farm culture of diverse crops in Paucartambo tells us that diversity need not be doomed by acculturation. Cultural change among the Quechua peasants of the region is selective and partial rather than uniform and absolute. They have retained and occasionally recast a triumvirate of cultural beliefs and practices—their Quechua language, the kawsay concept of a fit livelihood, Andean Christian faith—in their continued cultivation of the diverse crops. Their cultural capacity could be strengthened by programs and policies that support the tilling of diversity while recognizing the inevitability of change. Bilingual education, radio programs, and harvest fairs would help sustain the voices that are vital for in situ conservation. The recent reinvention and social sharing of diverse crops in the cuisine of well-off farmers suggests that the eating customs of regional societies beyond rural dwellers could be enthusiastically marshalled for their foodstuffs.

The broad-based environmental adaptation of most diverse crops conveys other key insights for in situ conservation. Since the large majority of diverse species and landraces are not limited to narrowly defined microenvironments, this diversity's survival does not depend on the protection of each and every farm habitat. The advocates of sustainable development would be poorly advised to assume that the mere protection of a multitude of microenvironments ensured in situ conservation. Their basic premise would be faulty and their program likely to incur prohibitive costs. The predominance of generalist style adaptation in the diverse crops does not mean, however, that the wealth of nature in these organisms is infinitely flexible. Variation of ripening period, for example, is a concomitant of diversity within most major crops such as Andean potatoes and maize. It imposes biological limits on the capacity of farmers to adjust livelihoods while still tilling their landrace-rich staples. Overlooking this complexity in the crops would be as disastrous for diversity as assuming the existence of fine-tuned adaptations.

When farmers innovate versatile techniques in order to take advantage of flexibility in their crops a potent combination is struck for in situ conservation. The Quechua farmers in Paucartambo have succeeded notably by intercropping their diverse minor crops such as quinoa, tarwi, and broad beans within small valley fields not far from their houses.[8] Their efforts appear one step removed from relocating them to the highly diverse gardens alongside their homes, often called house or dooryard gardens. Success of farmers thus far offers a useful example and a hopeful promise; however, it also urges the scrutiny of such innovations in terms of biological and other impacts. On the one hand, while the less complex cultivation of minor crops, evidenced for instance in a simple style of classification, might not be jarred, the more elaborate partitioning of diversity in the major crops could in many cases clash with the less ample spaces of the smaller fields and gardens. Minor crop diversity, on the other hand, could be imperiled if its cultivation were forced to be dovetailed even more closely with the staples.[9]

Not least, the transition to intercropped fields and gardens in Third World farming often implies the allocation of new work with the diverse crops to the women of peasant households. Women farmers in Paucartambo and other Andean regions are shouldering an ever greater share of farm work in general (Collins 1986, 1988; Deere 1982, 1990). While the men in peasant families migrate for short periods and labor in the new commercial crops, women farmers are the ones who are becoming the actual keepers of their biodiversity. The new intercropped Valley fields located close to Quechua homes in Paucartambo epitomize the further feminization of growing the diverse domesticates. Efforts to plan for in situ conservation and sustainable development must be especially aware of the gender specificity of altered labor costs implied by new diversity-saving innovations found among peasant farmers.[10]

The role of places in the peasant farming of Paucartambo is at the heart of still broader messages for in situ conservation and sustainable development. Quechua farmers gain their knowledge of diverse crops through the experience of work and living in particular places. Their economically driven decisions about agricultural change are likewise specific to certain places. The character of place-based distinctions thus shapes the fortunes of biodiversity, both creating and giving context to the influences of socioeconomic difference and ethnicity. The importance of place in biodiversity issues is gaining more weight due to the myriad nongovernmental organizations (NGOs) that currently attempt to foster sustainable development in rural regions like Paucartambo. Since the NGOs locate their operations in some places but not others, their place-based influences on agricultural change add to the salience of this feature in the actual course of conservation and development.

Quechua farming of diverse crops in Paucartambo makes clear that in situ conservation will depend on the enabling of certain development rather than attempting to constrain it altogether. Peasant farmers are eager to pursue the planting of improved high-yielding varieties and other income-generating activities. Their aspiration to improve their livelihoods indicates that efforts at sustainable development and in situ conservation will need to involve social, economic, and political sectors that reach well beyond the peasantries of rural regions. While Paucartambo farmers have selectively and ably combined the improved varieties with their diverse crops, they are evermore subject to the propagandistic promotion of agricultural modernization involving scant biodiversity. To moderate the effects of this governmental and private sector propaganda, information campaigns in favor of in situ conservation could be waged by respected sources such as a small but growing number of sustainable agriculture research organizations.

Involvement of Paucartambo's Quechua farmers in the modern agricultural economy also suggests a role for the international centers of agricultural research and development. The International Potato Center (CIP) based in Lima and funded through the CGIAR (Consultative Group on International Agricultural Research) has for decades sent collectors to Paucartambo for potato germplasm, while its high-yielding improved varieties have been among the first to be adopted by Paucartambo farmers. Given the long-term mission of the International Potato Center in germplasm conservation, they might logically join in programs aiming at in situ conservation (Huamán 1986). The

Center's development of improved crop types might also be designed to offer the diverse landraces as much protection as possible.

The skillful labors of the Quechua farmers in Paucartambo deserve to be rewarded with benefits when lucrative improved cultigens are created from their diverse crops. The peasant cultivators are producing a fair share of the raw material utilized in conventional plant breeding and increasingly in biotechnological engineering at sites far from the Paucartambo Andes. Abundant detail in the preceding chapters indicates that the Paucartambo farmers are also incurring a not inconsequential cost in their farming. That cost could be partly compensated by the companies that profit from crop and seed sales. Whether through proprietary rights or low-cost access to the "improved" end products they desire, the Quechua in Paucartambo deserve just compensation. They ought to share in financial profits when the hard-earned fortunes of their diverse crops enrich others (Brush 1992, 1993; Fowler and Mooney 1990; Kloppenburg 1988; Kloppenburg and Kleinman 1987).[11]

The saga of diversity among the Quechua in Paucartambo alerts us also to many effects of national government. The scissors effect that has pressured peasant farmers and led many to forego their diverse crops is a consequence of Peruvian government policies, some being reinforced at the international level. Social and economic policies that worsen the poverty of peasants without offering viable alternatives have not formed in a context of well-established civil and democratic institutions.[12] Yet the problems of rural poverty and environmental decay, including the loss of diverse crops, are being addressed in various ways by broad-based social movements that are working to strengthen and renew civil institutions in Peruvian society (Rengifo 1988; Repo-Carrasco 1988; Salis 1985). The dilemma of the diversity-growing Quechua in Paucartambo suggests that concerns about their fortunes will be necessary among broader sectors of civil society for the full success of in situ conservation and sustainable development.

Finally, the country of Peru as a whole faces an issue of food security that is manifest in the eating habits of the Quechua people in Paucartambo. The disappearance of crop diversity there is being accelerated by the import of inexpensive foodstuffs as low-cost aid from other countries, principally the United States and Western Europe.[13] Wheat products like noodles in particular are used by many poor farmers to underwrite their curtailment of the diverse crops and their adoption of greater commerce. The adoption of this lowest-cost diet by peasant farmers offers a curious inversion of staple crops when compared to the Western European experience. There, in the nineteenth century, the potato supplanted wheat in the diets of commoners due to the effect of government policies promoting cheap food (Salaman 1985; Thompson 1963).[14] Peru's growing dependence on wheat imports at present must raise the issue of national food security. Tragically ironic, many citizens in the country whose predecessors deeded the world its greatest heritage of diverse farm plants are surviving on floods of the cheapest imported foodstuffs.

Appendix A
Common Explanations of Genetic Erosion

Key to the Explanation	*Author(s), Date*
"Meanwhile, extension of commercial agriculture is causing a rapid extinction of the primitive domestic forms."	Sauer 1938, 769
"The Mexican wheats have washed over Asia with astonishing speed, replacing major centers of diversity almost overnight."	Harlan 1972, 214
"They [native cultivars] are also in the most immediate danger of extinction through replacement by modern cultivars."	Frankel 1974, 56
"Warnings of a greatly accelerated rate of displacement of primitive crop varieties by locally selected or introduced cultivars."	Frankel and Hawkes 1975, 1
"Modern varieties replace ancient populations that have provided genetic variability for plant breeding programs."	Harlan 1975a, 618
"The main forces behind the genetic erosion of the important food and cash crops in their areas of cultivation . . . include the displacement of landraces by modern varieties."	Plucknett et al. 1987, 9

Appendix B
Crop Biogeography and Vegetation in Paucartambo

Appendix B.1 The Biogeography and Evolution of Diverse Crops

Crop	*Main Elevations (in meters)*[1]	*Evolution in Eastern Andes*	*Authority*
Amaranth	2,800–3,500	*	J. Sauer 1967
Arracacha	2,800–3,000	*	Hodge 1954
Achira	2,800–3,000	*	León 1964; Gade 1975
Chile pepper	2,800–3,600	*	Pickersgill 1971
Quinoa	2,800–3,700	*	Wilson 1988
Winter squash	2,800–3,100	*	Whitaker and Cutler 1965
Crookneck squash	2,800–3,100		
Achocha	2,800–3,300		
Tarwi or bush lupine	2,800–3,900		
Oca	3,700–4,050		
Passionfruits	2,750–3,000		
Avocado	2,750–2,900		
Common or Kidney bean	2,750–2,950	*	Kaplan 1980
Yacon	2,750–3,100		
Floury Potatoes	3,700–4,050	*	Hawkes 1978, 1990
Precocious potato	2,900–3,300	*	Hawkes 1978, 1990
Bitter potatoes	3,900–4,100		Hawkes 1978, 1990
Mashua	3,700–4,050		
Ulluco	3,500–3,900	*	Cárdenas 1969
Maize	2,700–3,550	*	Bird 1970; Grobman et. al. 1961

1. Main elevations refers to the typical range within the Paucartambo Andes. *Indicates that the crop species either originated in the eastern Andean valleys or underwent significant evolutionary differentiation as a crop there.

Appendix B.2 Plants Characteristic of the Major Environments (Common names in parentheses)

Environment and Elevation Range	*Trees*	*Shrubs*	*Grasses, Herbs, Vines*
Thorn Savanna (8,850–11,150 feet 2,700–3,400 meters)	Trichocereus sp. (*hawanqollay*) Acacia sp. (*añanway*) Schinus molle (*molle*) Escallonia resinosa (*chachakoma*) Kageneckia lanceolata (*llok'e*)	Bernadesia horrida (*llawlli*) Fuchsia boliviana (*chimpu-chimpu*) Colletia sp. (*roqe*) Opuntia sp. (*tuna, chanki*) Puya sp. (*achupalla*)	Passiflora sp. (*qampaqway, tumbo*)
Shrub Savanna (10,850–12,880 feet 3,300–3,900 meters)	Polylepis racemosa (*q'euña*) Buddleia incana (*kishwar*)	Bernadesia horrida (*llawlli*) Baccharis sp. (*chilka*; *tayanka*) Senna birostris (*mutuy*)	Vulpa myuros (*soqllo pasto*) Stipa ichu (*ichu*)
Grassland Moor (12,500–13,600 feet 3,800–4,150 meters)	Polylepis racemosa (*q'euña*)		Stipa ichu (*ichu*) Festuca dolichophylla (*orqo ichu*) Calamagrostis sp. (*pampa ichu*) Bromus catharticus (*qachu*)
Cloud Forest (9,190–11,800 feet 2,800–3,600 meters)	Alnus jorullensis (*labran, aliso*) Hesperomeles lanuginosa (*linli*) Podocarpus sp. (*romerío*) Weinmannia auriculata	Vallea stipularis (*chinchilmay*) Gaultheria reticulata (*monte thumana*) Brachyotum grisebachii (*tiri* or *tili*) Miconia sp. (*tiri* or *tili*)	Cortaderia nitida (*ñiwa*) Chusquea scandens (*kurkur, bambu*) Bomarea sp. (*champa-champa*)

Appendix C The Human Geography of Agriculture in the Paucartambo Andes

Appendix C.1 Work Relations on Hacienda Estates

Category	*Obligations*	*Rights*
Tenants (*mañayruna* or *sayaqruna*)	Labor rent (one-half year or more)	Cultivation of fields, pasture rights (five to fifteen *topos*/household; *topo*=land for planting three hundred lbs. of potatoes)
Tenant sharecroppers (*yanapaq* or *puchuruna*)	Labor rent (one-half requirement of tenants)	Cultivation of fields (one-half area allotted tenants)
Servants (*pongos* or *pongo-chacras*)	Labor rent (five to seven days per week)	Shelter, food, small field (in some cases)
Renters (*arrendires*)	Monetary payment	Cultivation of fields
Sharecroppers (*partidarios*)	Seed, labor	One-half harvest

Appendix C.2 Birthplace of the Members of Peasant Communities in Paucartambo 1969–75[1]

Community	*Total Population*	*Same Community*	*Paucartambo (Different Community)*	*Cuzco*	*Puno*
Majopata (middle valley)	81	45	17	14	5
Pasto Grande (lower valley)	63	5	24	15	16*
Mollomarca (upper valley)	70	43	10	14	3
Ccotatoclla (interior)	49	45	3	0	1

* Three persons were born in still other departments

Appendix C.3 Population of Paucartambo, 1961–81

Area	*1961*	*1972*	*1981*	*Population Density (Hectares/person, 1972)*
Paucartambo Province	26,455	29,983	32,149	
Paucartambo District	8,053		9,390	3.80
Colquepata District	5,327		6,873	2.94
Challabamba District[2]	6,829		5,592	10.0
Paucartambo Village	1,928		1,620	
Colquepata Village	484		425	
Challabamba Village	201		325	

Sources: DNEC 1966, INE 1983, ONEC 1975

1. *Paucartambo* refers to Paucartambo Province, *Cuzco* to the provinces of Cuzco Department other than Paucartambo, and *Puno* to Puno Department.

2. Redistricting caused the decline of population in Challabamba between 1961 and 1981. Without this loss the growth of population in the district exceeded the other areas.

Appendix C.4 Livestock Holdings per Peasant Household, 1970–87

	Sheep		*Cattle*		*Pigs*		*Horses*		*Camelids*	
Community	*1971*	*1987*	*1971*	*1987*	*1971*	*1987*	*1971*	*1987*	*1971*	*1987*
Mollomarca (upper valley)	15	27	2	4	1	4	2	4	0	0
Majopata (middle valley)	9	12	3	6	3	5	1	2	1	4
Colquepata (interior)	20	11	2	1	3	3	2	2	1	1
Ninamarca* (interior)	12		3		2		1		0	
Southern Peruvian sierra*	7		2		1		1		0	

* (1976–79; from Figueroa 1984, 19)

Appendix C.5 The Fraction of Families *Not* Producing the Diverse Crops, 1987

Crop	*Not Producing (n=30)*	*Lacking Suitable Field Sites*
Boiling Potatoes	14/30	11/14
Maize	10/30	6/10
Ulluco	24/30	3/24
Quinoa	23/30	3/23

Appendix C.6 Common Farm Tools in Paucartambo

Task	*Tools*
Plowing and Tillage	foot-driven hoe, or foot-plow (*chakitaklla*) oxen-drawn traction, or scratch plow (*arado*) pick (*pico*) crusher (*qorana*)
Mounding	wide-blade hand hoe (*lampa*)
Harvest	narrow-blade hand hoes (*qasuna, rawkana, kuti, allachua, q'achapa*) sickle (*ichuña*)
Postharvest	maize husker (*tipina*)

Appendix C.7 The Number of Families Receiving Unequal-Size Land Allotments[1]

Community	*0–3 hectares*	*3–6 hectares*	*6–12 hectares*
Mollomarca	36	28	6
Pasto Grande	33	23	4
Majopata	38	40	4
Ccotatoclla	35	10	2

1. Between one third and one fifth of the land areas were arable. The data are compiled from agrarian reform records of the early 1970s.

Appendix D Production Techniques and Farm Spaces in Paucartambo

Appendix D.1 Techniques for the Storage of Tuber and Seed Crops

Crop	*Semiarid Climate (Southern Paucartambo and Quencomayo Valley)*	*Humid Climate (Northern Paucartambo Valley)*
Potatoes and Ulluco	straw cylinders (*taqe*) straw-bed piles	elevated bin (*troje*) straw-bed piles
Maize	straw cylinders (*taqe*)	bamboo platforms (*marka*)

Appendix D.2 The Apportionment of Yield from the Diverse Crops[1]

Crop	*Number of Fields Studied*	*Subs. only*	*Subs.- Seed*	*Comm. only*	*Comm.- Seed*	*Subs.- Comm.*	*Subs.- Comm.-Sd.*
Floury Potatoes	30	2 (7%)	23 (78%)		1 (3%)	1 (3%)	3 (9%)
One Cultivar	5			1 (20%)	1 (20%)	1 (20%)	2 (40%)
Multiple Cultivars	25	2 (8%)	23 (92%)				
Maize	61	6 (10%)	10 (16%)				45 (74%)
One Cultivar	12						12 (100%)
Multiple Cultivars	49	6 (12%)	10 (20%)				33 (68%)
Ulluco	33	2 (6%)	15 (45%)	2 (6%)	2 (6%)	4 (12%)	8 (24%)
One Cultivar	14			2 (14%)	2 (14%)	4 (28%)	6 (43%)
Multiple Cultivars	19	2 (11%)	15 (79%)				2 (11%)
Quinoa	22	0	12 (55%)			1 (5%)	9 (41%)
Multiple Cultivars	22	0	12 (55%)		0	1 (5%)	9 (41%)
Total	146	14 (10%)	46 (31%)	2 (1%)	12 (8%)	8 (5%)	52 (36%)

1. Abbreviations stand for the uses of yield from the diverse crops. "Subs." abbreviates subsistence, "Comm." is an abbreviation for commerce, and "Sd." stands for seed.

Appendix D.3 Range in the Elevation of Production Spaces in Paucartambo

Production Space	*Semiarid Climate Southern Paucartambo Valley and the Quencomayo Valley*[1]	*Subhumid Climate Middle Paucartambo Valley*[2]	*Humid Climate Northern Paucartambo Valley*[3]
Valley (*kheshwar*)	9,850–11,650 feet (3,000–3,550 meters)	9,190–11,150 feet (2,800–3,400 meters)	8,850–10,150 feet (2,700–3,100 meters)
Oxen Area (*yunlla*)	9,850–12,800 feet (3,000–3,900 meters)	9,190–12,500 feet (2,800–3,800 meters)	8,850–11,150 feet (2,700–3,400 meters)
Early Planting (*maway*)	9,850–12,100 feet (3,100–3,700 meters)	10,850–12,100 feet (3,300–3,700 meters)	8,850–10,850 feet (2,700–3,300 meters)
Hill (*loma*)	12,470–13,450 feet (3,800–4,100 meters)	12,140–13,450 feet (3,700–4,100 meters)	not used

1. Based on observations in the mid-1980s in the Peasant Communities of Mollomarca, Huaynapata, Payajana, Colquepata, Roquechiri, and Chocopía.
2. Based on observations in the mid-1980s in the Peasant Communities of Inquilpata, Majopata, and Jajahuana.
3. Based on observations in the mid-1980s in the Peasant Communities of Huaqanqa, Acobamba, Totora, and Pilco.

Appendix D.4 The Three Principal Maize Plantings

	Early Planting Big Seed (hatun muhu)	*Intermediate Planting Medium Seed* (chawpi muhu)	*Late Planting Small Seed* (uch'uy muhu)
Planting Month (Feast Day Marker)	August Saint Rose of Lima Saint Mary the Virgin (*Virgen de la Asención*)	October Mama Chanca Virgen del Rosario	November All Saints' Day Saint Andrew
Harvest Date (Associated Feast Day)	June–July Saint John the Baptist	May–June Saint Elena (*Cruz Velacuy*)	May–June Corpus Christi
Main Elevation Range	8,850–9,850 feet 2,700–3,000 meters	9,190–11,150 feet 2,800–3,400 meters	9,850–11,650 feet 3,000–3,550 meters
Ripening Time	9–12 months	6–8 months	5–7 months
Major Races	Huancavelicano Morocho Ancashino Paro	Chullpi Cusco Gigante Kulli Pisccoruntu San Geronimo	Confite Puntiagudo Cusco Cristalino Uchuquilla
Number of Landraces	6	10	11

Appendix E
Cultural Attributes of the Diverse Crops in Paucartambo

Appendix E.1 Principal Maize Uses and the Preferred Landraces

Preparation	*Local Name*	*Preferred Landrace(s)*
Boiling	*mot'e* or *mut'i*	*k'ellu, qosñiy, puka, fallcha*
Parching	*hank'a*	*chullpi, ch'uspi, qoqotoway*
Beer-making	*aqha, chicha*	*kulli, k'ellu*
Hominy	*fata*	*k'ellu, yuraq*
Finely ground mush	*api*	*k'ellu*
Fine gruel	*saqha*	*k'ellu*
Coarse gruel	*chaqe*	*perlas*
Corn on the cob	*choklo*	*yuraq, k'ellu*
Silage	*challa*	*k'ellu*
Pudding	*mazamora*	*k'ellu*
Ritual offerings	*chayasqa*	*misa, k'uti, taqe*
Soup thickener	*lawa*	*k'ellu*
Steamed	*huminta*	*k'ellu*
Fried	*tortilla*	*k'ellu*
Flour or Snack Flour	*sara hak'u* or *kakow*	*ch'uspi, qoqotoway*

Appendix E.2 The Advantages of Ulluco Cultivation

Moderate labor, little plowing

Moderate soil fertility

Disease resistant (especially important in perennially humid climate)

Semitolerant of drought

Frost tolerant

Dietary esteem

Appendix E.3 Chief Drawbacks Perceived in Quinoa (in order of relative importance)

High demand for soil nitrogen

Subpar response to chemical fertilizers

Lodges if inadequately fertilized

Labor intensive cultivation (transplanted by hand)

Labor intensive postharvest tasks (threshed with feet)

Stubble injurious to livestock (if not sickled close to ground)

Appendix E.4 Production Rationales for Adopting Barley

Low preharvest labor requirements

Low demand for soil nutrients

Moderate to high response to chemical fertilizers

Existing threshing floors (*eras*) and horses

Little tillage required

Appendix E.5* Cropping Calendars of Production Systems[1]

	A	*S*	*O*	*N*	*D*	*J*	*F*	*M*	*A*	*M*	*J*	*J*
Valley Big Seed Maize	X	X	X	X	X	X	X	X	X	X	X	–
	Saint Mary the Virgin (August 15) Saint Rose of Lima (August 30)									Saint Isidoro (May 25) Saint John the Baptist (June 24)		
Valley Medium Seed Maize	–	–	X	X	X	X	X	X	X	X	–	–
			Mama Chanca (September 16) Virgen del Rosario (October 8)							Corpus Christi (May) Saint Elena (May 2)		
Valley Small Seed Maize	–	–	–	X	X	X	X	X	X	X	–	–
				All Saints' (November 1) Saint Andrew (November 30)						Corpus Christi (May) Saint Elena Cruz Velacuy (May 2)		
Oxen Area	–	–	–	X	X	X	X	X	X	X	X	–
				All Saints' (November 1)								
Early Planting	X	X	X	X	X	X	–	–	–	–	–	X
												Pentecost (July) Virgen del Carmen (July 16)
	Mosoq Wata (August 1)											
Middle Early Planting	–	X	X	X	X	X	X	–	–	–	–	–
	Saint Mary the Virgin (August 15)											
Hill	–	–	X	X	X	X	X	X	X	X	X	–
			Saint Jerome (September 30) Virgen del Rosario (October 8)									Señor de Qollorit'i (July) Saint James (July 25)

* Initials in the top column of this table stand for months of the year, beginning with August.

1. For the most part, Quechua farmers in Paucartambo think of their cropping calendars in terms of Saints' Days rather than calendar dates.

Appendix F
Farm Management of the Diverse Crops

Appendix F.1 The Use of Human Anatomy in Crop Terminology

Potatoes and Ulluco
Eyes (*ñawi*): eyes
Babies (*wawa*): protrusions from eyes
Face (*uya*): general aspect
Skin (*qarallan*): skin or peel
Eyebrow (*qhechipra*): ridge above eye
Rib cage (*waqtan*): sides of tubers
Head (*uma*): apical section
Heart (*sonqo*): tuber flesh or pulp
Veins (*vena*): streaks in flesh
Bone (*tullu*): shoot of plant
Maize
Nose (*senqa*): tip of kernel
Head (*uma*): tip of ear

Appendix F.2 Preferred Traits in the Potato and Ulluco Crops

Eyes	
Favorable	*Unfavorable*
Beautiful eyes (*t'ika ñawi*)	Protruding eyes (*pita ñawi, nunupaasaq ñawi*)
Many eyes (*ñawi cape* or *ñawi ñawi*)	Deep eye (*t'oqo ñawi*)
Apical eye (*uman ñawi*)	Liplike eyes (*wak'a ñawi*)
Well-formed eye (*allin ruhusqa ñawiyoq*)	Elongate eyes (*wasqa ñawi*)
Round eyes (*muyu ñawi*)	Eyes about to sprout (*ch'iqchi ñawi*)
Diseaseless eyes (*ch'uya ñawi, lluska ñawi*)	
Good eyes (general) (*chawfra ñawi*)	
Plant	
Much-branching (*rafra-rafrachamanta*)	
Good form (*yuranmi allinta*)	

Appendix F.3 Culinary Traits Preferred in the Diverse Boiling Potatoes

Favorable	*Unfavorable*
Mealy (*hak'u*)	Watery (*uñutaq*)
Very mealy (*kaqho*)	Bitter (*qoymi*)
Dry and mealy (*chaki chaki hak'u*)	
Sweet (*miskhi saborchayoq*)	
Cooks quickly (*ratulla wayk'usqa*)	
Skin bursts when cooked (*fatan wayk'umpi*)	

Appendix F.4 Maize Types in Religious Ritual

Local Name	*Morphology*	*Preferred Landrace(s)*	*Ceremonial or Ritual Use*
Taqe (Store)	deformed ear with three spikelets	White (*yuraq*) or Yellow (*k'ellu*)	protect crop stores; fertility ritual
Misa (Mass)	bicolor	from Yellow (*k'ellu*) and Red (*puka*); also *yawar chasqo* and *waqankillay*	fertility rituals at planting and harvest, divination, casting spells
Kuti (Upside-down)	kernels "locked"	none	fertility rituals; ritual healing
Haha sara	seed coat ruptured at distal end	Red (*puka*)	ritual healing (especially for lightning victims)

Notes

Notes to Chapter 1

1. Faustino Condori Castillo, Umamarca, May 18, 1987. I use pseudonyms for all the living Quechua peasants cited in order to protect their anonymity, although the conversations did occur as reported. Even the most seemingly innocuous comment discussed conceivably might jeopardize the individual at some later day. The peasants' lack of secure civil and human rights makes this precaution a necessity.

2. The main period of my fieldwork in Paucartambo occurred between March 1986 and August 1987. Reports of this fieldwork focused on particular themes, among them the ecology of agricultural biodiversity (Zimmerer 1991a; Zimmerer and Douches 1991), ethnobotany and resource management (Zimmerer 1989, 1991b), and the changing fates of crop diversity (Zimmerer 1991c, 1992a, 1992b, 1992c, 1993a, n.d.). My major research interest as an environmental geographer has always been to integrate findings on environmental resources, social change, and economic development. Between 1987 and 1994 I broadened my historical research utilizing archival documents, the published records of residents and travelers, and the verbal accounts of Paucartambo inhabitants. I also added a detailed analysis of genetic variation in the potato and maize crops to the results of field sampling and experiments. At the outset my study may have resembled Artillicus's specializing hedgehog, but the project definitely metamorphosed into his omnivorous fox.

3. The tendency to attribute Andean crops to Inca agricultural prowess is repeated elsewhere in Nova's *Seeds of Tomorrow* and in publications such as the National Research Council's *Lost Crops of the Inca.* The latter, a book designed to inform a wide public, asserts without warrant that "it was the Incas who, by the time of the Spanish Conquest, had brought these plants to their highest state of development" (NRC 1989). Such statements hazard the assumption that crop evolution reached a zenith during the Inca period. They also advance the untenable claim that the Inca government, rather

than its subjects, were the ones responsible for agricultural biodiversity (Gade 1993; Zimmerer 1993a).

4. Notable variation in the estimates hinges partly on delimitation of "domesticated plants," which, strictly speaking, are dependent entirely on farmers for survival. "Cultivated plants," following many definitions, are planted and harvested but may survive and reproduce in the absence of farmers' interventions (Harlan 1975b, 63–64). The figure of forty agricultural species in the highlands refers mostly to domesticates whereas the estimate of two hundred encompasses many cultivated but undomesticated plants. Since dispersals were likely common among Andean regions during the process of plant evolution, I desist from using the descriptor "native," which generally indicates "a pedigree that goes back forever in association with a particular geographical region" (Crosby 1991, 84).

5. Names and spellings of crops in the Andes vary conspicuously. I adopt the names and spellings standardized in English in the widely distributed *Lost Crops of the Incas* (NRC 1989). The names are left without italics except in the case of local versions that differ from the standard ones.

6. This region was also studied in *Changes in Andean Agriculture,* a project of principal investigators Stephen B. Brush and Enrique Mayer. Steve Brush, César Fonseca Martel, and I chose the Paucartambo Andes as a study region during a one-week visit there in 1985. César Fonseca subsequently supervised the project with its five full-time field assistants from September 1985 until March 1986. I arrived in March 1986, and after the tragedy of César's accidental death in late March, I supervised the project until its completion in June. Steve Brush visited the field project in mid-April of 1986 and joined Enrique Mayer in visiting and terminating the project in June of that year. At that time I began my main period of field research, which lasted through July 1987.

7. I use estimates of elevation in feet and meters to make the role of topography and elevation-related environment readily intelligible to readers familiar with either measurement system. Since the figures are estimates, the equivalencies in each measurement system are rounded off for easier reading.

8. To this day Paucartambo villagers recount Crazy Yabar's memorable claim of having hybridized a cabbage and a pear tree.

9. It is unlikely that a single unified theory would adequately explain the complex nature of environmental modification. Regional political ecology recommends the creation of theoretical substructures and particular concepts that correspond to and elucidate the character of specific "under-determined" conditions (Blaikie 1985, 35). The less grand role does not diminish theory but rather situates it more usefully for human environmental study.

10. Ives and Messerli (1989) expound more fully on the overlooked importance of establishing definitions for key terms in accounts of environmental change.

11. The actual loss of genes is undocumented in all but a few cases due in part to the difficulty of sufficient sampling for cross-historical and cross-geographical comparison.

12. On the role of multifaced motives for resource management, see Denevan (1980, 1983), Ellen (1982), Gade (1981), Ingold (1992), A. W. Johnson (1972), and Orlove (1980).

13. Among Andean people this ethical norm was described by Cieza de León ([1553] 1959) as a "comfortable subsistence." Salomon defined it as "the absolute minimum of comfort which divided acceptable subsistence from deprivation" (Salomon 1986a, 89).

The social, historical, and normative dimensions embedded in the idea of subsistence were fused by Karl Marx: "The number and extent of his so-called necessary wants, as also the modes of satisfying them, are themselves the product of historical development and depend therefore . . . on the habits and degree of comfort in which the class of free laborers has been formed." Although writing on capitalism rather than colonialism, the statements by Marx reflect the same "historical and moral element" that had entered into the definition of subsistence rights under the Inca and the Spaniards (Marx [1867] 1987).

14. On the idea of the cultural or moral aesthetic, see Tuan (1993). Generally similar ideas were applied to farming people by Berry (1977), Sauer (1956), and Scott (1976).

15. A growing literature on agriculture and the micropolitics and politics of peasant resistance includes volumes on Southeast Asia (Peluso 1992; Scott 1985), applications to the Andes (Stern 1987; Thurner 1993), and suggestive re-readings of certain ethnobotany literature (Nabhan 1985, 1989).

16. An impressive set of studies on the biological evolution of the crop quartet examined here—potatoes, maize, ulluco, and quinoa—with respect to environments in the central Andes (along with Mexico and Central America in the case of maize), and, in some cases, to Andean farming techniques as well, includes the following: Bird (1970), Brandolini (1970), Grobman et al. (1961), Hawkes (1978, 1990), Hawkes and Hjerting (1989), Iltis (1983), Johns and Keen (1986), Pickersgill and Heiser (1978), and H. Wilson (1978, 1988).

17. Brush's initial study contains the kernel of this critique. He concluded that the elevation-stacked tiers of land use in Uchucmarca community result from the combination of "environmental features (topography, exposure, altitude, precipitation) and the characteristics of the particular crop or crops which are cultivated" (Brush 1976, 150, 1977, 82). The study data shows, however, that crops occur in either two or three zones and in as many as four (Brush 1976, 158). Thus, even the early findings of environment-driven adaptation hinted that the spatial concept of nonoverlapping zones in local land use was a considerable abstraction.

18. Like other livelihood norms expressed by peasants in the Andes and elsewhere, the *kawsay* concept should be approached as an ideal that is subject to historical change and social contestation (Larson 1988; Mallon 1983; Marx [1867] 1987; Scott 1976; Thurner 1993; Weismantel 1988; see also footnote 13 of this text). My use of the concept of "fit livelihood" is informed by the work of Toribio Mejía Xesspe, a Quechua peasant who became the chief assistant and ethnographic source for the renowned Peruvian archaeologist Julio Tello (Mejía Xesspe 1978). I am grateful to Frank Salomon who pointed me to the work of Mejía Xesspe and supplied the insight on his role as Tello's assistant. Although the expression *kawsay* used in Paucartambo might be taken to mean the fitness of subsistence, I believe it refers more broadly to livelihood, since peasant economies in the Andes have typically involved exchange (including barter) and commerce (since at least the 1500s), rather than narrowly defined subsistence (compare Brush 1977, 8; Larson 1988; Murra 1985a, 60). In adopting the term *fit livelihood* I am following Wolf's appraisal of a peasant's "hold on his source of livelihood" as a fundamental ideal in peasant lifeways and politics at present and historically (Wolf 1969, xiv).

19. The unevenness of recent change in peasant economies, especially those of the

central Andes, is described in Collins (1988), Deere (1982, 1990), Deere and de Janvry (1981), Godoy (1990), Guillet (1992), Larson (1988), Lehman (1982a, 1982b), Mitchell (1991), Orlove and Custred (1980), Reinhardt (1988), Sheridan (1988), and Zimmerer (1991c, 1992a).

20. Recent historical-geographical studies on the resource use of peasant and indigenous peasants in developing countries recounts the lengthy, lurching, and nonlinear course of the major anthropogenic modification of environments (Blaikie and Brookfield 1987; Butzer 1993; Ives and Messerli 1989; Lewis 1992, 1993; Stevens 1993; Turner et al. 1990; see also Geertz 1966; C. Sauer 1938, 1956, 1958).

21. On the historical reinvention of African subsistence customs and environment use, see Richards (1985, 1986).

22. The distinct and often understated use of ecological concepts in the Sauerian school or the geographical approach known as "cultural-historical ecology" is worthy of a lengthy discussion that appears elsewhere (Zimmerer 1995; see also Zimmerer and Langstroth 1994, for a brief review of ecological research on Latin America by the Sauerian school).

23. Modernization Theory, it should be noted, took the growth of marketing as proof of development's benefits reaching the rural poor. The Quechua in Paucartambo, however, like their counterparts in other Andean regions, have long suffered the consequences of policies and sociopolitical power biased against the peasant economy under economic programs of free-trade liberalism as well as state regulation (de Janvry et al. 1989; Thorp and Bertram 1978). A wealth of case studies assess the process and problems, both recent and historical, of regional underdevelopment in the Peruvian Andes. In addition to the citations contained in the section in this volume entitled "Social Theory and Biodiversity," see, for example, Alvarez (1983), Baca Tupayachi (1985), Brisseau (1978), Figueroa (1981, 1984), Flores Galindo (1977, 1988), Glave and Remy (1983), González de Olarte (1987), Guillén Marroquín (1989), Hopkins (1981), and Mörner (1978).

24. It would be erroneous, however, to conclude that farming of diverse crops hangs simply on the chronic underdevelopment of a region. If that were true, one would similarly expect the least developed sector of the farm population to till the most diverse crops. As described earlier, the wealthiest peasant farmers in the Paucartambo Andes are disproportionately represented in the ranks of diversity's growers.

25. With respect to regional political ecology, the emphasis on historical concepts is urged in order to prevent the sorts of gross mistakes that have been broadcast in at least a few exaggerated claims of impending biospheric devastation (for this critique, see Blaikie and Brookfield 1987; Ives and Messerli 1989; and Turner et al. 1990).

Notes to Chapter 2

1. On the basis of topography, the ranges of the Western Cordillera, the Central Cordillera, and the Eastern Cordillera make up the central Andes mountains. Geographers Pulgar Vidal of Peru and Ismael Montes de Oca of Bolivia have written valuable overviews of the physical and human geography in their countries (Montes de Oca 1989; Pulgar Vidal 1946).

2. Distributions of the wild descendents of presumed crop progenitors have been counted as a major proof of probable domestication since the pioneering analysis of Alphonse de Candolle (1908) and through the more recent summaries by C. Sauer (1952), Harlan (1975b), and J. G. Hawkes (1983). Crop evolutionist Harlan observed that much crop domestication in South America occurred "over a broad range northward along the eastern slopes of the Andean range" (Harlan 1975b, 232).

3. The extremely dry western ranges and Pacific coastal lowlands preserve plant remains much better than the humid Eastern Slopes.

4. Validity of the "natural life-zones" identified in these classifications is dubious with respect to vegetation (Tosi 1960). Classification of climate types on the basis of rainfall, temperature, and evapotranspiration is, however, sufficiently sound.

5. The grassland moor at present drops below 11,500 feet (3,500 meters) near Challabamba, for example, while in the interior of the Quencomayo watershed near Colquepata the same limit descends only to 13,125 feet (4,000 meters).

6. It is important to note that although farmers could re-create diversity through subsequent selection, they could not guarantee quantumlike leaps of evolution.

7. Upslope protraction during the course of crop evolution was localized within the highlands. Highland crops did not originate from lowland progenitors nor did Andean agriculture appear as "derivative" following diffusion from outside the region (compare to C. Sauer 1952, 44).

8. A handful of extant Aymara toponyms also suggest altiplanic influences although the interpretation of linguistic and settlement evidence still remains a debated issue (for example, Bird et al. 1983–84).

9. Garcilaso's exceptional familiarity with the region probably derived from his family's ownership of a coca field in Avisca located in the Paucartambo montaña, or foothills (Garcilaso de la Vega [1609] 1987, 222).

10. Recent ethnohistorical research on the Eastern Andes underlines the long-lived frontier quality of Paucartambo (Renard-Casevitz, Saignes, and Taylor 1988). The authors of the book write, "Paucartambo, Challabamba, the site of Tres Cruces and its fantastic panorama about the lowlands, the famous ridge of Cañac-Huay. No other lowland area is so close to Cuzco or coincides better with the eastern frontier of the Imperial City. Tres Cruces, Cañac-Huay and the coca-growing region of Avisca along the Tono river marked the true border of *Antisuyo*" (Renard-Casevitz, Saignes, and Taylor 1988, 117).

11. The success of the Inca in naming the landscape was apparent in colonial documents and in later republican sources. Among the people within the region, however, at least some renaming efforts failed. To this day, the Quechua peasants in Paucartambo refer to the river as the Mapacho, retaining a local tradition of toponymy that presumably predated the Inca imposition of Quechua names.

12. The significance of this insight nonetheless escaped various authors on biodiversity and was overlooked altogether in the recent *Lost Crops of the Inca* (NRC 1989).

13. Paucartambo landraces belonging to these maize races are shown in table 9. The principal landrace of Kulli in the region is *kulli* (Dark Purple) and the main Chullpi landrace is *chullpi.*

14. Like terracing and irrigation works, most if not all maize cultivars in state and cult agriculture were technological innovations that preceded Inca rule. Even Cusco

Gigante is thought by the Grobman team to have originated prior to the consolidation of Inca power (Grobman 1961, 298). The Cusco Gigante type of maize is labeled in English as the Giant Corn of the Incas and is incorporated as the genetic basis for the commercial product of the California-based Corn Nuts company.

15. I adopt the spelling *chuño* because of its fairly wide usage, rather than use *ch'uñu*. Otherwise orthography mostly follows the style of *Diccionario Quechua Cuzco-Collao* by Antonio Cusihuaman (1976). Place names are rendered in standard rather than orthographically correct form: Cuzco, for instance, is not spelled Qosqo.

16. Garcilaso de la Vega ([1609] 1987, 250) wrote: "All the potatoes produced on the lands of the Sun [the Inca cult] and of the Inca were treated in this way [freeze-dried into chuño] and kept in the storehouses," and Guamán Poma de Ayala listed only freeze-dried forms among tubers maintained in the storage depots of the Inca state (Guamán Poma de Ayala [1613] 1980, vol. 1:338). This point is made also by Murra (1980, 13). He wrote that of seventy-seven references to stored foods in the chronicles that he reviewed (twenty-eight chroniclers), no less than thirty-six specified maize, seven indicated chuño, and most of the remainder were general in character. There is little doubt, however, that several types of storehouses were managed by the Inca rulers and that some, such as the ones mentioned in this section, were likely to have contained more diverse crops than others (Garcilaso de la Vega [1609] 1987, 255).

17. Aggregations of local kinship groups, the state, and the cult received land in each region under Inca rule (de Acosta [1590] 1940; Polo de Ondegardo [1561] 1940; Cobo [1653] 1979). The state's share of territory was usually the largest, while the amount allotted to local kinship groups varied according to the number and size of households. Each family received one *topo* of land (a measure of the quantity of seed that could be planted rather than field area per se) with the grant of an additional topo for each male child and one-half topo for female offspring.

18. Although the Inca rulers Pachacuti and Topa Inca had gained control over the territory of the Paucartambo foothills through subduing its inhabitants (the Antis to the Inca), military hold of the montaña and subjugation of the Antis were difficult to retain. Susceptibility of Incan highlanders to the ravages of endemic disease, particularly mucocutaneous leishmaniasis, may have limited the empire's rule over lowland peoples and terrain, although the settlement pattern and political economy of the Antis were likely to have posed impediments of equal or greater importance (Gade 1979; Le Moine and Raymond 1987).

19. Supply of Indian peasants in the colonial labor draft depended on the power of local authorities, the kurakas, who recruited the mita workers for the crown economy. The principal authorities of indigenous communities in the central Andes were referred to as kurakas, although the Spaniards often used the term *caciques* in legal documents and other writing. Adoption of the mita corvée demonstrated how the colonial Spaniards built many economic institutions by grafting Inca arrangements (Rowe 1957).

20. Los Andes de Paucartambo was one of numerous provinces established in Cuzco. The province initially consisted of six districts (Paucartambo, Pitomarca, Chinchos, Amparaes, Caycay, and Obay), although by 1630 acts of subdivision resulted in a total of thirteen districts, which included the core districts of present-day Paucartambo, such as Paucartambo, Challabamba, and Colquepata, as well as ones that have since been ceded to neighboring provinces (for example, Amparaes and Ccatca).

21. The exact location of the earliest seat of government in Paucartambo remains

unclear. Notwithstanding evidence for the founding of a settlement named Paucartambo by the late sixteenth century (Maurtúa 1906), a document from the late seventeenth century indicates that San Antonio de Llaullipata served as capital of the province (Villanueva Urteaga [1693] 1982). San Antonio de Llaullipata was sited in present-day Callipata on a river terrace perched roughly one hundred meters above the confluence of the Paucartambo and Quencomayo rivers. At an unknown later date it was moved downhill to a slightly elevated portion of the floodplain and became widely known as Paucartambo (Villasante Ortiz 1975, vol. 1). The powerful Paucartambo River, which periodically ravages floodplain sites, may have erased the remnants of a first settlement.

22. Until 1559 encomienda grants entitled grantees to recruit Indian labor without pay as well (Rowe 1957a).

23. In other regions, by contrast, European crops were adopted for subsistence within a few decades after the arrival, settlement, and conquest of the Spaniards. Traveling through Quito (in present-day Ecuador) during the 1550s, Cieza de Léon observed that "now that they know the value and utility of wheat and barley, many of the natives around this city of Quito plant them both, and eat them and make drinks of barley" (Cieza de León [1553] 1959). Such adoption increased over time; during an eleven-year sojourn in Peru, Ecuador, and Colombia between 1735 and 1746, the Spaniards Jorge Juan and Antonio de Ulloa listed ground barley (*machca*) as one of the staple foods among Quito inhabitants (Juan and Ulloa [1772] 1975, 157).

24. Don García Pancorva was the main community authority, or *Cacique Principal,* and Don Gerónimo Compe was the second-ranking community authority, or *Segunda Persona.*

25. Use of the topo, a unit based on seed quantity, demonstrated another instance how the Spaniards adopted Inca customs while also altering them.

26. Under reducción policy the new villagers could not hold land farther than one league or about three miles from their nucleated settlements (Spalding 1984, 179).

27. Another obvious neglect of Maravier's orders and his reordering of the farm landscape was manifest in 1658, when a group of nine Quechua peasants purchased territory near Colquepata village from a Dominican priest and founded the ayllu community known as Sonqo ("Heart"; LAS 1658). The sectoral fallow system of Colquepata community with twelve well-defined sectors remained in practice until roughly 1950, when it was abandoned (Zimmerer 1991e).

28. It is reasonable to assume that endemic diseases and debilitating conditions of the coca-growing lands caused mortality rates in the labor-providing Paucartambo Andes to exceed the southern Peruvian average of fifty percent (N. Cook 1981, 227). Mucocutaneous leishmaniasis, or *espundia,* devastated highlanders laboring in the foothills, so that one of every two migrant workers probably died (Gade 1979, 268). Labor abuses, such as the illegal use of human porters to transport coca bundles from the lowlands, worsened the effects of all diseases and escalated mortality (Polo de Ondegardo [1561] 1940, 190). During more than two centuries of population decline, tribute quotas were rarely adjusted; at times even the cauacachiqqueyoc "haves" must have felt their accustomed livelihoods to be in jeopardy (N. Cook 1981, 246).

29. The term *forastero* contrasted original, or *originario,* which referred to someone born in a community.

30. *Forasteros sin tierra* contrasted with *forasteros con tierra,* the latter fiscal category being established in the 1730s to refer to Indian migrants who had been granted

land and who were held accountable for tribute (Larson 1988, 112–13). This predominance of forasteros sin tierras in Paucartambo continued through the nineteenth century; in 1845 seventy percent of the Indian population fell into this category (Peralta Ruiz 1990).

31. Numerous owners of prosperous coca estates in the montaña foothills purchased highland counterparts in Paucartambo to serve as storage areas for coca leaf and as suppliers of staple foodstuffs. They located leaf storage at coca depots, or *despachos,* along the main Cañac-Huay coca trail in the northern Paucartambo Valley, in estates such as Jajahuana, Pasto, Pilco, Totora, and Acobamba (map 3). Estates designed to serve as staple-provisioning bread-producing estates, or *haciendas de panllevar,* clustered in the central and southern Paucartambo Valley in sites such as Espingone, Cusipata, Guacomoco, La Soledad, Llaychu, Runtu Runtu, and Sayllapata (ADC 1784–85a, ADC 1784–85b, ADC 1785–87, ADC 1800, ADC 1807–8).

32. In the Colquepata repartimiento colonial officials ordered quotas of freeze-dried chuño potatoes, fresh potatoes, and maize (Escobedo Mansilla 1979). Once obtained, the surplus of staple crops was sold and used directly by the encomienda holders and the state to feed servants and workers in their agricultural enterprises such as coca growing and in other economic activities, especially mining. Sixteenth-century annalist Pedro de Cieza León reported that several tons of freeze-dried potatoes were imported each year for the miners at Potosí (de Acosta [1590] 1940, 241; Cieza de León [1553] 1959). The pressure of such demands on Indian tribute-payers worsened when drought, frost, or other environmental vagaries struck their fields, which, according to one observer, occurred in three of every five years (Polo de Ondegardo [1561] 1940, 168). Although official policy of the Peruvian viceroyalty prescribed the waiving or commutation of tribute requirements if crops failed, such adjustments were in fact uncommon and, as a consequence, excessive demands frequently burdened Indian farmers and especially besieged those occupying the unenviable category of "have-nots."

33. Rendered *kawsaychiqayoq* in contemporary Quechua orthography (Cusihuaman 1976).

34. The widespread expression of comida del indio in government reports can be taken to suggest also that the symbolism of diverse crops helped permit the rulers to view the desperation of poorer peasants as atypical in the sense of not being customary of the group. See Gade (1975) and Weismantel (1988) on contemporary use of comida del indio as a derogatory term in the Peruvian and Ecuadorian Andes, respectively.

35. Following his visit to southern Peru in 1911 as a member of the Yale Expedition, Isaiah Bowman noted similarly that "Paucartambo town, itself, once important for its commerce in coca is now in a sadly decadent condition" (Bowman 1916, 77). Although accurate, the description by Bowman was not based on personal observation but instead probably relied on Antonio Raimondi, whose accounts he had read.

36. Paucartambo offered less suitable expanses for grazing than the central Andes, especially in comparison with the intermontane High Plain, or *altiplano,* that stretched south from Cuzco (Orlove 1977a).

37. Gristmill remains dot the landscape at Mollomarca, Payajana, Cusipata, Molinopata, Chusu (above Jajahuana), Acobamba, Mandorpuquio (opposite Acobamba and downstream), and Paucona (two mills) (map 3). Three others, named Keskes, Chaupi Molino (Middle Mill), and Yawar Molino (Blood Mill), lined the Quencomayo River in Paucartambo village.

38. These epiphytotic outbreaks resembled those of the nearby Urubamba Valley that occurred in 1932 and 1933 (Gade 1975, 138). Barberry shrubs, or *Berberis lutea,* common in Paucartambo, hosted the fungal pathogen. Small farmers in other world regions have combatted the black stem rust by the regionwide destruction of barberry shrubs (see, for example, Cronon 1983). Paucartambo farmers, and those in the nearby Urubamba Valley for that matter, did not take concerted action, perhaps because wheat was already a diminishing item in their crop repertoires.

39. Coca workers were provisioned with a diversity of customary foodstuffs (maize, potatoes, beans, quinoa), as well as a few introduced crops (peas, onions, wheat) and dried beef, or *sesina,* as it was known. Production of the highland estates that articulated closely with the lowland economy was distinguished also by alfalfa fields *(alfalfares)* for feeding mule teams used to transport the foodstuffs and coca. Alfalfa parcels similarly greened the irrigated sites near other major trails that connected the highlands to the upper Amazon prior to road links (Brisseau 1978).

40. While the distribution of the precocious *S. phureja* species draped across the eastern Andean valleys from Venezuela as far south as Bolivia, unique intraspecific landraces lined each of the individual pockets, such as the Paucartambo Andes (Brücher 1969; Camino et al. 1981; Hawkes 1990).

41. The sector people tenants and helper tenant sharecroppers were not, however, guaranteed their suites of land and labor resources, notwithstanding the emphasis that Palacio Pimental placed on the contractual nature of peasant-estate rights and obligations. Comparisons among estates of the labor-time demanded from sector people tenants with similar land areas demonstrated a twofold variation during the 1950s (Palacio Pimental 1957a, 205–6). This one-hundred percent variation revealed the dissimilar and often contested nature of the unwritten agreements between estate peasants and owners. For estate peasants, it must be remembered, the motivations for controlling land and a portion of their own labor included opportunities for both commerce and subsistence.

42. In some cases the Quechua of Colquepata supported their claims to private property with the painstakingly saved land titles issued to communities by colonial officials hundreds of years earlier. Using these documents and the concept of property rights, Colquepata peasant leaders, such as Daniel Hurca and Miguel Quispe, protested the ruthless theft of tools, livestock, and land by appealing to the proper conduct of laws governing property in the "civilized nation that is Peru" (ARA 1923). In one deposition Miguel Quispe along with his brother Manuel and four other members of the Sayllapata community argued that landowners and nonindigenous villagers "intended to erode our liberty and legal rights that were inherited from our ancestors in our properties, known as *parcialidades* or *aillos*" (ARA 1922, 1). Although the official land titles granted in 1927 to Indigenous Communities in the Colquepata area, such as Chocopía, Colquepata, Sayllapata, Sipascancha, and Sonqo, did not eliminate conflicts over resource control, the government's titling did strengthen the legal recourse of community members.

43. Continued use of the subsistence rights idea is documented in many other sources as well. For example, a judicial document housed in highland Bolivia reports the protest filed by an indigenous peasant in Cochabamba Department in 1881. Challenging the legal validity under which the Bolivian government sought to privatize rural land holdings through the law of *Ex-Vinculación,* the claimant argued that traditionally

"each family was assigned a portion of land, the product of which was enough to furnish a *comfortable subsistence.*" This claim over the right to land was followed by a citation to "Garcilazo," a reference to the chronicler Garcilaso de la Vega who had interpreted Inca history for the Spaniards more than three centuries earlier (Zimmerer 1993a).

Notes to Chapter 3

1. Santusa Castillo Quispe, Umamarca, June 5, 1987.

2. General Velasco's Revolutionary Government of the Armed Forces planned land reform in order to modernize peasant production, end the skewed concentration of land ownership (eighty-seven percent of the country's agricultural land was held by four percent of its population), and defuse political opposition that included a variety of pro-Communist parties, peasant and trade unions, and popular social movements. The military junta discouraged and repressed popular participation as part of its public concern "to prevent Communism." While fervently anti-Communist, the military government was equally opposed to Peru's ruling civilian oligarchy, which was thwarting the country's development by resisting much-needed economic and social reform (Bourque and Palmer 1975; Cotler 1975).

3. The macroscale policies of Velasco's military government resembled the "functional dualism" model of economic development, whereby the peasant sector served "both as a source of primitive accumulation through cheap semiproletarian labor and cheap food and a contradictory process that leads to the destruction of the peasantry" (de Janvry 1981, 4). Peru's military rulers and their economic planners did indeed envision that the peasant sector would produce inexpensive foodstuffs for urban populations while, at the same time, lessening the need for costly imports (Cotler 1975). Furthermore, like earlier Peruvian governments, they also planned for the transfer of capital from the countryside to the city and mines. Although "functional dualism" was initially defined partly on the basis of the peasant sector eventually being eliminated, economic circumstances in Paucartambo and other highland Peruvian regions during the postreform period neither impoverished peasants so much nor offered sufficiently compelling opportunities for migration that peasant farmers and their economic sector disappeared (de Janvry et al. 1989). Macroeconomic policies in Peru during the postreform period could also be understood as turning from an earlier orientation based on the largely separate sectors of exports (sugar cane, cotton) and national production (wheat, potatoes) toward one where national commerce was integrated with growing agroindustries and urban markets under the umbrella of import-substitution policies (Alvarez 1983; Hopkins 1981; Thorp and Bertram 1978).

4. Faustino Condori Castillo, Umamarca, March 8, 1986.

5. The preference for community units could also be seen in the unquestioned devolution of various Paucartambo haciendas that were consolidated by single owners into separate Peasant Communities following the Land Reform of 1969.

6. Section IV of the Special Statute of Peasant Communities of 1970, entitled "About Community Members," had stipulated that community membership could be conferred only due to birth there or marriage with a member (La Dirección y Promoción de la Reforma Agraria. n.d., 7).

7. Substantial immigration from Puno Department belonged to a long-term histori-

cal flow of outmigrants that was present as early as the sixteenth century and that probably peaked during the waves of forastero migration to Paucartambo in the 1600s and 1700s (chapter two; see Mörner 1978; Polo de Ondegardo [1561] 1940, 168; Wightman 1990).

8. Terms of trade for highland agriculture remained negative throughout most of the period from 1969 to 1987 (Alvarez 1983; Hopkins 1981; Thorp and Bertram 1978). Among items considered in the terms of trade were agricultural inputs such as seed, fertilizer and pesticide—inputs that had become commonplace in Paucartambo during the postreform period (Figueroa 1984, 33). In Ninamarca, ninety-three percent of families purchased pesticides during a three-year study in the late 1970s.

9. Unfavorable market conditions for potato agriculture in the Peruvian Andes stemmed from chronic overproduction and the biases of government policy following 1969. Postreform governments in Peru steered credit, technical assistance, and research and development toward large-scale commercial potato farms in the irrigated coastal valleys (Alvarez 1983; Wilson and Wise 1986).

10. Founded in Arequipa in 1898 by a pair of German-born Peruvians, Ernesto Gunther and Traservo Rehder, the predominantly Swiss-owned company continued as one of the country's three main beer producers through the 1980s (Hopkins 1978, 14).

11. A similar arrangement whereby peasant producers shoulder the bulk of economic risk operates in small-scale mining operations in Bolivia (Godoy 1990, 59). One major injustice that the Quechua in Paucartambo perceived was that the Beer Company of Southern Peru set barley prices according to three "grades" that were assigned to a lot at the time of sale. Peasant growers complained that their product was never awarded the top grade with the highest buying price (grade one) and that it was frequently assessed grade three when it merited the second level. Clapp (1988) discusses how the growing contracts appeared on the surface to be fair but were in fact biased against peasant growers. It bears mention that the real market prices of barley fell during the period from 1950 to 1969 and continued this trend during the postreform epoch, which further favored the beer company's capacity to contract peasant growers (Fano and Benavides 1992).

12. Although coca fields covered less area of the Pilcopata-Qosñipata lowlands than during the colonial epoch, they were still substantial. Their production was destined mainly for regional consumption rather than export in the coca-cocaine trade. Migration from Paucartambo for longer periods frequently involved ninety-day contracts that sent workers to the Quince Mil placer mines. A secondary destination for temporary migrants from Paucartambo was the La Convención lowland in northern Cuzco Department.

13. Improved potato varieties were introduced in the Peruvian Andes during the 1950s (Gade 1975, 207).

14. Faustino Condori Castillo, Umamarca, November 13, 1986.

15. The Quechua farmers in Paucartambo managed a special flora of spinescent shrubs for use as fence plants. These shrub taxa included *paqpa* (agave, *Agave americana*), *cheqche* (*Berberis lutea*), *roqe* (*Colletia spinosissima*), *chanki* (prickly pear cactus, *Opuntia sp.*), *llawlli* (*Bernadesia horrida*), and *tankar* (*Solanum sp.*) (Zimmerer 1989). A few of the dooryard garden plants are listed in table 1.

16. Landlessness remained rare in the Paucartambo countryside throughout the 1970s and the 1980s due in part to provisions in the Special Statute of Peasant Communities

that prohibited community members from selling land. Although some community members violated the reform law, the transfer of land ownership among peasants was generally uncommon.

17. The existence of labor-time shortages in peasant agriculture in the Peruvian highlands following the Agrarian Reform of 1969 has been widely documented, although the environmental consequences of the shortages have been assessed only with respect to frontier colonization sites. Other studies assess the impact on general social and economic activities in the highlands (Baca Tupayachi 1985; Brush 1977, 107; Caballero 1981; Collins 1988, 137; Kervyn 1989; Orlove and Custred 1980, 36; compare with Zimmerer 1991c, 1992b).

18. Studies on the labor demands of postharvest processing and other time-consuming tasks such as intercropping in the peasant growing of diverse crops include Brush et al. (1981), Clawson (1985), Collins (1988), Gade (1972b), Johannessen (1970), Richards (1985, 1986), Weismantel (1988), and Zimmerer (1991b, 1991c).

19. On the use of social kinship for labor recruitment in other Andean regions, see Brush (1977), Collins (1988), Guillet (1992), and Mitchell (1991).

20. In similar fashion an analogous term *masa* defined the field area that two ploughmen with the foot-levered hoes known as chakitaklla, or foot-plows, could furrow in a single day.

21. While the vast majority of Paucartambo cultivators commanded an adequate and frequently sophisticated knowledge of landrace work, such knowledge did vary among individuals and groups within the peasant society, which is discussed in chapters five and six.

22. Palacio Pimental (1957a, 1957b) wrote that some wealthy tenants of Paucartambo estates in effect operated "small haciendas" of their own. He also pointed out that the sector people tended to be the most influential of estate peasants in everyday affairs as well as in occasional disputes with the hacienda.

23. Differences among the three socioeconomic groups were readily apparent. Many Paucartambo peasants could sit on a hillside above the scattered huts of their community and discuss the economic resources of most or all inhabitants. In Peasant Communities with emphasis on a certain resource that endowment weighed heavily in the assessment of wealth. In sheep-raising Colquepata, for instance, the well-off typically owned more than thirty head of sheep, while the poor owned less than fifteen. Additional factors such as dress and the style of one's home (for example, whether or not it was painted) might also be considered.

24. Eufemia Champi Amao, Umamarca, January 5, 1987.

25. Líbano Yapa Flores, Umamarca, May 4, 1986.

26. According to law, former owners could retain the estate houses and upward of thirteen acres. The thirteen acres in many cases comprised prime lands. With the resettlement of owners into villagers, the estate houses became rarely used and slipped into disrepair and decay, poignant reminders of the downfall of the haciendas.

27. Patrons and clients usually cemented ties through the customs of social kinship. Much Andeanist scholarship has investigated the symbolic properties of this relationship (Allen 1988; Isbell 1978; Mayer 1977). Cultural ecology research in the Andes has emphasized the important role of social kinship in economic exchanges for agricultural production given the labor costs, technology, and social organization of farming (for example, Brush 1977; Golte 1980).

28. Examples of market villages included Pisac, Calca, Ocongate, Urubamba, Urcos, Santo Tomas, and Sicuani, and literally scores of others.

29. The claim could just as easily be inverted to argue that their domination of the countryside made the region ripe for activity by the Shining Path, or *Sendero Luminoso.* In any case the Shining Path took only a few actions in Paucartambo, due in part to the fact that its activity near Cuzco gained greater publicity from attacks in the city and well-known places, such as the Machu Picchu Railroad.

30. Government recognition of Peasant Communities in Peru slowed after climbing from roughly 1,600 to 2,337 during the early 1970s (Bourque and Palmer 1975; Isbell 1978, 29; La Dirección de Promoción de Reforma Agraria n.d.). By 1977 that number had grown to 2,837, and by the late 1980s it reached 3,500 (Flores Galindo 1988, 7; Kervyn 1989).

31. Villagers' continued economic advantage was illustrated by the terms of sharecropping. Prior to the Agrarian Reform of 1969, estate peasants and especially the relatively wealthy tenants (the mañay runa, or "sector people") typically sharecropped with landless villagers through the latter supplying seed together with some labor inputs in order to gain an even share of yield from a field (Palacio Pimental 1957a, 197). By the 1980s, the sharecropping villagers supplied seed and field inputs such as fertilizer and pesticide, but no longer furnished any labor in the joint-venture fields. The balance of economic power between village and countryside, as evidenced in sharecropping, thus shifted even further toward the villagers after the Land Reform of 1969.

32. By way of cultural geographic comparison, it is interesting to note that Quechua-speaking peasants in the Ayacucho region north of Paucartambo deployed the term *naked one,* or *qala,* in reference to nonindigenous villagers (Isbell 1978).

33. Size of the diversity-poor fraction of Quechua farmers in the years from 1969 to 1990 likely surpassed that of preceding eras. During earlier epochs many of the diversity-poor peasants tended to desert their farms altogether due to death or desperate migration.

34. Líbano later explained that the feast concocted for Señor Juancito merely intimated at the countless gunny sacks of potato landraces and fresh corn that he and Natividad gifted each year after harvest to a few principal villagers including the mayor and the priest. All the expenditures were worthwhile, he added.

35. In this sense the macroscale dictates of new policies such as the Peruvian government's land reform and its economic programs and the microscale imperatives of resource access and cultural meaning were accurate first-order estimates of the environmental outcomes that issued forth from the transitions of farm nature and society between 1969 and 1990. The political economic and socioeconomic forces could not, however, account fully for the unevenness of diversity's fortunes at the regional scale, where the characters of farm spaces and farming places were strong forces.

Notes to Chapter 4

1. Líbano Yapa Flores, Umamarca, October 18, 1986.

2. Influence of elevation on agriculture and livestock-raising in the Andes was the subject of a pioneering tradition of geographical study beginning with Alexander von

Humboldt and Aimé Bonpland ([1802] 1959) through Carl Troll (1958, 1968). They emphasized elevation-related climate as the factor that most molds the distribution and boundaries of Andean land use. Subsequent studies added a cultural historical dimension comprised of subsistence needs and political-economic demands to further explain the spatial differentiation of land use (for example, see Mayer 1985, Murra 1964, 1972, 1980, 1985a, 1985b, Salomon 1985).

3. This handling of landraces by groups also set the stage for the management of individual landraces and the determination of diversity within landrace groups.

4. The farmers especially took note of resistant landraces in fields stricken by disease. The Quechua in Paucartambo, like other Andean farmers, were also keen to find varieties that withstood attacks of the fungal parasite known as Potato Late Blight, or *Phytophtera infestans* (Brush et al. 1981; Jackson et al. 1980).

5. The production spaces are designated as proper nouns, since among the Quechua of Paucartambo they are widely recognized and well-defined.

6. Líbano Yapa Flores, Umamarca, October 18, 1986.

7. Ecologically specialized communities, by definition, were restricted to elevations above 11,500 feet (3,500 meters). They lacked Valley fields and, in many cases, Early Planting fields as well.

8. Mayer's definition of "production zone" emphasizes the presence of multifaceted distinctiveness: "Specific productive resources in which crops are grown in distinctive ways. It includes infrastructural features, a particular system of rationing resources (such as irrigation water and natural grasses), and the existence of rule-making mechanisms that regulate how these resources are to be used" (Mayer 1985, 50–51).

9. Field owners ensured the completion of tasks through the careful allocation of spatial quotas such as rows and subsequently the arrangement of work itself into an informal competition among workers. Influence of the sociospatial organization of work routines also shaped the character of cultivation technologies (Zimmerer 1994b).

10. A minor limitation stemmed from the moderate suite of farm tools (appendix C.6). Although varied, the farm tool kit did not offer a panoply that otherwise may have led tillers to define more divergent sorts of production spaces.

11. On the vividness of livelihood-related cognitive contrasts among Quechua peasants in the Andes, see Mayer (1985, 50) and Salomon (1985, 515).

12. O. Harris (1985) discusses the kitchen symbol applied to high-elevation potato agriculture among the inhabitants of Northern Potosí in Bolivia.

13. It also deviated from the model of a "smooth continuum," sketched by production zones in the Bolivian Andes (Godoy 1984). On a critique of optimality in assumptions about Andean land use, see Knapp (1991).

14. For this reason, I take pains to substitute "production system" for "production zone" in adopting the concept of farm space in Andean agriculture proposed by Mayer (1985).

15. The high-elevation grassland moor is referred to as *jalca* in the northern Peruvian sierra (Pulgar Vidal 1946; Weberbauer 1945). It should be noted that some of the original ecological definitions placed the puna above the limits of agriculture (Troll 1968; Weberbauer 1945).

16. In the ecologically extended, or specialized, communities, this Hill agriculture comprised nearly all cultivation. Even in most ecologically compressed, or generalized, communities, the area suitable for Hill farming was extensive.

17. They looked at the cover of fertility loving plants, such as ichu grass, as one measure of a fallowed field's readiness for farming.

18. The Quechua in Paucartambo produced miska maize for two quite different purposes. In the northern valley they grew it primarily as forage. In the southern valley they counted on its contribution to the long-awaited thimp'u dish in a Carnival feast during February.

19. The symbolic dualism of Valley and Hill agriculture was shaped historically by social and environmental forces as well. The continued prominence of Saints' Days in Valley farming resembled the religious role granted this farming by the Inca cult that lorded maize in ceremonial ritual (Mitchell 1980; Murra 1960). Such religious prominence also reflected the status of maize as the principal surplus crop of the Inca political economy and as a counterpoint to subsistence-oriented "autochthonous" tuber agriculture. It was further reinforced by the environmental and hence agricultural uncertainty of Valley farming (O. Harris 1985; Platt 1982).

20. In regions with weaker growth and less diversification of farm commerce than Paucartambo the use of sectoral fallow commons persisted in Oxen Area agriculture after the Agrarian Reform of 1969, a situation demonstrated about one hundred miles to the south in the Cuyo-Cuyo District of Puno Department (Camino et al. 1981).

21. Líbano held one or more Oxen Area plots within each of the pie-shaped sectors that formerly had reached downhill from Hill farming and stretched across the middle slopes.

22. Although landholdings and settlement in Colquepata were more dispersed than in other parts of the Paucartambo Andes, the social and environmental patterns did not preclude the concentration of bog properties among better-off families.

23. Long-term conspicuousness of the Oxen Area of Paucartambo gives cause to question the apparent simplicity of "bizonal cultivation" described in northern Potosí in Bolivia (O. Harris 1985; Platt 1982, 1986). Widespread expansion of the Oxen Area under colonial rule and present-day production strongly suggests a greater landscape complexity there.

24. The sum of adjustment was brought about through a complex and changing "manufacture" of farm space, as Mayer terms it (1985), although I believe this term is ill-chosen due to its inference that human agency alone creates production spaces. In reality the forces of nature are also at work in producing these spaces.

25. The convenient tagging of a field in terms of its production space did not obviate the detailed knowledge of individual fields that was held by most farmers. The point is that they conceived of these detailed attributes in relation to the production spaces and guided their decisions on this basis.

26. Many assessments of land use and agriculture in the central Andes have assigned primacy to either environmental or social factors in defining the geographic areas larger than land use units. Studies defining areas on the basis of watersheds and river valleys prioritized environmental ordering (Brush et al. 1981; Gade 1975; Mayer 1979; Mayer and Fonseca 1979). The choice of Peasant Communities and Cooperatives, pilgrimage networks, and trade or market regions emphasized social ordering (Brush 1976, 1977; Guillet 1979; Mitchell 1991; Orlove 1977a; Webster 1973). I believe that the unit of the geographical place and the spatial scale it represents are well-suited to the combined environmental and social ordering that operates in the changing role of diverse crops.

Notes to Chapter 5

1. They also transported the products of various economic booms in the moñtana, such as quinine bark or cascarilla. The descendents of Scottish immigrants continued to bear surnames such as MacDougal, while otherwise blending imperceptibly into the Quechua peasantry of the northern Paucartambo Valley.

2. Trees reached only as high as 11,150 feet (3,400 meters) in the northern valley, whereas they climbed above 12,950 feet (3,950 meters) in the much drier Quencomayo Valley. The depression of temperature in the northern valley due to cloud cover also shaped the distribution of subtropical crop plants. Inhabitants of Huaqanqa located at roughly 9,200 feet (2,800 meters) could not count on garden crops of figs and papaya that thrived at the same elevation in the drier and hence warmer Urubamba Valley to the west (Gade 1975).

3. Houses themselves in the northern Paucartambo Valley displayed a unique rain-shedding architecture of four-cornered, gableless roofs and walls of stone rather than adobe block. With respect to Manú National Park and local land use, it is worth noting that Quechua of the northern valley held ideas about rights to pasture in the grassland-moor that were different than the park management, which defined this territory as off-limits to livestock. The northern valley haciendas had grazed livestock there and even kept herders' quarters that later became park buildings. After the Land Reform of 1969, the Quechua farmers felt they deserved access to the former resources of the haciendas.

4. The highway was traversed daily by numerous large trucks that not only shipped freight (including much timber of rain forest trees) but also shuttled passengers. Tres Cruces, where the highway crested the Cordillera Paucartambo, lay about five miles south of the Cañac-Huay saddle that had seated the colonial coca trail.

5. Although frequent fungal outbreaks dissuaded barley cropping in the northern valley, the full story of its absence entailed a bitter barley-growers strike waged near Challabamba in the mid-1970s. The strike had galvanized opposition among northern valley growers and led the brewery to concentrate its efforts in other areas of Paucartambo.

6. Cultivation of swiddens distinguished the northern Paucartambo Valley; farmers applied the term *chaqo* to describe a recently cleared swidden.

7. There I offered them rides to Huaqanqa, and the couple, who I met one year earlier, discussed their farming situation (May 30, 1987).

8. The noteworthy resistance of *Solanum phureja* to Potato Late Blight and a potato-infecting bacterial wilt known as *Pseudomonas solanaceum* has drawn the attention and efforts of crop collectors and plant breeders since the 1950s.

9. Cattle-raising for commerce was a longstanding economic activity in the sparsely populated northern valley (Palacio Pimental 1957a, 1957b). Buyers from other Cuzco provinces periodically traveled to the northern Paucartambo Valley in order to purchase cattle herds in the years from 1969 to 1990; see appendix C.4.

10. In nearby Calca Province agricultural economist Bruno Kervyn and his research team documented the substantial labor costs of building rock fences similar to those constructed in the northern Paucartambo Valley (Kervyn 1989).

11. The Pitumarca traders claimed to hold coca and maize fields in the northern Paucartambo Valley, probably reflective of an earlier "vertical archipelago" of land use.

The Huaqanqa people referred to them as qolla in reference to their origins in southern Cuzco. In barter with the Pitumarca traders Huaqanqa agriculturalists offered 480 kilograms (4 *fanegas*) of unshelled maize for one horse or, alternatively, 120 to 180 kilograms of unshelled maize for the wool sheared from an adult alpaca. In barter with the nearby inhabitants of west-bank communities possessing high-elevation territory (Chaqllaybamba, Mandorpuquio, and Qachu-Qachu), Huaqanqa farmers traded maize for freeze-dried chuño and moraya and for fresh floury potatoes. Other common terms of local barter stipulated twenty-three kilograms of shelled maize for one sheep yearling.

12. This fate of *S. phureja* diversity in the Paucartambo Andes appears to be paralleling similar declines in other Andean regions. While the species covers the eastern Andean Cordillera from Venezuela to Bolivia, its widespread disappearance has already been noted (Brücher 1969, Camino et al. 1981, Hawkes 1990). Its status as an endangered potato in the Paucartambo Andes may thus apply to South America in general.

13. Similar bogs in the Cuzco Province of Sicuani were called *oqho* (Orlove 1977a, 81). On the approximate distribution of swale bogs in Colquepata, see map 9.

14. Drainage of the bogs was sufficient so that farmers could occasionally rotate their Early Planting potatoes with a crop of oats. They grew the hydrophilous oat crop for fodder, which, during the 1980s, was shipped to southern Cuzco provinces under the advice and arrangements made by the local Ministry of Agriculture office based in Huancarani (8 miles [5 kilometers] from Colquepata on the Cuzco-Paucartambo Highway).

15. The liver fluke as well as its snail host inhabit the standing water and soggy soils of bogs, marshes, and small lakes of the southern Peruvian sierra. The fluke also plagues the humid pastures of other Cuzco provinces, such as Anta and Sicuani (Orlove 1977a; Watters 1994). The history of its spread and the role of humans in this environmentally important diffusion remain virtually unknown. Colquepata farmers have unsuccessfully and usually unsafely applied chemical treatments including copper sulfate and arsenic to combat the liver flukes.

16. The numerically dominant presence of Quechua peasants in Colquepata did not negate the power of a small group of mestizo or white (misti) villagers who held the major political posts such as mayor (see also Allen 1988). Although cruel, the power of these white Colquepata villagers was less than their counterparts in Paucartambo and Challabamba.

17. See map 4.

18. Farming systems of the Quencomayo bogs made up one of only a few extant examples of wetland or hydraulic agriculture, also known as raised-field agriculture, which was widespread in the New World and especially South America during the pre-European period. Many details of wetland farming in the Quencomayo bogs did not, however, closely resemble the particulars of the widespread pre-European raised fields (Denevan 1980; Parsons 1971; Zimmerer 1991d, 1994b).

19. In field systems other than their Early Planting the Colquepata women broke new ground in the local division of labor. Some drove the oxen-drawn plow, for instance, to till rainfed Oxen Area parcels.

20. This generalist-style scattering of fieldholding and farming among Colquepata families was oddly incongruous with their dispersed pattern of settlement. Dispersed rather than nucleated settlement is more typically of specialist land users who in the

Andes are primarily herders (Brush 1976, Brush and Guillet 1985, Guillet 1981b, Mayer 1985, Orlove 1977b).

21. The few Colquepata farmers still growing quinoa in the 1980s all intercropped it within their maize fields.

22. Gonzalo Vargas, Paucartambo, March 15, 1986.

23. Juan Santos Yucra, Colquepata, December 3, 1986.

24. Some unterraced pass-through slope fields may have concealed a farmer's intent to rebuild the step-step parcels at a later date. The predominance of pass-through slope fields, however, affirmed the prevalence of decisions to abandon the agricultural terraces for good.

25. Wealthy peasants could afford to buy the bottled Cuzco Beer regularly, although they preferred chicha on many occasions.

26. In the case of potato landraces detailed study of genetic constituents in a sample of six widespread landraces from Paucartambo showed that the unique genes cluster locally within particular places and thus would become extinguished under a scenario of local loss (Quiroz et al. 1990; Zimmerer and Douches 1991). Similar research has not been conducted on the biogeography of genetic variation in other landrace crops.

27. Earlier descriptions enumerated landraces as well as subtypes, the latter being distinguished mostly on the basis of skin color. The 1986–90 samples yielded 183 sublandraces, although they were grouped into the 79 landraces identified in table 8.

28. Row tillage did secure the cropping of potatoes that sported shorter stolons and it thus sufficed for the high-yielding improved varieties.

29. The all-important consequences of this style of seed management for the diversity of landrace crops are discussed in the next chapter.

30. In fact, the synonym *mixed* potatoes was only slightly less common than the oft-heard *floury* potatoes, while being more frequent than the quaint *ancient* potatoes.

31. It was not uncommon for a farmer to explain how she or he had garnered an array of floury potato seeds by working for in-kind payment over one or more years for a variety of field owners.

32. The reduced diversity of floury potato commerce resembled the case of maize in the northern Paucartambo Valley.

33. The remaining eight landraces were cultivated only in downstream Peasant Communities near Challabamba and as far north as Huaqanqa; see table 9.

34. The useful flora of field balks included a large number of medicinal and esculent types—such as *salvia,* or *Lepechina meyenii*; *mayo manzanillu,* or *Epilobium denticulatum*; and *alqo yawar chunka,* or *Alternanthera caracasana* (Zimmerer 1989).

35. Field Notes, Notebook Seven, pages 81–82.

36. These actions showed how the seemingly simple acts of decision-making, as outlined in the framework regional political ecology (Blaikie 1985; Blaikie and Brookfield 1987), were thoroughly cultural practices.

Notes to Chapter 6

1. Natividad Puma Huaman, Umamarca, April 15, 1987.
2. Natividad Puma Huaman, Umamarca, June 15, 1987.
3. An example in the potato crop was the landrace variant *moro wayra,* whose skin

showed a mix of red and yellow splotches in lieu of the solid red characteristic of plain *wayra.*

4. References to the human form in the metaphors of Quechua farming were but one of a long series of cosmological, sociopolitical, and environmental associations with the body that were drawn by the Quechua peasants (Classen 1993; Fonseca 1988).

5. Similar beliefs may be widespread in the intercropping of other Andean peasants, although comparable evidence of cultural practices is partial thus far (Rhoades and Bebbington 1990; Yamamoto 1985).

6. Similarly, they held that the combination of landraces sown into a single hole was chosen on the basis of general differences between them rather than the particular properties of individual types.

7. By contrast, the cultivators of the eastern Andes in Puno erected eight use-categories in the classification of potato landraces, although the groups were not restrictive and farmers classified many landraces according to more than one use-category (Camino et al. 1981, 179; see also Patrón 1902).

8. Natividad Puma Huaman, Umamarca, April 15, 1987.

9. A cultural legacy could also be discerned here: "throughout most of the history of the human occupancy in the Andes, the pressure has been on taming the high *puna*" (Murra 1960, 394).

10. Ch'ilkas people in Pitumarca caravaned their llamas and alpacas yearly en route to an archipelago-like assortment of lands and trading partners at lower elevations in the far northern Paucartambo Valley. Their visits to Paucartambo and their significance for diverse crops are discussed in chapter five.

11. It is interesting that the boiling potato fields were usually at a safe distance from the maize-growing Valley lands that had been the ones most contested and usurped by outsiders.

12. The barbed names were not unknown among at least some villagers. Like many discourses of less powerful groups, the one on potato naming gained meanings through its expression within the group as well as through its occasional exposure to outsiders (Scott 1985); see table 8.

13. Even the names for the diseases of ulluco coincided with those of potatoes.

14. The most common ulluco dishes in Paucartambo entailed the following preparations: cutting into elongate slices for soup (*lisas almuerzo*); dicing for stew (*lisas uchu*); parboiling, pounding with a wooden spoon, and boiling (*lisas wakta*); partial flattening with the rocker stone followed by boiling (*lisas sakta*); and freeze-drying alternated with soaking in water (*lingle*). The last preparation, similar to the freeze-dried and leached form of bitter potatoes known as moraya, was noted by W. H. Hodge in his classic article on the ulluco (Hodge 1951).

15. A cultivator paid close attention also to the general quality of maize seed in terms of its likely viability. After the seed ears had been selected, she chose viable seed by shelling kernels from the middle portion, using those from the head and base only for consumption. Suitable seed possessed an intact tip (nose) and fully developed germ (embryo).

16. Early planting and late planting fields likewise consisted of some landraces slated primarily for production and others mostly for consumption.

17. The variety of maize dishes in the Paucartambo Andes was made possible by the underappreciated mastery of grinding techniques. In fact, several flours and gruels

prepared from Andean and introduced grains and legumes through grinding were consumed in a wide variety of dishes by the Quechua farmers. A primer and perhaps the most definitive statement on the ethnobotanical importance of grinding in the Peruvian Andes remains the work of Mejía Xesspe ([1931] 1978; see also Zimmerer 1992c).

18. The extreme biological importance of intermediate-level use-categories suggests additional insights for both general folk taxonomic models and the application of the models to crop diversity management by farmers (Berlin 1976; Berlin et al. 1973; Brunel 1969; Brush 1980; Brush et al. 1981; Zimmerer 1991b).

19. Segundino Condori Castillo, March 28, 1986.

20. For example, the farmers made ritual offerings when planting the improved or high-yielding varieties of potatoes. They also imbued them with the familiar anthropomorphic designations and Quechua terms, commenting that the HYVs "walked" (*purin*), "slept" (*punun*), "ate" (*miqhun*), and even "chose" particular soils (*hallp'ata aqlla*), much like their traditional crop varieties.

21. On Quechua ethnoastronomy in Cuzco, see Urton (1981).

Notes to Chapter 7

1. These studies elaborate on a common topic of agrarian discontent. The scissors effect is lamented poignantly in the *Northern Lights* documentary about Upper Midwestern grain growers. Paraphrasing the documentary, farming is a strange business, you sell wholesale and buy retail.

2. An example of a rigid assumption about subsistence farming is that of Scott's "moral economy of the peasant" (Scott 1976). His assumption that the adoption of new crops varies strictly as a function of competing labor requirements in subsistence fails to discuss whether either production system may be flexible enough to spread labor-time demands sufficiently for the adoption of one and the continuation of the other.

3. Vital cultural customs are also evident in the current management of diverse maize in the central Chiapas region of Mexico (Bellon 1991). Although expressions of intimacy between person and plant among the Quechua in Paucartambo may someday become undermined by feelings of being old-fashioned or boring, the bonds remained strong during the decades between 1969 and 1990.

4. The erroneous tenet often masks an unstated functionalism. Such reasoning figures that if well-to-do indigenous peasants by definition serve in the vanguard of modernization they must thoroughly reject those traditional customs thought to be incompatible with Westernization.

5. These studies interpret the environment as a background scene or a mere receptacle of human impacts rather than being credited with a recursive function. Granting the environment a role in social and economic dynamics means profiling it as a player and not merely as a prop.

6. One illuminating debate on this topic raises the diversity-rich forest islands known as *apêtê* that exist amid the savannas of the Kayopó Indians of eastern Brazil. Focusing on the issue of whether the forest islands are the creation of the Kayopó, the debate evidences some of the assumptions of the "ecologically noble savage" (Parker 1992, 1993; Posey 1983, 1992a, 1992b).

7. Although many counterexamples to the general assumption of the ecologically

noble savage are intended to be revisionary rather than reactionary, they have sometimes profitted the adversaries of conservation and indigenous people.

8. A conservation strategy based on highly diverse gardens resembles a number of in situ efforts in the United States through groups like Seed Saver's Exchange and Native Seeds/SEARCH (Fowler and Mooney 1990; Gillis 1993; Nabhan 1989). The transfer of useful information and experience between the United States-based groups and indigenous peasants in developing countries may prove mutually helpful, although the differences in settings will likely moderate commensurability.

9. Analysis of agriculture in the Paucartambo Andes suggests that the similar cultivation of minor and major domesticates is strongly motivated by pressures to standardize farming systems. The degree of biological similarity of the minor-major crop pairs may be less than has been assumed.

10. Altered work regimes of the women farmers in Paucartambo, including greater cultivation of crops other than the diverse ones, is likely to emerge as an increasingly important force prompting the abandonment of landrace agriculture (Deere 1982).

11. "Professional NGOs" working with grassroots NGOs may be able to address the issue of spatial scale that is currently one impediment in the issue of property rights to diverse crops. Since national ownership likely disfavors indigenous peasants and because other large-scale aggregations are plagued by the similar social problem of ownership (Brush 1992, 1993; Kloppenburg and Gonzales 1994; Zimmerer 1992c), organization of a more local scale may help in providing a workable resolution.

12. Absence and weakness of those institutions was one main factor in the rise of the Shining Path Movement and the devastation of its civil war with Peruvian government forces (Poole and Rénique 1992).

13. Myriad ties entwine the seemingly faraway region of the Paucartambo Andes and the places of the United States and other developed countries: economic, political, cultural, and, not least, environmental. One emphasis of second generation work on in situ conservation and sustainable development will be to examine and expand on the connections between interests concerned with environmentally sound agriculture in places such as the United States with those places such as Paucartambo (Berry 1977; Gillis 1993; Jackson 1984; Nabhan 1989).

14. The historical replacement of staple crops was summarized by the social historian E. P. Thompson: "Mr. Salaman, the historian of the potato, has given a convincing blow by blow account of the 'battle of the loaf,' by which landowners, farmers, parsons, manufacturers, and the Government itself sought to drive laborers from a wheaten to a potato diet" (Thompson 1963, 314). As both Salaman and Thompson discuss, the new reliance on potatoes simply maintained workers in the countryside and cities at less expense than the wheat-based diets, the inverse of today's force-field in the Paucartambo Andes. Subsidized imports not only can subvert a country's cultivation of its staples but even distort development by trapping governments into economic policies dependent on low-cost but uncertain food aid (Watts 1987).

Bibliography

Acosta, Joseph de. [1590] 1940. *Historia natural y moral de las indias.* Reprint, Mexico City: Fondo de Cultura Económica.

Adams, W. M. 1990. *Green Development: Environment and Sustainability in the Third World.* London: Routledge.

Aguirre, Lurecia, and Mario E. Tapia. 1982. "Estudios sobre quinuas del valle." In *Tercer Congreso Internacional de Cultivos Andinos—La Paz, Bolivia,* 55–61. La Paz: IBTA—CIID.

Alberti, Giorgio, and Enrique Mayer. 1974. *Reciprocidad e intercambio en los Andes peruanos.* Lima: Instituto de Estudios Peruanos.

Alcorn, Janis B. 1984. *Huastec Mayan Ethnobotany.* Austin: University of Texas Press.

Allen, Catherine J. 1988. *The Hold Life Has: Coca and Cultural Identity in an Andean Community.* Washington, D.C.: Smithsonian Institution Press.

Altieri, Miguel A. 1983. *Agroecology: The Scientific Basis of Alternative Agriculture.* Berkeley: Division of Biological Control, University of California.

Altieri, Miguel A., and Laura C. Merrick. 1987. "*In situ* Conservation of Crop Genetic Resources through Maintenance of Traditional Farming Systems." *Economic Botany* 41 (1): 86–96.

———. 1988. "Agroecology and in situ Conservation of Native Crop Diversity in the Third World." In *Biodiversity,* edited by E. O. Wilson, 361–369. Washington, D.C.: National Academy Press.

Alvarez, Elena. 1983. *Política económica y agricultura en el Perú, 1969–1979.* Lima: Instituto de Estudios Peruanos.

Anderson, Edgar. 1952. *Plants, Man, and Life.* Berkeley: University of California Press.

Baca Tupayachi, Epifanio. 1985. *Economía campesina y mercados de trabajo: el caso del sur oriente.* Cuzco: Centro de Estudios Rurales Bartolomé de las Casas.

Balée, William. 1992. "Indigenous History and Amazonian Biodiversity." In *Changing Tropical Forests: Historical Perspectives on Today's Challenges in Central and South*

America, edited by H. Steen and R. Tucker, 185–197. Durham: Duke University and Forest History Society.

Basile, David G. 1974. *Tillers of the Andes: Farmers and Farming in the Quito Basin.* Studies in Geography, no. 8. Chapel Hill: University of North Carolina, Department of Geography.

Bassett, Thomas J. 1988. "The Political Ecology of Peasant-Herder Conflicts in the Northern Ivory Coast." *Annals of the Association of American Geographers* 78 (3): 453–472.

Bayliss-Smith, T. P. 1982. *The Ecology of Agricultural Systems.* Cambridge: Cambridge University Press.

Bellon, Mauricio R. 1991. "The Ethnoecology of Maize Variety Management: A Case Study from Mexico." *Human Ecology* 19 (3): 389–418.

Bennett, Wendell C. 1947. "The Andean Highlands, an Introduction." In *Handbook of South American Indians,* edited by J. Steward, 1–60. Vol. 2. Washington, D.C.: Smithsonian Institution Bureau of American Ethnology.

Berlin, Brent. 1976. "The Concept of Rank in Ethnobiological Classification: Some Evidence from Aguaruna Folk Botany." *American Ethnologist* 3 (3): 381–410.

Berlin, Brent, Dennis E. Breedlove, and Peter H. Raven. 1973. "General Principles of Classification and Nomenclature in Folk Biology." *American Anthropologist* 75 (1): 214–242.

Bernstein, Henry. 1979. "African Peasantries: A Theoretical Framework." *Journal of Peasant Studies* 6 (4): 421–443.

Berry, Wendell. 1977. *The Unsettling of America: Culture and Agriculture.* New York: Avon Books.

Bertonio, Ludovico. 1956. *Vocabulario de la lengua aymara.* La Paz: Don Bosco.

de Betanzos, Juan. [1551] 1968. *Suma y naración de los Incas.* Reprint, Madrid: Imprenta M. G. Herández.

Beyersdorf, Margot. 1984. *Léxico agropecuario quechua.* Cuzco: Centro de Estudios Rurales Bartolomé de las Casas.

Bird, Robert M. 1970. "Maize and its Cultural and Natural Environment in the Sierra of Huánuco, Peru." Ph.D dissertation, University of California, Berkeley.

———. 1980. "Maize Evolution from 500 B.C. to the Present." *Biotrópica* 12 (1): 30–41.

Bird, Robert M., David L. Browman, and Marshall E. Durbin. 1983–84. "Quechua and Maize: Mirrors of Central Andean Culture History." *Journal of the Steward Anthropological Society* 15 (1–2): 187–239.

Black, Richard. 1990. "Regional Political Ecology in Theory and Practice: A Case Study from Northern Portugal." *Transactions of the Institute of British Geographers* 15: 35–47.

Blaikie, Piers. 1985. *The Political Economy of Soil Erosion in Developing Countries.* London: Longman.

———. 1988. "The Explanation of Land Degradation in Nepal." In *The Himalayan Dilemma: Reconciling Development and Conservation,* edited by Jack D. Ives and Bruno Messerli, 132–158. London: Routledge.

Blaikie, Piers, and Harold Brookfield, eds. 1987. *Land Degradation and Society.* London: Methuen.

Blakewell, Peter. 1984. *Miners of the Red Mountain: Indian Labor in Potosí, 1545–1650.* Albuquerque: University of New Mexico Press.

Blanco Galdos, Oscar. 1981. "Recursos genéticos y tecnología de los andes altos." In *Agricultura de ladera en América tropical,* edited by Andrés R. Novoa and Joshua L. Posner, 297–303. Turrialba, Costa Rica: CATIE.

Boster, James S. 1985. "Selection for Perceptual Distinctiveness: Evidence from Aguaruna Cultivars of *Manihot esculenta.*" *Economic Botany* 39 (3): 310–325.

Botkin, Daniel B. 1990. *Discordant Harmonies.* Oxford: Oxford University Press.

Bourque, Susan C., and David S. Palmer. 1975. "Transforming the Rural Sector: Government Policy and Peasant Response." In *The Peruvian Experiment: Continuity and Change under Military Rule,* edited by Abraham F. Lowenthal, 179–219. Princeton: Princeton University Press.

Bowman, Isaiah. 1916. *The Andes of Southern Peru.* New York: The American Geographical Society.

Braatz, Susan. 1992. *Conserving Biological Diversity.* Washington, D.C.: The World Bank.

Bradby, Barbara. 1975. "The Destruction of Natural Economy in Peru: A Problem of the Articulation of Modes of Production." *Economy and Society* 4 (2): 127–161.

Brandolini, Aureliano. 1970. "Maize." In *Genetic Resources in Plants—Their Exploration and Conservation,* edited by O. H. Frankel and E. Bennet, 273–309. Philadelphia: F. A. Davis Company.

Brisseau, Jeanine. 1978. "El rol histórico del Cuzco como centro regional." In *Cuzco: geografía e historia: documentos y apuntes de interpretación,* edited by J. Brisseau, M. Burga, A. Giesecke, and C. A. Ugarte, 1–36. Lima: Universidad Nacional Agraria La Molina.

Brown, J. H. 1981. "Two Decades of Homage to Santa Rosalia: Toward a General Theory of Diversity." *American Zoologist* 21 (4): 877–888.

Brücher, Heinz. 1969. "Venezuelas Primitiv-Kartoffeln." *Angewandte Botanik* 42 (5/6): 179–188.

Brunel, G. 1969. "Variation in Quechua Folk Biology." Ph.D diss., Department of Anthropology, University of California, Berkeley.

Brush, Stephen B. 1976. "Man's Use of an Andean Ecosystem." *Human Ecology* 4 (2): 147–166.

———. 1977. *Mountain, Field, and Family: The Economy and Human Ecology of an Andean Valley.* Philadelphia: University of Pennsylvania Press.

———. 1980. "Potato Taxonomies in Andean Agriculture." In *Indigenous Knowledge Systems and Development,* edited by David W. Brokensha, D. M. Warren, and Oswald Werner, 37–47. New York: The University Press of America.

———. 1986. "Genetic Diversity and Conservation in Traditional Farming Systems." *Journal of Ethnobiology* 6 (1): 151–167.

———. 1987. "Diversity and Change in Andean Agriculture." In *Lands at Risk in the Third World,* edited by Peter D. Little, Michael M. Horowitz, and A. Endres Nyerges, 290–315. Boulder: Westview Press.

———. 1989. "Rethinking Crop Genetic Resource Conservation." *Conservation Biology* 3 (1): 19–29.

———. 1992. "Farmers' Rights and Genetic Conservation in Traditional Farming Systems." *World Development* 20 (11): 1617–1630.

———. 1993. "Indigenous Knowledge of Biological Resources and Intellectual Property Rights: The Role of Anthropology." *American Anthropologist* 95 (3): 653–671.

Brush, Stephen B., Heath J. Carney, and Zósimo Huamán. 1981. "Dynamics of Andean Potato Agriculture." *Economic Botany* 35 (1): 70–88.

Brush, Stephen B., Mauricio Bellon Corrales, and Ella Schmidt. 1988. "Agricultural Development and Maize Diversity in Mexico." *Human Ecology* 16 (3): 307–327.

Brush, Stephen B., and David W. Guillet. 1985. "Small-Scale Agro-Pastoral Production in the Central Andes." *Mountain Research and Development* 5 (1): 19–30.

Brush, Stephen B., and J. Edward Taylor. 1992. "Technology Adoption and Biological Diversity in Andean Potato Agriculture." *Journal of Development Economics* 39: 365–387.

Bryant, Raymond L. 1992. "Political Ecology: An Emerging Research Agenda in Third-World Studies." *Political Geography* 11 (1): 12–36.

Bueno, Cosme. 1951. *Geografía del Perú virreinal.* Lima: Daniel Valcarcel.

Bukasov, S. M. 1981. *Las plantas cultivadas de México, Guatemala y Colombia.* Turrialba, Costa Rica: Agencia Alemana de Ayuda Técnica, GTZ.

Buttel, Frederick H. n.d. "Sustainable Agriculture as Self-Fulfilling Marginality." In *Varieties of Sustainability,* edited by Paul B. Thompson and D. Bruce Dickson. College Station: Texas A & M Press.

Butzer, Karl W. 1993. "No Eden in the New World." *Nature* 362: 15–17.

Caballero, José María. 1981. *Economía agraria de la serra peruana: antes de la reforma agraria de 1969.* Lima: Instituto de Estudios Peruanos.

Camino, Alejandro, Jorge Recharte, and Pedro Bidegaray. 1981. "Flexibilidad calendárica en la agricultura tradicional de las vertientes orientales de los Andes." In *La tecnología en el mundo andino: Runakunap Kawayninkupaq Rurasqankunaqa,* edited by Heather Lechtman and Ana María Soldi, 169–194. Mexico: Universidad Nacional Autónoma de México.

Candolle, Alphonse de. 1908. *Origin of Cultivated Plants.* New York: D. Appleton & Company.

Cárdenas, Martín. 1966. "The South American Potential Germ-Plasm." *American Potato Journal* 43 (10): 367–370.

———. 1969. *Manual de plantas económicas de Bolivia.* Cochabamba: Ichtus.

Carney, Judith, and Michael Watts. 1991. "Disciplining Women? Rice, Mechanization, and the Evolution of Mandinka Gender Relations in Senegambia." *Signs: Journal of Women in Culture and Society* 16 (41): 651–681.

Chang, J. 1977. "Tropical Agriculture: Crop Diversity and Crop Yields." *Economic Geography* 53: 241–254.

Cieza de León, Pedro de. [1553] 1853. "La crónica del Perú." In *Historiadores primitivos de Indias,* edited by Enrique de Vedia, 349–458. Biblioteca de Autores Españoles, no. 26. Reprint, Madrid: M. Rivadeneyra.

———. [1553] 1959. *The Incas.* Translated by Harriet de Onis. Reprint, Norman: University of Oklahoma Press.

Clapp, Roger A. J. 1988. "Representing Reciprocity, Reproducing Domination: Ideology and the Labour Process in Latin American Contract Farming." *The Journal of Peasant Studies* 16 (1): 5–39.

Classen, Constance. 1993. *Inca Cosmology and the Human Body.* Salt Lake City: University of Utah Press.

Clawson, David L. 1985. "Harvest Security and Intraspecific Diversity in Traditional Tropical Agriculture." *Economic Botany* 39 (1): 56–67.

Cleveland, David A., Daniela Soleri, and Steven F. Smith. 1994. "Do Folk Crop Varieties Have a Role in Sustainable Agriculture?" *BioScience* 44 (11): 740–751.

Cobo, Father Bernabé. [1653] 1956. *Obras,* 2 vols. Biblioteca de Autores Españoles, nos. 91–92. Reprint, Madrid: Ediciones Atlas.

———. [1653] 1979. *History of the Inca Empire.* Translated by Roland Hamilton. Reprint, Austin: University of Texas Press.

Collins, Jane L. 1986. "The Household and Relations of Production in Southern Peru." *Comparative Studies in Society and History* 28: 651–671.

———. 1988. *Unseasonal Migrations: The Effects of Rural Labor Scarcity in Peru.* Princeton: Princeton University Press.

Cook, Noble D. 1981. *Demographic Collapse: Indian Peru, 1520–1620.* Cambridge: Cambridge University Press.

Cook, O. F. 1925a. "Peru as a Center of Domestication, Part I." *Journal of Heredity* 16 (2): 33–46.

———. 1925b. "Peru as a Center of Domestication, Part II." *Journal of Heredity* 16 (3): 95–110.

Cook, Sherburne F. 1949. *The Historical Demography and Ecology of the Teotlalpan.* Ibero-Americana, no. 33. Berkeley: University of California Press.

Cotler, Julio. 1975. "The New Mode of Political Domination in Peru." In *The Peruvian Experiment: Continuity and Change under Military Rule,* edited by Abraham F. Lowenthal, 44–78. Princeton: Princeton University Press.

Cox, George W., and Michael D. Atkins. 1979. *Agricultural Ecology: An Analysis of World Food Production Systems.* San Francisco: W. H. Freeman & Company.

Cronon, William. 1983. *Changes in the Land: Indians, Colonists, and the Ecology of New England.* New York: Hill & Wang.

Crosby, Alfred W. 1986. *Ecological Imperialism: The Biological Expansion of Europe, 900–1900.* Cambridge: Cambridge University Press.

———. 1991. "Metamorphosis of the Americas." In *Seeds of Change: Five Hundred Years Since Columbus,* edited by Herman J. Viola and Carolyn Margolis, 70–89. Washington, D.C.: Smithsonian Institution Press.

Cusihuaman, Antonio. 1976. *Diccionario quechua: Cuzco-Collao.* Lima: Instituto de Estudios Peruanos.

Darwin, Charles R. 1868. *The Variation of Animals and Plants under Domestication.* London: J. Murray.

Deere, Carmen Diana. 1982. "The Division of Labor by Sex in Agriculture: A Peruvian Case Study." *Economic Development and Cultural Change* 31 (4): 795–911.

———. 1990. *Household and Class Relations: Peasants and Livelihoods in Northern Peru.* Berkeley: University of California Press.

Deere, Carmen D., and Alain de Janvry. 1981. "Demographic and Social Differentiation among Northern Peruvian Peasants." *The Journal of Peasant Studies* 8 (3): 335–366.

Denevan, William M. 1980. "Latin America." In *World Systems of Traditional Resource Management,* edited by Gary A. Klee, 217–244. New York: John Wiley.

———. 1983. "Adaptation, Variation, and Cultural Geography." *Professional Geographer* 35 (4): 399–407.

———. 1986. "The Cultural Ecology, Archaeology, and History of Terracing and Terrace Abandonment in the Colca Valley of Southern Peru." Technical Report to the National Science Foundation and the National Geographic Society. Madison, Wisconsin: University of Wisconsin, Department of Geography.

———. 1992. "The Pristine Myth: The Landscape of the Americas in 1492." *Annals of the Association of American Geographers* 82 (3): 369–385.

Deustua, José, and José Luis Rénique. 1984. *Intelectuales, indigenismo y descentralismo en el Perú, 1897–1931.* Cuzco: Centro de Estudios Rurales Andinos Bartolomé de las Casas.

Diamond, Jared M. 1986. "The Environmentalist Myth." *Nature* 324: 19–20.

DNEC (Dirección Nacional de Estadística y Censos). 1966. *Centros Poblados, Censo de 1961.* Tomo II. Lima: Dirección Nacional de Estadística y Censos.

Dobyns, Henry F. 1963. "An Outline of Andean Epidemic History to 1720." *Bulletin of the History of Medicine* 37 (6): 493–515.

Dollfus, Olivier. 1981. *El reto del espacio andino.* Lima: Instituto de Estudios Peruanos.

———. 1991. *Territorios andinos.* Lima: Instituto de Estudios Peruanos.

Donkin, R. A. 1979. *Agricultural Terracing in the Aboriginal New World.* Tucson: University of Arizona Press.

Doolittle, William E. 1984. "Agricultural Change as an Incremental Process." *Annals of the Association of American Geographers* 74: 124–137.

Ehrenfeld, David. 1987. "Sustainable Agriculture and the Challenge of Place." *American Journal of Alternative Agriculture* 2: 184–187.

———. 1993. *Beginning Again: People and Nature in the New Millennium.* Oxford: Oxford University Press.

El Comercio (Lima newspaper). 1987. "Hay 6 mil variedades de maíz en el país." Lima.

Ellen, Roy. 1982. *Environment, Subsistence, and System.* Cambridge: Cambridge University Press.

Engel, Frederic. 1970. "Exploration of the Chilca Canyon, Peru." *Current Anthropology* 11 (1): 55–58.

Entrikin, J. Nicholas. 1991. *The Betweenness of Place.* Baltimore: The Johns Hopkins University Press.

Escobedo Mansilla, Renaldo. 1979. *El tributo indígena en el Perú (siglos XVI y XVII).* Pamplona, Spain: Ediciones Universidad de Navarra.

Evans, L. T. 1980. "The Natural History of Crop Yield." *American Scientist* 68: 388–397.

Fano, Hugo, and Marisela Benavides. 1992. *Los cultivos andinos en perspectiva.* Cuzco and Lima: Centro de Estudios Regionales Andinos and International Potato Center.

Figueroa, Adolfo. 1981. *La economía campesina de la sierra del Perú.* Lima: Pontificia Universidad Católica del Perú.

———. 1984. *Capitalist Development and the Peasant Economy in Peru.* Cambridge: Cambridge University Press.

Flores Galindo, Alberto. 1977. *Arequipa y el sur andino: ensayo de historia regional (siglos XVIII–XX).* Lima: Editorial Horizonte.

———. 1988. "Presentación." In *Comunidades campesinas: cambios y permanencias,* edited by Alberto Flores Galindo, 7–9. Lima: Centro de Estudios Sociales SOLIDARIDAD.

Flores Ochoa, Jorge A., and Juan V. Nuñez del Prado B. n.d. *Q'ero: El ultimo ayllu inka.* Cuzco: Centro de Estudios Andinos Cuzco.

Fonseca, César. 1988. "Diferenciación campesina en los Andes peruanos." In *Comunidad y producción en la agricultura andina,* edited by César Fonseca and Enrique Mayer, 165–196. Lima: FOMCIENCIAS.

Fonseca, César, and Enrique Mayer. 1988. *Comunidad y producción en la agricultura andina.* Lima: FOMCIENCIAS.

Fowler, Cary, and Pat Mooney. 1990. *Shattering: Food, Politics, and the Loss of Genetic Diversity.* Tucson: University of Arizona Press.

Frankel, O. H. 1974. "Genetic Conservation: Our Evolutionary Responsibility." *Genetics* 78: 53–65.

Frankel, O. H., and E. Bennett, eds. 1970. *Genetic Resources in Plants: Their Exploration and Conservation.* Oxford: Blackwell Scientific.

Frankel, O. H., and J. G. Hawkes, eds. 1975. *Crop Genetic Resources for Today and Tomorrow.* Cambridge: Cambridge University Press.

Franquemont, Christine R. 1988. "Chinchero Plant Categories: An Andean Logic of Observation." Ph.D diss., Cornell University, Ithaca, New York.

Fuenzalida, Fernando. 1970. "La estructura de la comunidad de indígenas tradicional." In *La hacienda, la comunidad y el campesino en el Perú,* edited by José Matos Mar, 61–104. Lima: Moncloa-Campodónico.

Gade, Daniel W. 1969. "Vanishing Crops of Traditional Agriculture: The Case of Tarwi (*Lupinus mutabilis*) in the Andes." *Proceedings of the Association of American Geographers* 1: 47–51.

———. 1970. "Ethnobotany of Cañihua (*Chenopodium pallidicaule*), Rustic Seed Crop of the Altiplano." *Economic Botany* 24 (1): 55–61.

———. 1972a. "Setting the Stage for Domestication: Brassica Weeds in Andean Peasant Ecology." *Proceedings of the Association of American Geographers* 4 (1): 38–41.

———. 1972b. "South American Lupine and the Process of Decline in the World Cultigen Inventory." *Journal d'Agriculture Tropicale et de Botanique Appliquée* 19 (4–5): 85–92.

———. 1975. *Plants, Man, and the Land in the Vilcanota Valley of Peru.* The Hague: Dr. W. Junk.

———. 1979. "Inca and Colonial Settlement, Coca Cultivation and Endemic Disease in the Tropical Forest." *Journal of Historical Geography* 5 (3): 263–279.

———. 1981. "Some Research Themes in the Cultural Geography of the Central Andean Highlands." In *L'homme et Son Environment á Haute Altitude,* edited by P. Baker and C. Jest, 123–128. Paris: C.N.R.S.

———. 1992. Review of *Lost Crops of the Incas: Little-Known Plants of the Andes with Promise for Worldwide Cultivation, by National Research Council. Mountain Research and Development* 12 (1): 97–98.

———. 1993. "Landscape, System, and Identity in the Post-Conquest Andes." *Annals of the Association of American Geographers* 82 (3): 460–477.

Gade, Daniel W., and Mario Escobar. 1982. "Village Settlement and the Colonial Legacy in Southern Peru." *Geographical Review* 72 (4): 430–449.

Garcilaso de la Vega, El Inca. [1609] 1987. *Royal Commentaries of the Incas and General History of Peru.* Translated by Harold V. Livermore. Reprint, Austin: University of Texas Press.

Geertz, Clifford. 1966. *Agricultural Involution: The Process of Ecological Change in Indonesia.* Berkeley: University of California Press.

Gillis, Anna Maria. 1993. "Keeping Traditions on the Menu." *BioScience* 43 (7): 425–429.

Glave, Luis Miguel. 1983. "Trajines: un capítulo en la formación del mercado interno colonial." *Revista Andina* 1 (1): 9–67.

Glave, Luis Miguel, and María Isabel Remy. 1980. "La producción de maíz en Ollantaytambo durante el siglo XVIII." *Allpanchis* 14: 109–132.

———. 1983. *Estructura agraria y vida rural en una región andina: Ollantaytambo entre los siglos XVI y XIX.* Cuzco: Centro de Estudios Rurales Andinos Bartolomé de las Casas.

Godoy, Ricardo A. 1984. "Ecological Degradation and Agricultural Intensification in the Andean Highlands." *Human Ecology* 12 (4): 359–383.

———. 1990. *Mining and Agriculture in Highland Bolivia: Ecology, History, and Commerce among the Jukumanis.* Tucson: The University of Arizona Press.

Goland, Carol. 1992. *Cultivating Diversity: Field Scattering as Agricultural Risk Management in Cuyo Cuyo, Department of Puno, Peru.* Working Paper, no. 4, Production, Storage, and Exchange in a Terraced Environment on the Eastern Andean Escarpment. Edited by Bruce Winterhalder. Chapel Hill, N.C.: University of North Carolina, Department of Anthropology.

Golte, Jürgen. 1980. *La racionalidad de la organización andina.* Lima: Instituto de Estudios Peruanos.

Gómez Molina, Eduardo, and Adrienne V. Little. 1981. "Geoecology of the Andes: The Natural Science Basis for Resource Planning." *Mountain Research and Development* 1 (2): 115–144.

Gómez-Pompa, Arturo, and Andrea Klaus. 1992. "Taming the Wilderness Myth." *BioScience* 42 (4): 271–279.

González Holguín, Diego. [1608] 1952. *Vocabulario de la lengua general de todo el Perú llamada lengua qquichua o del Inca.* Lima: Imprenta Santa María.

González de Olarte, Efraín. 1987. *Inflación y campesinado: comunidades y microregiones frente a la crisis.* Lima: Institution de Estudios Peruanos.

Gould, Stephen J. 1982. "Darwinism and the Expansion of Evolutionary Theory." *Science* 216: 380–387.

Gould, Stephen J., and Richard Lewontin. 1979. "The Spandrels of San Marco and the Panglossian Paradigm: A Critique of the Adaptationist Programme." *Proceedings of the Royal Society of London B, Biological Science* 205: 581–598.

Grobman, Alexander, Wilfredo Salhuana, and Ricardo Sevilla. 1961. *Races of Maize in Peru: Their Origins, Evolution, and Classification.* Washington, D.C.: National Academy of Sciences.

Grossman, Lawrence S. 1984. *Peasants, Subsistence Ecology, and Development in the Highlands of Papua New Guinea.* Princeton: Princeton University Press.

Grubb, Michael. 1993. *The Earth Summit Agreements.* London: Earthscan Publications.

Grumbine, R. Edward. 1992. *Ghost Bears: Exploring the Biodiversity Crisis.* Covelo, California: Island Press.

Guamán Poma de Ayala, Felipe. [1613] 1980. *El primer nueva crónica y buen gobierno,* 3 vols. Edited by John V. Murra and Rolena Adorno. Translated by Jorge L. Urioste. Reprint, Mexico: Siglo XXI.

Guillén Marroquín, Jesús. 1980. *El capital transnacional en la industria cervecera.* Cuzco: Centro de Estudios Rurales Andinos Bartolomé de las Casas.

———. 1989. *La economía agraria del Cusco, 1900–1980.* Cuzco: Centro de Estudios Rurales Andinos Bartolomé de las Casas.

Guillet, David W. 1979. *Agrarian Reform and Peasant Economy in Southern Peru.* Columbia: University of Missouri Press.

———. 1981a. "Land Tenure, Agricultural Regime, and Ecological Zone in the Central Andes." *American Ethnologist* 8 (1): 139–158.

———. 1981b. "Agrarian Ecology and Peasant Production in the Central Andes." *Mountain Research and Development* 1 (1): 19–28.

———. 1992. *Covering Ground: Communal Water Management and the State in the Peruvian Andes.* Ann Arbor: University of Michigan Press.

Gutiérrez, Ramón. 1984. *Notas sobre las haciendas del Cusco.* Lima: Fundación para la Educación, la Ciencia y la Cultura.

Harlan, J. R. 1951. "Anatomy of Gene Centers." *The American Naturalist* 85 (821): 97–103.

———. 1971. "Agricultural Origins: Centers and Noncenters." *Science* 174: 468–474.

———. 1972. "Genetics of Disaster." *Journal of Environmental Quality* 1 (3): 212–215.

———. 1975a. "Our Vanishing Genetic Resources." *Science* 188: 618–621.

———. 1975b. *Crops and Man.* Madison: American Society of Agronomy and Crop Science Society of America.

Harlan, J. R., and J. M. J. de Wet. 1971. "Toward a Rational Classification of Cultivated Plants." *Taxon* 20 (4): 509–517.

Harris, Olivia. 1985. "Ecological Duality and the Role of the Center: Northern Potosí." In *Andean Ecology and Civilization: An Interdisciplinary Perspective on Andean Ecological Complementarity,* edited by Shozo Masuda, Izumi Shimada, and Craig Morris, 311–335. Tokyo: University of Tokyo Press.

Harris, P. M. 1978. *The Potato Crop: The Scientific Basis for Improvement.* London: Chapman & Hall.

Hastorf, Christine A. 1990. "The Effect of the Inka State on Sausa Agricultural Production and Crop Consumption." *American Antiquity* 55 (2): 262–290.

Hawkes, J. G. 1941. *Potato Collecting Expeditions in Mexico and South America.* Cambridge: Imperial Bureau of Plant Breeding and Genetics.

———. 1944. *Potato Collecting Expeditions in Mexico and South America. II: Systematic Classification of the Collections.* Cambridge: Imperial Bureau of Plant Breeding and Genetics.

———. 1947. "On the Origin and Meaning of South American Potato Names." *Journal of the Linnaean Society, Botany* 53: 205–250.

———. 1978. "Biosystematics of the Potato." In *The Potato Crop: The Scientific Basis for Improvement,* edited by P. M. Harris, 15–69. London: Chapman & Hall.

———. 1983. *The Diversity of Crop Plants.* Cambridge: Harvard University Press.

———. 1990. *The Potato: Evolution, Biodiversity, and Genetic Resources.* Washington, D.C.: Smithsonian Institution Press.

Hawkes, J. G., and J. P. Hjerting. 1989. *The Potatoes of Bolivia: Their Breeding Value and Evolutionary Relationships.* Oxford: Clarendon Press.

Hecht, Susanna, and Alexander Cockburn. 1990. *The Fate of the Forest: Developers, Destroyers, and Defenders of the Amazon.* New York: Harper Collins.

Hobsbawm, Eric, and Terence Ranger. 1992. *The Invention of Tradition.* Cambridge: Cambridge University Press.

Hodge, W. H. 1951. "Three Native Tuber Food Plants of the High Andes." *Economic Botany* 5 (3): 185–201.

———. 1954. "The Edible Arracacha: A Little Known Root Crop of the Andes." *Economic Botany* 8 (3): 195–221.

Hopkins, Raúl. 1978. "La industria cervecera y la agricultura de cebada en el súr del Perú." Unpublished paper, Departamento de Economía, Pontificia Universidad Católica del Perú.

———. 1981. *Desarrollo desigual y crisis en la agricultura peruana, 1944–1969.* Lima: Instituto de Estudios Peruanos.

Horkheimer, Hans. [1960] 1990. *Alimentación y obtención de alimentos en los andes prehispánicos.* Translated by Ernesto More. Reprint, La Paz: HISBOL.

Horton, Douglas Earl. 1976. *Haciendas and Cooperatives: A Study of Estate Organization, Land Reform, and New Reform Enterprises in Peru.* Dissertation Series, no. 67. Ithaca: Cornell University Press.

———. 1987. *Potatoes: Production, Marketing, and Programs for Developing Countries.* Boulder: Westview Press.

Huamán, Zósimo. 1986. "Conservación de recursos genéticos de papa en el CIP." *Circular* (Centro Internacional de la Papa) 14 (2): 1–7.

Huamán, Z., J. T. Williams, W. Salhuana, and L. Vincent. 1977. *Descriptors for the Cultivated Potato.* Rome: International Board for Plant Genetic Resources.

von Humboldt, Alexander, and Aimé Bonpland. [1802] 1959. *Essai sur la geographie des plantes.* Reprint, London: Society for the Bibliography of Natural History.

Iltis, Hugh H. 1983. "From Teosinte to Maize: The Catastrophic Sexual Transmutation." *Science* 222: 886–894.

Instituto Nacional de Estadística (INE). 1983. *Censos nacionales: VIII de Población—III de Vivienda, 12 de julio de 1981.* Lima: Imprenta Nacional.

Ingold, Tim. 1992. "Culture and the Perception of the Environment." In *Bush Base: Forest Farm: Culture, Environment, and Development,* edited by Elisabeth Croll and David Parkin, 39–57. London: Routledge.

International Bureau for Plant Taxonomy and Nomenclature. 1980. *International Code of Nomenclature for Cultivated Plants—1980.* The Hague: W. Junk.

Isbell, Billie Jean. 1978. *To Defend Ourselves: Ecology and Ritual in an Andean Village.* Austin: University of Texas Press.

Ives, Jack D., and Bruno Messerli. 1989. *The Himalayan Dilemma: Reconciling Development and Conservation.* London: Routledge.

Jackson, M. T., J. G. Hawkes, and P. R. Rowe. 1980. "An Ethnobotanical Field Study of Primitive Potato Varieties in Peru." *Euphytica* 29 (1): 107–113.

Jackson, Wes. 1984. "A Search for the Unifying Concept for Sustainable Agriculture." In *Meeting the Expectations of the Land,* edited by Wes Jackson, Wendell Berry, and Bruce Coleman, 208–230. San Francisco: North Point Press.

de Janvry, Alain. 1981. *The Agrarian Question and Reformism in Latin America.* Baltimore: The Johns Hopkins University Press.

de Janvry, Alain, Elisabeth Sadoulet, and Linda Wilcox Young. 1989. "Land and Labour in Latin American Agriculture from the 1950s to the 1980s." *Journal of Peasant Studies* 16 (3): 396–424.

Jennings, Bruce H. 1988. *Foundations of International Agricultural Research: Science and Politics in Mexican Agriculture.* Boulder: Westview Press.

Johannessen, Carl L. 1963. *Savannas of Interior Honduras.* Ibero-Americana 46. Berkeley: University of California Press.

———. 1970. "The Dispersal of Musa in Central America: The Domestication Process in Action." *Geographical Review* 60 (4): 689–699.

Johannessen, Carl L., Michael R. Wilson, and William A. Davenport. 1970. "The Domestication of Maize: Process or Event?" *Geographical Review* 60 (3): 393–413.

Johns, Timothy, and Susan L. Keen. 1986. "Ongoing Evolution of the Potato on the Altiplano of Western Bolivia." *Economic Botany* 40 (4): 409–424.

Johnson, Allan W. 1972. "Individuality and Experimentation in Traditional Agriculture." *Human Ecology* 1 (2): 149–160.

Johnson, A. M. 1976. "The Climate of Peru, Bolivia, and Ecuador." In *Climates of Central and South America,* edited by Werner Schwerdtfeger, 147–218. Amsterdam: Elsevier.

Juan, Jorge, and Antonio de Ulloa. [1772] 1975. *A Voyage to South America.* Translated by John Adams. Reprint, Tucson, Arizona: Center for Latin American Studies, University of Arizona Press.

Kaplan, Lawrence. 1980. "Variation in the Cultivated Beans." In *Guitarrero Cave: Early Man in the Andes,* edited by Thomas F. Lynch, 145–148. New York: Academic Press.

Kelly, Kenneth. 1965. "Land-Use Regions in the Central and Northern Portions of the Inca Empire." *Annals of the Association of American Geographers* 55 (2): 327–338.

Kervyn, Bruno. 1989. "Campesinos y acción colectiva: la organización del espacio en comunidades de la sierra sur del Perú." *Revista Andina* 7 (1): 7–60.

Kloppenburg, Jack Jr. 1988. *First the Seed: The Political Economy of Plant Biotechnology, 1492–2000.* Cambridge: Cambridge University Press.

———. 1991. "Social Theory and the De/Reconstruction of Agricultural Science: Local Knowledge for an Alternative Agriculture." *Rural Sociology* 56 (4): 519–548.

Kloppenburg, Jack Jr., and Daniel L. Kleinman. 1987. "The Plant Germplasm Controversy: Analyzing Empirically the Distribution of the World's Plant Genetic Resources." *BioScience* 37 (3): 190–198.

Kloppenburg, Jack Jr., and Tirso Gonzales. 1994. "Between State and Capital: NGOs as Allies of Indigenous Peoples." In *Intellectual Property Rights for Indigenous Peoples, A Sourcebook,* edited by Tom Greaves, 163–178. Oklahoma City: Society for Applied Anthropology.

Knapp, Gregory. 1991. *Andean Ecology: Adaptive Dynamics in Ecuador.* Boulder, Colorado: Westview Press.

La Barre, Weston. 1947. "Potato Taxonomy among the Aymara Indians of Bolivia." *Acta Americana* 5: 83–103.

La Dirección y Promoción de la Reforma Agraria. n.d. *Estatuto Especial de Comunidades Campesinas: Texto intégro y comentarios.* Lima: La Dirección y Promoción de la Reforma Agraria.

La Prensa. 21 August 1970. Texto de la Ley de la Reforma Agraria (Decreto Supremo 17716). Lima: La Prensa.

Larmie, Anne C. 1993. *Work, Reproduction, and Health in Two Andean Communities (Department of Puno, Peru).* Working Paper, no. 5. Production, Storage, and Exchange in a Terraced Environment on the Eastern Andean Escarpment. Edited by Bruce Winterhalder. Chapel Hill, N.C.: University of North Carolina, Department of Anthropology.

Larson, Brooke. 1988. *Colonialism and Agrarian Transformation in Bolivia: Cochabamba, 1550–1900.* Princeton: Princeton University Press.

Lehman, David. 1982a. *Ecology and Exchange in the Andes.* Cambridge: Cambridge University Press.

———. 1982b. "Beyond Lenin and Chayanov: New Paths of Agrarian Capitalism." *Journal of Development Economics* 11: 133–161.

Le Moine, Genevieve, and J. Scott Raymond. 1987. "Leishmaniasis and Inca Settlement in the Peruvian Jungle." *Journal of Historical Geography* 13 (2): 113–129.

León, Jorge. 1964. *Plantas alimenticias andinas.* Lima: Instituto Interamericano de Ciencias Agrícolas Zona Andina.

Leopold, Aldo. 1949. *A Sand County Almanac.* Oxford: Oxford University Press.

Levillier, Roberto, ed. 1925. *Gobernantes del Perú: Cartas y papeles siglo XVI.* Madrid: Imprenta de Juan Pueyo.

Lewis, Martin W. 1992. *Wagering the Land: Ritual, Capital, and Environmental Degradation in the Cordillera of Northern Luzon, 1900–1986.* Berkeley: University of California Press.

———. 1993. *Green Delusions: An Environmentalist Critique of Radical Environmentalism.* Durham: Duke University Press.

Lira, Jorge. 1945. *Diccionario kechuwa-español.* Tucumán: Universidad Nacional de Tucumán.

Llona, S. E. 1904. *Mapa histórico geográfico de los valles de Paucartambo* (map scale 1: 200,000). Lima: Sociedad Geográfica de Lima.

Loucks, Orie L. 1977. "Emergence of Research on Agro-Ecosystems." *Annual Review of Ecology and Systematics* 8: 173–192.

Lyon, Patricia J. 1984. "The Attackers or the Attacked? The Invention of 'Hostile Savages' in the Valleys of Paucartambo, Cuzco, Peru." Unpublished manuscript, Institute of Andean Studies, Berkeley, California.

Lynch, Thomas F., ed. 1980. *Guitarrero Cave: Early Man in the Andes.* New York: Academic Press.

Lynch, Thomas F., R. Gillespie, John A. J. Gowlett, and R. E. M. Hedges. 1985. "Chronology of Guitarrero Cave, Peru." *Science* 229: 864–867.

MacBride, J. Francis. 1936. *Flora of Peru.* Vol. 13, pt. 2. Chicago: Field Museum of Natural History.

Macera, Pablo. 1968. *Mapas coloniales de haciendas cuzqueñas.* Lima: Universidad. Nacional Mayor de San Marcos, Seminario de Historia Rural Andina.

MacNeish, Richard S. 1977. "The Beginnings of Agriculture in Central Peru." In

Origins of Agriculture, edited by Charles A. Reed, 753–802. The Hague: Mouton Publishers.

———. 1992. *The Origins of Agriculture and Settled Life.* Norman: University of Oklahoma Press.

Málaga Medina, Alejandro. 1974. "Las reducciones en el Perú." *Historia y Cultura* (Lima) 8: 141–172.

Mallon, Florencia E. 1983. *The Defense of Community in Peru's Central Highlands: Peasant Struggle and Capitalist Transition, 1860–1940.* Princeton: Princeton University Press.

Manglesdorf, Paul C. 1974. *Corn: Its Origin, Evolution, and Improvement.* Cambridge: Harvard University Press.

Martínez Alier, Juan. 1973. *Los huacchilleros del Perú: Dos estudios de formaciones sociales agrarios.* Lima: Instituto de Etudios Peruanos.

———. 1992. "Ecology and the Poor: A Neglected Dimension of Latin American History." *Journal of Latin American Studies* 23: 621–639.

Marx, Karl. [1867] 1987. *Capital: A Critique of Political Economy.* Vol. 1. Translated by Samuel Moore and Edward Aveling. Reprint, New York: International Publishers.

Maurtúa, Víctor M., ed. 1906. *Juicio de límites entre Perú y Bolivia.* Barcelona: Imprenta Heinrich.

Mayer, Enrique. 1977. "Beyond the Nuclear Family." In *Andean Kinship and Marriage,* edited by Ralph Bolton and Enrique Mayer, 60–80. Washington, D.C.: American Anthropological Association.

———. 1979. *Land Use in the Andes: Ecology and Agriculture in the Mantaro Valley of Peru with Special Reference to Potatoes.* Lima: Centro Internacional de la Papa.

———. 1985. "Production Zones." In *Andean Ecology and Civilization,* edited by Shozo Masuda, Izumi Shimada, and C. Morris, 45–84. Tokyo: University of Tokyo Press.

———. 1988. "De hacienda a comunidad: el impacto de la reforma agraria en la provincia de Paucartambo, Cusco." In *Sociedad andina: pasado y presente: contribuciones en homenaje a la memoria de César Fonseca Martel,* edited by Ramiro Matos Mendieta, 60–99. Lima: FOMCIENCIAS.

Mayer, Enrique, and Marisol de la Cadena. 1989. *Cooperación y conflicto en la comunidad andina: zonas de producción y organización social.* Lima: Instituto de Estudios Peruanos.

Mayer, Enrique, and César Fonseca. 1979. *Sistemas agrarios en la cuenca del río Cañete.* Lima: Impresos ONERN.

Mayer, Enrique, and Manuel Glave. 1990. "Papas regaladas y papas regalo: rentabilidad, costos en inversión." In *Perú: el problema agrario en debate,* edited by Alberto Chirif, Nelson Manrique, and Benjamín Quijandría, 87–101. Lima: SEPIA.

———. 1992. "Rentabilidad en la producción compesina de papas." In *La chacra de papa: economía y ecología,* edited by Enrique Mayer, 29–188. Lima: CEPES.

Mejía Xesspe, M. T. [1931] 1978. "Kausay: alimentación de los Indios." In *Tecnología andina,* edited by Rogger Ravines, 205–226. Reprint, Lima: Instituto de Estudios Peruanos.

Milstead, Harley P. 1928. "Distribution of Crops in Peru." *Economic Geography* 4 (1): 88–106.

Mishkin, Bernard. 1947. "The Contemporary Quechua." In *Handbook of South American Indians.* Vol. 2. Edited by J. Steward, 411–470. Washington, D.C.: Smithsonian Institution Bureau of American Ethnology.

Mitchell, William P. 1980. "Local Ecology and the State: Implications of Contemporary Quechua Land Use for Inca Sequence of Agricultural Work." In *Beyond the Myths of Culture: Essays in Cultural Materialism,* edited by Eric B. Ross, 139–154. New York: Academic Press.

———. 1991. *Peasants on the Edge: Crop, Cult, and Crisis in the Andes.* Austin: University of Texas Press.

Montes de Oca, Ismael. 1989. *Geografía y recursos naturales de Bolivia.* La Paz: Editorial Educacional.

Mörner, Magnus. 1967. *Race Mixture in the History of Latin America.* Boston: Little Brown.

———. 1978. *Perfil de la sociedad rural del Cuzco a fines de la colonia.* Lima: Universidad del Pacífico.

———. 1985. *The Andean Past: Land, Societies, and Conflicts.* New York: Columbia University Press.

———. 1990. "Alcances y límites del cambio estructural: Cusco, Perú, 1895–1920." In *Perú: el problema agrario en debate,* edited by Alberto Chirif, Nelson Manrique, and Benjamín Quijandría, 137–156. Lima: SEPIA.

———. N.d. "Distribution of Property and Income in a Small Andean Town in the Early Nineteenth Century."

Morris, Craig. 1981. "Tecnología organización inca del almacenamiento de viveres en la sierra." In *La tecnología en el mundo andino,* edited by Heather Lechtman and Ana María Soldi, 327–376. Mexico: Universidad Nacional Autónoma de México.

Murra, John V. 1960. "Rite and Crop in the Inca State." In *Culture in History: Essays in Honor of Paul Radis,* edited by Stanley Diamond, 393–407. New York: Columbia University Press.

———. 1964. "Una apreción etnológica de la visita." In *Visita hecha a la provincia de Chucuito por Garci Diez de San Miguel en el año 1567,* edited by Waldemar Espinoza Soriano, 421–442. Lima: La Casa de la Cultura del Perú.

———. 1972. "El 'control vertical' de un máximo de pisos ecológicos en la economía de las sociedades andinas." In *Visita de la provincia de León de Huánuco (1562).* Vol. 2. Huánuco: Universidad Hermilio Valdizán.

———. 1980. *The Economic Organization of the Inka State.* Greenwich, Connecticut: Jai Press.

———. 1985a. "'El Archipiélago Vertical' Revisited." In *Andean Ecology and Civilization: An Interdisciplinary Perspective on Andean Ecological Complementarity,* edited by Shozo Masuda, Izumi Shimada, and Craig Morris, 3–13. Tokyo: University of Tokyo Press.

———. 1985b. "The Limits and Limitations of the 'Vertical Archipelago' in the Andes." In *Andean Ecology and Civilization: An Interdisciplinary Perspective on Andean Ecological Complementarity,* edited by Shozo Masuda, Izumi Shimada, and Craig Morris, 15–20. Tokyo: University of Tokyo Press.

Nabhan, Gary Paul. 1985. "Native Crop Diversity in Aridoamerica: Conservation of Regional Gene Pools." *Economic Botany* 39 (4): 387–399.

———. 1989. *Enduring Seeds: Native American Agriculture and Wild Plant Conservation.* San Francisco: North Point Press.

———. 1992. "Epilogue: Native Crops of the Americas: Passing Novelties or Lasting Contributions to Diversity?" In *Chiles to Chocolate: Food the Americas Gave the World,* edited by Nelson Foster and Linda S. Cordell, 143–162. Tucson: University of Arizona Press.

National Academy of Sciences (NAS). 1972. *Genetic Vulnerability of Major Crops.* Washington, D.C.: National Academy of Sciences.

National Research Council (NRC). 1989. *Lost Crops of the Incas: Little-Known Plants of the Andes with Promise for Worldwide Cultivation.* Washington, D.C.: National Academy Press.

———. 1991. *Managing Global Genetic Resources: The U.S. National Plant Germplasm System.* Washington, D.C.: National Academy Press.

Netting, Robert McC. 1974. "Agrarian Ecology." *Annual Review of Anthropology* 3: 21–56.

———. 1993. *Smallholders, Householders: Farm Families and the Ecology of Intensive Sustainable Agriculture.* Palo Alto: Stanford University Press.

Nietschmann, Bernard Q. 1973. *Between Land and Water: The Subsistence Ecology of the Miskito Indians, Eastern Nicaragua.* New York: Seminar Press.

Ochoa, Carlos M. 1975. "Potato Collecting Expeditions in Chile, Bolivia and Peru, and the Genetic Erosion of Indigenous Cultivars." In *Crop Genetic Resources for Today and Tomorrow,* edited by O. H. Frankel and J. G. Hawkes, 167–173. Cambridge: Cambridge University Press.

———. 1990. *The Potatoes of South America: Bolivia.* Translated by Donald Ugent. Cambridge: Cambridge University Press.

Oldfield, Margery L., and Janis B. Alcorn. 1991. "Conservation of Traditional Agroecosystems." In *Biodiversity: Culture, Conservation, and Ecodevelopment.* Ed. Margery L. Oldfield and Janis B. Alcorn, 37–58. Boulder: Westview Press.

Oficina Nacional de Estadística y Censos (ONEC). 1975. *Censos Nacionales. VII de Población—II de Vivienda, 4 de junio de 1972.* Tomo I. Lima: Oficina Nacional de Estadística y Censos.

Oficina Nacional de Evaluación de Recursos Naturales (ONERN). 1976. *Mapa Ecológico del Perú: guía explicativa.* Lima: ONERN.

Orlove, Benjamin S. 1977a. *Alpacas, Sheep, and Men: The Wool Export Economy and Regional Society in Southern Peru.* New York: Academic Press.

———. 1977b. "Integration through Production: The Use of Zonation in Espinar." *American Ethnologist* 4: 84–101.

———. 1979. "The Breaking of Patron-Client Ties: The Case of Surimana in Southern Peru." *Nova Americana* 2: 83–104.

———. 1980. "Ecological Anthropology." *Annual Review of Anthropology* 9: 235–273.

Orlove, Benjamin S., and Glynn Custred. 1980. *Land and Power in Latin America: Agrarian Economies and Social Processes in the Andes.* New York: Holmes & Meier Publishing.

Orlove, Benjamin S., and Ricardo Godoy. 1986. "Sectoral Fallowing Systems in the Central Andes." *Journal of Ethnobiology* 6 (1): 169–204.

Orwell, George. 1958. *The Road to Wigen Pier.* New York: Harcourt Brace Jovanovich.

Palacio Pimental, H. Gustavo. 1957a. "Relaciones de trabajo entre el patrón y los colonos en los fundos de la provincia de Paucartambo." *Revista Universitaria* (Cuzco) 112: 173–221.

———. 1957b. "Relaciones de trabajo entre el patrón y los colonos en los fundos de la provincia de Paucartambo." *Revista Universitaria* (Cuzco) 113: 45–72.

Parker, Eugene. 1992. "Forest Islands and the Kayopó Resource Management in Amazonia: A Reappraisal of the *Apêtê.*" *American Anthropologist* 94 (2): 406–428.

———. 1993. "Fact and Fiction in Amazonia: The Case of the *Apêtê.*" *American Anthropologist* 95 (3): 715–723.

Parsons, James J. 1971. "Ecological Problems and Approaches in Latin American Geography." In *Geographic Research on Latin America, Benchmark 1970,* edited by B. Lenteck, R. L. Carmin, and T. L. Martinson, 13–32. Muncie, Indiana: Ball State University.

Patrón, Pablo. 1902. "La papa en el Perú primitivo." *Boletín de la Sociedad Geográfica de Lima* (11): 316–324.

Pearsall, Deborah. 1978. *Paleoethnobotany in Western South America: Progress and Problems.* University of Michigan Anthropological Papers, no. 67. Ann Arbor: University of Michigan Press.

———. 1992. "The Origins of Plant Cultivation in South America." In *The Origins of Agriculture,* edited by C. Wesly Cowan and Patty Jo Watson, 173–205. Washington, D.C.: Smithsonian Institution Press.

Peluso, Nancy. 1992. *Rich Forests, Poor People: Resource Control and Resistance in Java.* Berkeley: University of California Press.

Peralta Ruiz, Víctor. 1990. "El estado repúblicano y los campesinos cusqueños en los inicios de la República (1821–1854)." In *Perú: el problema agrario en debate,* edited by Alberto Chirif, Nelson Manrique, and Benjamín Quijandría, 157–170. Lima: SEPIA.

Pickersgill, Barbara. 1971. "Relationships Between Weedy and Cultivated Forms in Some Species of Chili Peppers (Genus *Capsicum*)." *Evolution* 25 (4): 683–691.

Pickersgill, Barbara, and Charles B. Heiser, Jr. 1978. "Origins and Distribution of Plants Domesticated in the New World Tropics." In *Origins of Agriculture,* edited by Charles A. Reed, 803–835. The Hague: Mouton.

Platt, Tristan. 1982. "The Role of the Andean *ayllu* in the Reproduction of the Petty Commodity Regime in Northern Potosí (Bolivia)." In *Ecology and Exchange in the Andes,* edited by David Lehman, 27–69. Cambridge: Cambridge University Press.

———. 1986. "Mirrors and Maize: The Concept of *yanantin* Among the Macha of Bolivia." In *Anthropological History of Andean Polities,* ed. J. V. Murra, N. Wachtel, and J. Revel, 228–259. Cambridge: Cambridge University Press.

Plucknett, D. L., N. J. H. Smith, J. T. Williams, and N. Murthi Anishetty. 1983. "Crop Germplasm Conservation and Developing Countries." *Science* 220: 163–169.

———. 1987. *Gene Banks and the World's Food.* Princeton, N.J.: Princeton University Press.

Polanyi, Karl. 1957. *The Great Transformation.* Boston: Beacon Press.

Polo de Ondegardo, Juan. [1561] 1940. "Informe del Licenciado Juan Polo de Ondegardo al Licenciado Briviesca de Muñatones sobre la perpetuidad de las encomiendas en el Perú." Reprint, *Revista Histórica* (Lima) 40: 125–196.

Poole, Deborah. 1992. "Figueroa Aznar and the Cusco Indigenistas: Photography and Modernism in Early Twentieth-Century Peru." *Representations* 38: 39–75.
Poole, Deborah, and Gerardo Rénique. 1992. *Peru: Time of Fear.* New York: Latin American Bureau.
Posey, Darrell Addison. 1983. "Indigenous Ecological Knowledge and Development of the Amazon." In *The Dilemma of Amazonian Development,* edited by Emilio Moran, 225–257. Boulder: Westview Press.
———. 1992a. "Reply to Parker." *American Anthropologist* 94 (2): 441–443.
———. 1992b. "Interpreting and Applying the 'Reality' of Indigenous Concepts: What is Necessary to Learn from the Natives?" In *Conservation of Neotropical Forests: Working from Traditional Resource Use,* edited by Kent H. Redford and Christine Padoch, 21–34. New York: Columbia University Press.
Pred, Allan. 1984. "Place as Historically Contingent Process: Structuration and the Time-Geography of Becoming Places." *Annals of the Association of American Geographers* 74 (2): 279–297.
Prescott-Allen, R., and C. Prescott-Allen. 1982. "The Case for *in situ* Conservation of Crop Genetic Resources." *Nature and Resources* 23: 15–20.
Pulgar Vidal, Javier. 1946. *Historia y geografía del Perú: las ocho regiones naturales del Perú.* Lima: Universidad Nacional Mayor de San Marcos.
Querol, Daniel. 1993. *Genetic Resources: A Practical Guide to Their Conservation.* London: Zed Books.
Quiros, C. F., S. B. Brush, D. S. Douches, G. Hewstes, and K. S. Zimmerer. 1990. "Biochemical and Folk Assessment of Variability of Andean Cultivated Potatoes." *Economic Botany* 44 (2): 254–266.
Radcliffe, Sarah A. 1986. "Gender Relations, Peasant Livelihood Strategies and Migration: A Case Study from Cuzco, Peru." *Bulletin of Latin American Research* 5 (2): 29–47.
Raimondi, Antonio. [1874–1913] 1965. *El Perú.* 3 vols. Lima: Imprenta del Estado.
Ravines, Rogger, ed. 1978. *Tecnología andina.* Lima: Instituto de Estudios Peruanos.
Redford, Kent H. 1990. "The Ecologically Noble Savage." *Orion Nature Quarterly* 9: 25–29.
———. 1992. "The Empty Forest." *BioScience* 42 (6): 412–422.
Reinhardt, Nola. 1988. *Our Daily Bread: The Peasant Question and Family Farming in the Colombian Andes.* Berkeley: University of California Press.
Renard-Casevitz, F. M., T. Saignes, and A. C. Taylor. 1988. *Al este de los Andes: relaciones entre las sociedades amazónicas y andinas entre los siglos XV y XVII.* Quito: IFEA.
Rengifo, Grimaldo. 1988. *Recursos fitogenéticos andinos.* Cajamarca: Proyecto Piloto de Ecosistemas Andinos.
Repo-Carrasco, Ritva. 1988. *Cultivos andinos.* Cuzco: Centro de Estudios Rurales Bartolomé de las Casas.
Rhoades, Robert E. 1982. "The Incredible Potato." *National Geographic Magazine* (May): 668–694.
———. 1990. "Potatoes: Genetic Resources and Farmer Strategies. Comparison of the Peruvian Andes and Nepali Himalayas." In *Mountain Agriculture and Crop Genetic Resources,* edited by K. W. Rily, N. Mateo, G. C. Hawtin, and R. Yadav, 293–304. New Delhi: Oxford Publishing.

Rhoades, Robert, and Anthony J. Bebbington. 1990. "Mixing It Up: Variations in Andean Farmers' Rationales for Intercropping of Potatoes." *Field Crops Research* 25: 1–12.

Rhoades, Robert, and Stephen I. Thompson. 1975. "Adaptive Strategies in Alpine Environments: Beyond Ecological Particularism." *American Ethnologist* 2: 535–551.

Richards, Paul. 1985. *Indigenous Agricultural Revolution: Ecology and Food Production in West Africa.* London: Unwin Hyman.

———. 1986. *Coping with Hunger: Hazard and Experiment in an African Rice-Farming System.* London: Allen & Unwin.

Roseberry, William. 1993. "Beyond the Agrarian Question in Latin America." In *Confronting Historical Paradigms: Peasants, Labor, and the Capitalist World System in Africa and Latin America,* edited by Frederick Cooper, Allen F. Isaacman, Florencia E. Mallon, William Roseberry, and Steve J. Stern, 318–367. Madison: The University of Wisconsin Press.

Rowe, John Howland. 1945. "Absolute Chronology in the Andean Area." *American Antiquity* 10 (3): 265–284.

———. 1947a. "Inca Culture at the Time of the Spanish Conquest." In *Handbook of South American Indians.* Vol. 2. Edited by J. Steward, 198–330. Washington, D.C.: Smithsonian Institution Bureau of American Ethnology.

———. 1947b. "The Distribution of Indians and Indian Language in Peru." *Geographical Review* 37 (2): 205–215.

———. 1957. "The Incas under Spanish Colonial Institutions." *Hispanic American Historical Review* 37 (2): 155–199.

Ruiz, Hipólito, and Joseph Pavón. [1794] 1965. *Florae peruvianae et chilensis.* Reprint, New York: Stechert-Hafner.

Sack, Robert David. 1980. *Conceptions of Space in Social Thought: A Geographic Perspective.* Minneapolis: University of Minnesota Press.

———. 1986. *Human Territoriality: Its Theory and History.* Cambridge: Cambridge University Press.

Salaman, Redcliffe N. 1985. *The History and Social Influence of the Potato.* Cambridge: Cambridge University Press.

Salis, Annette. 1985. *Cultivos andinos: alternativa alimentaria popular?* Cuzco: Centro de Estudios Rurales Bartolomé de las Casas.

Salomon, Frank L. 1985. "The Dynamic Potential of the Complementarity Concept." In *Andean Ecology and Civilization: An Interdisciplinary Perspective on Andean Ecological Complementarity,* edited by Shozo Masuda, Izumi Shimada, and Craig Morris, 511–531. Tokyo: University of Tokyo Press.

———. 1986a. *Native Lords of Quito in the Age of the Incas: The Political Economy of North Andean Chiefdoms.* Cambridge: Cambridge University Press.

———. 1986b. "Vertical Politics on the Inka Frontier." In *Anthropological History of Andean Polities,* edited by John V. Murra, Nathan Wachtel, and Jacques Revel, 89–117. Cambridge: Cambridge University Press.

Sauer, Carl O. 1938. "Theme of Plant and Animal Destruction in Economic History." *Journal of Farm Economics* 20 (4): 765–775.

———. 1950. "Cultivated Plants of South and Central America." In *Handbook of South American Indians.* Vol. 6. Edited by J. Steward, 487–543. Washington, D.C.: Smithsonian Institution Bureau of American Ethnology.

———. 1952. *Agricultural Origins and Dispersals.* New York: American Geographical Society.

———. 1956. "The Agency of Man on the Earth." In *Man's Role in Changing the Face of the Earth,* edited by William L. Thomas, Jr., 49–69. Chicago: University of Chicago Press.

———. 1958. "Man in the Ecology of Tropical America." *Proceedings of the Ninth Pacific Science Congress, Bangkok* (1957) 20: 104–110. Reprinted in 1967 in *Land and Life: A Selection from the Writings of Carl Ortwin Sauer,* edited by John Leighly, 182–193. Berkeley: University of California Press.

Sauer, Jonathan D. 1967. "The Grain Amaranths and Their Relatives: A Revised Taxonomic and Geographic Survey." *Annals of the Missouri Botanical Garden* 54 (2): 103–137.

———. 1993. *Historical Geography of Crop Plants: A Select Roster.* Boca Raton: CRC Press.

Sauer, Jonathan D., and Lawrence Kaplan. 1969. "*Canavalia* Beans in American Prehistory." *American Antiquity* 34 (4): 417–424.

Scott, James C. 1976. *The Moral Economy of the Peasant: Rebellion and Subsistence in Southeast Asia.* New Haven: Yale University Press.

———. 1985. *Weapons of the Weak: Everyday Forms of Peasant Resistance.* New Haven: Yale University Press.

Sheridan, Thomas E. 1988. *Where the Dove Calls: The Political Ecology of a Peasant Corporate Community in Northwestern Mexico.* Tucson: University of Arizona Press.

Skeldon, Ronald. 1985. "Circulation: A Transition in Mobility in Peru." In *Circulation in Third World Countries,* edited by R. Mansell Prothero and Murray Chapman, 100–119. New York: Routledge & Kegan Paul.

Smith, C. Earle. 1980. "Plant Remains from Guitarrero Cave." In *Guitarrero Cave: Early Man in the Andes,* edited by Thomas F. Lynch, 87–120. New York: Academic Press.

Soulé, M. E., and B. A. Wilcox, eds. 1980. *Conservation Biology: An Evolutionary-Ecological Perspective.* Sundeland, Massachusetts: Sinauer Associates.

Spalding, Karen. 1974. *De indio a campesino.* Lima: Instituto de Estudios Peruanos.

———. 1984. *Huarochirí: An Andean Society under Inca and Spanish Rule.* Stanford: Stanford University Press.

Stern, Steve J. 1982. *Peru's Indian Peoples and the Challenge of Spanish Conquest: Huamanga to 1640.* Madison: University of Wisconsin Press.

———, ed. 1987. *Resistance, Rebellion, and Consciousness in the Andean Peasant World: 18th to 20th Centuries.* Madison: University of Wisconsin Press.

Stevens, Stanley F. 1993. *Claiming the High Ground: Sherpas, Subsistence, and Environmental Change in the Highest Himalaya.* Berkeley: University of California Press.

Tapia, M., and N. Mateo. 1992. "Andean Phytogenetic and Zoogenetic Resources." In *Mountain Agriculture and Crop Genetic Resources,* edited by K. W. Rily, N. Mateo, G. C. Hawtin, and R. Yadav, 235–254. New Delhi: Oxford Publishing.

Thompson, E. P. 1963. *The Making of the English Working Class.* New York: Vintage Books.

Thoreau, Henry D. 1993. *Faith in a Seed.* Covelo, California: Island Press.

Thorp, Rosemary, and Geoffrey Bertram. 1978. *Peru 1890–1977: Growth and Policy in an Open Economy.* New York: Columbia University Press.

Thurner, Mark. 1993. "Peasant Politics and Andean Haciendas in the Transition to Capitalism: An Ethnographic History." *Latin American Research Review* 28 (3): 41–82.

Tivy, Joy. 1990. *Agricultural Ecology.* London: Longman.

Todaro, Michael P. 1989. *Economic Development in the Third World.* 4th ed. New York: Longman.

Tosi, Joseph A. 1960. *Zonas de vida natural en el Perú: memoria explicativa sobre el mapa ecológico del Perú.* Lima: Instituto Interamericano de Ciencias Agricolas de la OEA, Zona Andina.

Towle, Margaret A. 1961. *The Ethnobotany of Pre-Columbian Peru.* Washington, D.C.: Werner-Gren Foundation.

Treacy, John M. 1994. *Las chacras de Coporaque: andenería y riego en el valle del Colca.* Lima: Instituto de Estudios Peruanos.

Troll, Carl. 1958. *Las culturas superiores andinas y el medio geográfico.* Lima: Instituto de Geografía, Universidad Nacional Mayor de San Marcos.

———. 1968. "The Cordilleras of the Tropical Americas: Aspects of Climatic, Phytogeographical and Agrarian Ecology." In *Geo-ecology of the Mountainous Regions of the Tropical Americas.* Proceedings of the UNESCO (Mexico) Symposium, 1966, 15–56. Bonn: Ferd Dummlers Verlag.

Tuan, Yi-Fu. 1974. *Topophilia: A Study of Environmental Perception, Attitudes, and Values.* New York: Prentice Hall.

———. 1977. *Space and Place: The Perspectives of Experience.* Minneapolis: University of Minnesota.

———. 1993. *Passing Strange and Wonderful: Aesthetics, Nature, and Culture.* Washington, D.C.: Shearwater Books.

Turner, B. L. II, William C. Clark, Robert W. Kates, John F. Richards, Jessica T. Matthews, and William B. Meyer. 1990. *The Earth as Transformed by Human Action.* Cambridge: Cambridge University Press.

Ugent, Donald. 1970. "The Potato." *Science* 170: 1161–1166.

Urton, Gary. 1981. *At the Crossroads of the Earth and Sky: An Andean Cosmology.* Austin: University of Texas Press.

———. 1984. "Chuta: el espacio de la práctica social en Paqariqtambo." *Revista Andina* 2 (1): 7–44.

van den Berghe, Pierre L., and George P. Primov. 1977. *Inequality in the Peruvian Andes: Class and Ethnicity in Cuzco.* Columbia: University of Missouri Press.

Vargas C., César. 1948. *Las papas sudperuanas. Parte I.* Cuzco: Universidad Nacional de San Antonio de Abad del Cusco.

———. 1954. *Las papas sudperuanas. Parte II.* Cuzco: Universidad Nacional de San Antonio de Abad del Cusco.

Vavilov, N. I. 1949–50. "The Origin, Variation, Immunity, and Breeding of Cultivated Plants." Translated by K. Starr Chester. *Chronica Botanica* 13 (1/6): 1–364.

———. 1957. *World Resources of Cereals, Leguminous Seed Crops and Flax, and Their Utilization in Plant Breeding.* Moscow: The Academy of Sciences of the USSR.

Vickers, William T. 1983. "Tropical Forest Mimicry in Swiddens: A Reassessment of Geertz's Model with Amazonian Data." *Human Ecology* 11 (1): 13–34.

Vietmeyer, Noel D. 1981. "Rediscovering America's Forgotten Crops." *National Geographic* 159 (5): 702–712.
———. 1985. "The Lost Crops of the Incas." *Ceres* 99: 37–40.
———. 1986. "Lesser-Known Plants of Potential Use in Agriculture and Forestry." *Science* 232: 1379–1384.
Villanueva Urteaga, Horacio. [1693] 1982. *Cusco 1689: informes de los párrocos al obispo Mollinedo.* Reprint, Cuzco: Centro de Estudios Rurales Andinos Bartolomé de las Casas.
Villasante Ortiz, Segundo. 1952. "Expediciones realizadas a la provincia de Paucartambo." *Revista Universitaria* 102: 126–150.
———. 1955. "Apuntes para un estudio fitogeográfico de la provincia de Paucartambo." *Revista Universitaria* 109: 127–145.
———. 1975. *Paucartambo: visión monográfica.* 4 vols. Cuzco: Editorial León.
Viola, Herman J., and Carolyn Margolis, eds. 1991. *Seeds of Change: Five Hundred Years since Columbus.* Washington, D.C.: Smithsonian Institution Press.
Watters, R. F. 1994. *Poverty and Peasantry in Peru's Southern Andes, 1963–1990.* Pittsburgh: University of Pittsburgh Press.
Watts, Michael J. 1983. *Silent Violence: Food, Famine, and Peasantry in Northern Nigeria.* Berkeley: University of California Press.
———. 1987. "Powers of Production—Geographers among the Peasants." *Environment and Planning D, Society and Space* 5: 215–230.
———. 1993. "Development I: Power, Knowledge, and Discursive Practice." *Progress in Human Geography* 17 (2): 257–272.
Weberbauer, A. 1945. *El mundo vegetal de los Andes peruanos.* 2d ed. Lima: Ministerio de Agricultura.
Webster, Steven. 1973. "Native Pastoralism in the South Andes." *Ethnology* 12 (2): 115–134.
Weismantel, Mary J. 1988. *Food, Gender, and Poverty in the Ecadorian Andes.* Philadelphia: University of Pennsylvania Press.
West, Robert C., ed. 1982. *Andean Reflections: Letters from Carl O. Sauer While on a South American Trip Under Grant from the Rockefeller Foundation, 1942.* Boulder: Westview Press.
Whitaker, Thomas W., and Hugh C. Cutler. 1965. "Cucurbits and Cultures in the Americas." *Economic Botany* 19 (4): 344–349.
Whyte, William Foote, and Giorgio Alberti. 1976. *Power, Politics, and Progress: Social Change in Rural Peru.* Amsterdam: Elsevier.
Wightman, Ann M. 1990. *Indigenous Migration and Social Change.* Durham: Duke University Press.
Wilkes, Garrison. 1983. "Current Status of Crop Plant Germplasm." *CRC Critical Reviews in Plant Science* 1: 133–181.
———. 1991. "In Situ Conservation of Agricultural Systems." In *Biodiversity: Culture, Conservation, and Ecodevelopment,* edited by Margery L. Oldfield and Janis B. Alcorn, 86–101. Boulder: Westview Press.
Wilkes, Garrison, and Susan Wilkes. 1972. "The Green." *Environment* 14 (8): 32–39.
Williams, J. Trevor. 1984. "A Decade of Crop Genetic Resource Research." In *Crop Genetic Resources: Conservation and Evaluation,* edited by J. H. W. Holden and J. T. Williams, 1–17. London: George Allen & Unwin.

———. 1988. "Identifying and Protecting the Origins of Our Food Plants." In *Biodiversity,* edited by E. O. Wilson, 240–247. Washington, D.C.: National Academy Press.

Wilson, E. O., ed. 1988. *Biodiversity*. Washington, D.C.: National Academy Press.

Wilson, Hugh D. 1978. "*Chenopodium quinoa* Willd.: Variation and Relationships in Southern South America." *National Geographic Society Research Reports* 19: 711–721.

———. 1988. "Quinoa Biosystematics I: Domesticated Populations." *Economic Botany* 42 (4): 461–477.

Wilson, Patricia A., and Carol Wise. 1986. "The Regional Implications of Public Investment in Peru." *Latin American Research Review* 21 (2): 93–112.

Winterhalder, Bruce. 1994. "Rainfall Predictability and Water Management in the Central Andes of Southern Peru." In *Irrigation at High Altitudes,* edited by David Guillet and William P. Mitchell, 21–67. Arlington, VA: American Anthropological Association.

Wolf, Eric R. 1966. *Peasants.* Englewood Cliffs: Prentice Hall.

———. 1969. *Peasant Wars of the Twentieth Century.* New York: Harper & Row.

———. 1982. *Europe and the People without History.* Berkeley: University of California Press.

World Commission on Environment and Development (WCED). 1987. *Our Common Future.* Oxford: Oxford University Press.

World Resources Institute (WRI). 1992. *Global Biodiversity Strategy: Guidelines for Action to Save, Study, and Use Earth's Biotic Wealth Sustainably and Equitably.* Washington, D.C.: World Resources Institute, World Conservation Union, United Nations Environment Programme.

Worster, Donald, ed. 1988. *The Ends of the Earth: Perspectives on Modern Environmental History.* Cambridge: Cambridge University Press.

———. 1993. *The Wealth of Nature: Environmental History and the Ecological Imagination.* Oxford: Oxford University Press.

Yacovleff, E., and F. L. Herrera. 1934. "El mundo vegetal de los antiguos peruanos." *Revista del Museo Nacional* 3: 241–322.

———. 1935. "El mundo vegetal de los antiguos peruanos (conclusión)." *Revista del Museo Nacional* 1: 31–102.

Yamamoto, Norio. 1985. "The Ecological Complementarity of Agro-Pastoralism: Some Comments." *Andean Ecology and Civilization: An Interdisciplinary Perspective on Andean Ecological Complementarity,* edited by Shozo Masuda, Izumi Shimada, and Craig Morris, 85–99. Tokyo: University of Tokyo Press.

Young, Ken. 1987. Personal Communication. October 17.

Zevallos, C. M., W. C. Galinat, D. W. Lathrap, E. R. Leng, J. G. Marcos, and K. M. Klumpp. 1977. "The San Pablo Corn Kernel and Its Friends." *Science* 196: 385–389.

Zimmerer, Karl S. 1985. "Agricultural Inheritances: Peasant Management of Common Bean (*Phaseolus Vulgaris*) Variation in Northern Peru." Master's thesis, Department of Geography, University of California, Berkeley.

———. 1988. "Seeds of Peasant Subsistence: Agrarian Structure, Crop Ecology, and Quechua Agricultural Knowledge with Reference to the Loss of Biological Diversity in the Southern Peruvian Andes." Ph.D diss., Department of Geography, University of California, Berkeley.

———. 1989. "Plants of Paucartambo." Five hundred and fifty botanical accessions with herbarium labels and detailed ethnobotanical notes submitted to the Herbarium of the Field Museum, Chicago.

———. 1991a. "The Regional Biogeography of Native Potato Cultivars in Highland Peru." *Journal of Biogeography* 18: 165–178.

———. 1991b. "Managing Diversity in Potato and Maize Fields of the Peruvian Andes." *Journal of Ethnobiology* 11 (1): 23–49.

———. 1991c. "Labor Shortages and Crop Diversity in the Southern Peruvian Sierra." *The Geographical Review* 82 (4): 414–432.

———. 1991d. "Wetland Production and Smallholder Persistence: Agricultural Change in a Highland Peruvian Region." *Annals of the Association of American Geographers* 81 (3): 443–463.

———. 1991e. "Agricultura de barbecho sectorizada en las alturas de Paucartambo: luchas sobre la ecología del espacio productivo durante los siglos XVI y XX." *Allpanchis* 23: 189–226.

———. 1992a. "The Loss and Maintenance of Native Crops in Mountain Agriculture." *GeoJournal* 27 (1): 61–72.

———. 1992b. "Land-Use Modification and Labour Shortage Impacts on the Loss of Native Crop Diversity in the Andean Highlands." In *Sustainable Mountain Agriculture,* edited by N. S. Jodha, M. Banskota, and Tej Partap, 415–421. New Delhi: Oxford Publishing.

———. 1992c. "Biological Diversity and Local Development: 'Popping Beans' in the Central Andes." *Mountain Research and Development* 12 (1): 47–61.

———. 1993a. "Agricultural Biodiversity and Peasant Rights to Subsistence in the Central Andes during Inca Rule." *Journal of Historical Geography* 19 (1): 15–32.

———. 1993b. "Soil Erosion and Social Discourses: Perceiving the Nature of Environmental Degradation." *Economic Geography* 69 (3): 312–327.

———. 1994a. "Integrating the 'New Ecology' and Human Geography: Promise and Prospects." *Annals of the Association of American Geographers* 84 (1): 108–125.

———. 1994b. "Transforming Colquepata Wetlands: Landscapes of Knowledge and Practice in Andean Agriculture." In *Irrigation at High Altitudes,* edited by David Guillet and William P. Mitchell, 115–140. Arlington, VA: American Anthropological Association.

———. 1995. "Ecology as Cornerstone and Chimera in Human Geography." In *Concepts in Human Geography,* edited by Carville Earle and Kent Mathewson, 161–188. London: Rowman & Littlefield.

———. N.d. "Geographical Place and Resource Ethics among Non-Western Peoples: Their Implications for Sustainability." In *Varieties of Sustainability,* edited by Paul B. Thompson and D. Bruce Dickson. College Station: Texas A & M Press.

Zimmerer, Karl S., and David S. Douches. 1991. "Geographical Approaches to Crop Conservation: The Patterning of Genetic Diversity in Andean Potatoes." *Economic Botany* 45 (2): 176–189.

Zimmerer, Karl S., and Robert Langstroth. 1994. "Physical Geography of Tropical Latin America: The Spatial and Temporal Heterogeneity of Environments." *Journal of Tropical Geography* 14 (2): 157–172.

Archival Documents

Key to Abbreviations: AAC, Archivo del Arzobispado del Cusco; ADC, Archivo Departamental del Cusco; ARA, Archivo de la Reforma Agraria (Cusco); BN, Biblioteca Nacional, Lima; LACH, Libro de Actas de la Comunidad de Chocopía (Paucartambo Province); LAS, Libro de Actas de la Comunidad de Sonqo (Paucartambo Province).

AAC 1595, G.6.74.1. "Visita de Diego Maravier en el Pueblo de San Geronimo de Colquepata, 1595." Transcription by Henrique Urbano and Laura Hurtado.

AAC 1746, Paquete 8, Legajo 206, "Chamayro, Paucartambo—Concurso de Acreedores."

AAC 1777, Caja 87, Paquete 5, Expediente 87, "Don Melchor Gutierrez vecino y hacendado del Real Asiento de Paucartambo."

ADC 1639, Escribanos, Joseph Navarro, "Venta: Doña Juana de Vargas a Pedro Gonzales."

ADC 1784–85a, Intendencia, Provincias, Causas Ordinarias, Legajo 89, "De autos seguidos por el intestado fallecimiento del Dr. Don Juan Gonzalez."

ADC 1784–85b, Intendencia, Causas Ordinarias, Legajo 89, "Doña Estefania Gonzales, una de las herederas."

ADC 1785, Intendencia, Real Hacienda, Legajo 165, "Expediente relativo a el arrendamiento de unas tierras nombradas Tocoguailla."

ADC 1785–87, Intendencia, Provincias, Causas Ordinarias, Legajo 90, "Los Autos del Concurso que sea formado a la Hda. de San Ildefonso."

ADC 1791–93, Intendencia, Provincias, Causas Ordinarias, Legajo 92, "Rason del Chuño y Sesina que estoy dando para las Requas de Chamayro."

ADC 1794–96, Intendencia, Provincias, Causas Ordinarias, Legajo 93, "Sobre el remate de las haciendas de Ayre y Guatocto."

ADC 1793, Intendencia, Causas Ordinarias, Legajo 29, "Copia certificada a la cuenta de tributos y hospital del Partido de Paucartambo."

ADC 1794, Real Audiencia, Causas Ordinarias, Legajo 15, "Expediente de Nicolas Callisana i Carlos Yapo."

ADC 1800, Real Audiencia, Causas Ordinarias, Legajo 37, "Expendiente en que el Procurador Jose Gregorio Tinoco."

ADC 1807–8, Diezmos, Legajo 47, "Quaderno segundo del concurso del Acreedores formado a las Haziendas de Yapunccuni, Espingone, Cacayoc, Runtu Runtu, y otros nombres sitas en Paucartambo."

ADC 1845, Matriculas, "Extracto de Contribuyentes de Indigenas de la Provincia de Paucartambo, Año de 1845."

ADC 1850, Matriculas, "Extracto de Contribuyentes de Indigenas de la Provincia de Paucartambo, Año de 1850."

ADC 1890, Matriculas, "Extracto de Contribuyentes de Indigenas de la Provincia de Paucartambo, Año de 1890."

ADC 1962a, Prefectura, Letra H, Número 62.

ADC 1962b, Sub-Prefectura, Letra J, Número 7.

ARA 1922, Expediente 78, "Relatio a quejas diversas de las comunidades del distrito de Ccolqquepata (sic)."

ARA 1923, Expediente 234, "Expediente relativo a quejas diversas de las communidades del distrito 'COLQUEPATA' de la Provinvia de Paucartambo del departmento del Cuzco. Año 1923."

ARA 1935, Chocopía, "Queja de los indígenas de la comunidad de indigenes de Chocopía contra del Sr. Victor LaTorre, del distrito de Colquepata."

ARA 1972, Expediente 11–7–4, "Pasto Grande."

BN 1587, Odinario, Expediente A370, "Juan Iñiquez de Bermeo contra Pablo de Gamboa sobre que le pague 250 pesos."

LACH 1658, "Los linderos de Chocopía." Document transcribed by Leonidas Concha.

LACH 1950, "Litigio de tierras." Document transcribed by Leonidas Concha.

LAS 1658, "Título de propiedad de Soncco." Document transcribed by Leonidas Concha.

Index

/Acculturation. *See* Cultural/religious traditions
Acobamba. *See* Northern Paucartambo Valley
Agave, 132, 261 n.15
Agrarian Reform Law (1969). *See* Land Reform (1969)
Agricultural innovation. *See* Peasant innovation
Agroecology, 15–22, 111–112, 113–115, 221. *See also* Ecological adaptation theory
Aji chile pepper. *See* Capsicum
Alcoholic beverages. *See* Beer; Chicha
Alfalfa, 52, 259 n.39
Alikuya. See Liver fluke
Allen, Catherine, 78
All-Soviet Institute for Applied Botany, 13
Altiplano peoples, 34, 197, 255 n.8
Amazon Rubber Boom, 13, 57
Ancient agriculture, 3, 12, 26–34, 65, 255 n.3; and climate, 28, 255 n.4; crop evolution, 30–31, 33–34, 65, 253 n.16, 255 nn.6,7; crop types, 31–33; Eastern Andes, 26–30, 32–33, 254 n.1, 255 nn.3,10; interregional crop exchange, 33–34, 255 n.8
Andean Weevil, 112
Antisuyo Road, 35
Añu. See Mashua
Aqha beer, 38, 68, 171, 209, 268 n.25
Astronomy, 216, 270 n.21
Ayacucho, 32, 33
Aymara Kingdoms, 34, 255 n.8
Ayni labor. *See* Reciprocal labor exchange
Barley, 66, 116, 266 n.5; Colquepata area, 162, 168–169, 170; late colonial/republican cultivation, 59, 61, 258 n.37; Northern Paucartambo Valley, 154, 266 n.5; Oxen Area farm spaces, 139, 140; postreform cultivation, 79–80, 95–96; Southern Paucartambo Valley, 173, 178–181; Spanish colonial cultivation, 49, 52, 257 n.23
Barter, 95, 160, 177, 253 n.18, 266–267 n.11; as payment-in-kind, 7, 93, 96, 268 n.31
Beans, common, 39, 136
Beer, 59, 79–80, 261 nn.10,11; *aqha,* 38, 171, 209, 268 n.25; peasant consumption, 90, 96. *See also* Barley
Beer Company of Southern Peru, 79, 95, 261 n.10; and Colquepata area, 162, 170; exploitative practices, 80, 220, 261 n.11, 266 n.5; late colonial/republican period, 59; and Southern Paucartambo Valley, 178
Belaúnde, Fernando Terry, 78
Benavides, María, 14
Bennett, Erna, 5, 7–8
Berry, Wendell, 17
Betanzos, Juan de, 39
Biodiversity: extent of, 10, 12, 13–14, 252 n.4; geographical unevenness, 149; historical overview, 65–67; international study of, 10, 14–15, 148–149, 172, 225–232; property rights, 9, 13, 232; as status indicator, 23, 24, 96–97, 103–105, 214, 225, 254 n.24; terminology, 16, 17. *See also* Landrace loss; *specific topic and theories*

Bitter potatoes, 32, 39, 110, 111, 128
Blaikie, Piers, 21
Bowman, Isaiah, 27, 28, 30, 258 n.35
Broad beans, 60, 66, 86; and farm spaces, 135, 138; Southern Paucartambo Valley, 178, 180, 181
Bruntland Report, 226
Brush, Stephen, 15
Bukasov, Sergei, 13, 148
Butzer, Karl W., 227

Callacancha. *See* Southern Paucartambo Valley
Callisana, Nicolas, 53
Callispuquio. *See* Southern Paucartambo Valley
Campesinos. See Peasants
Cañac-Huay Pass, 45, 47
Candolle, Alphonse de, 10
Capsicum, 36, 39, 49, 95, 104
Cárdenas, Martín, 15, 33
Carpapampa. *See* Southern Paucartambo Valley
Catholicism. *See* Cultural/religious traditions
Cattle. *See* Livestock-raising
Cauacachiqqueyoc. See Fit livelihood ethic
Ccotatoclla. *See* Colquepata area
Center for the Investigation of Andean Crops, 137
Challabamba. *See* Northern Paucartambo Valley
Chawcha potatoes. *See* Precocious potatoes
Chawpi qhata. See Oxen Area farm spaces
Chenopodium hircinum, 33
Chenopodium pallidicaule. See Kañiwa
Chenopodium quinoa. See Quinoa
Chicha. See Aqha beer
Chile pepper. *See* Capsicum
Ch'ilkas people, 203, 269 n.10
Chocopía. *See* Colquepata area
Cieza de León, Pedro de, 40–41, 45, 252 n.13, 257 n.23, 258 n.32
CIP (International Potato Center), 14, 69, 148, 231–232
Cities and villages: as administrative centers, 56, 97, 98, 101; and diverse crops, 116–117, 214–215; founding of, 34, 47, 53, 256–257 n.21; as marketplaces, 58, 105, 166, 263 n.28; social functions, 1, 13, 48, 54, 97–104, 164. *See also* Cuzco City; *Reducción*; Villager-peasant relations
Climate, 28, 90, 115, 118–119, 128, 136, 152, 175, 255 n.4; Northern Paucartambo Valley, 152–153, 266 nn.2,3
Cloud forest. *See* Vegetation, non-crop
Cobo, Bernabé, 40–41, 42, 43, 45, 55
Coca: Inca cultivation, 36; late colonial/republican cultivation, 56, 59–60, 259 n.39; postreform cultivation, 153, 261 n.12; Spanish colonial cultivation, 45, 46–49, 52, 151, 257 n.28, 258 n.31
Colonialism, 44–45, 55–56, 141; administration, 45–46, 256–257 nn.20,21; coca cultivation, 46–49; conquest, 44–45; and cultural change, 23; *encomienda* system, 49, 257 n.22; and farm spaces, 124, 139; fit livelihood ethic under, 41, 53–54; *mit'a* labor under, 45, 47, 48, 49, 52, 256 n.19; and naming customs, 190, 203–204, 215; peasant demographic collapse under, 48, 49, 52–53, 151, 257 n.28, 258 n.35; and peasant innovation, 224; and peasant socioeconomic differentiation, 54–55, 258 nn.32, 34; *reducción,* 49–51, 151, 257 nn.25–27; tribute system, 49, 61, 124, 257 n.28, 258 nn.30,32; twin republics ideal, 51–52. *See also* Late colonial/republican period
Colquepata area, 161–172, 267 n.19; barley cultivation, 162, 168–169, 170, 268 n.25; consumption practices, 171; Early Planting farm spaces, 162, 164, 165–166, 167, 170, 265 n.22, 267 nn.14,18; field terracing, 170–171, 268 n.24; Hill farm spaces, 121; Inca period, 35, 37; and Land Reform, 73; land tenure, 167, 267–268 n.20; late colonial/republican period, 57, 60, 61, 62–64; postreform period, 78, 81, 82; potato cultivation, 162, 268 n.26; under Spanish colonialism, 50, 53–54; Valley farm spaces, 167–168, 268 n.21; villager-peasant relations, 99, 103, 164–165, 267 n.16
Comida del indio, 23, 66, 103, 258 n.34; and Spanish colonialism, 23, 44, 45, 55, 258 n.34
Commerce: Colquepata area, 165–166, 170; and farm spaces, 120, 134, 137–138, 139, 145; and fit livelihood ethic, 253 n.18; late colonial/republican period, 56, 58–59, 60; markets and diverse crops, 24, 49, 58–59, 65–67, 80, 82, 83, 116–117, 214–215, 221; and Modernization Theory, 24–25, 254 n.23; Northern Paucartambo Valley, 151, 154, 155, 159, 160–161, 266 n.9; postreform, 78–79, 83–84, 94–95, 96, 106; and seed management, 116–117; Southern Paucartambo Valley, 172–173; unreliability of, 67
Communities. *See* Peasant communities
Compadrazgo, 87, 262 n.19
Conservation: and ecologically noble savage assumptions, 2, 226, 227–228, 270–271 nn.6,7; and farm spaces, 145–147; in situ, 9–10, 228–231, 271 nn.8–11,13; and international organizations, 231–232, 271 nn.11,

13; and peasant innovation, 221–225; peasant perceptions, 181–185; and sustainable development, 226–227, 228–229. *See also* Biodiversity; Landrace loss; *specific topics*
Consumption practices, 18, 20–21, 171; cultural change, 214; and diversity rationales, 187, 200–201; and extrafamily labor, 96–97; and Inca agriculture, 38, 41; Northern Paucartambo Valley, 157, 160; post-reform, 78, 89–90, 95–97; preparation techniques, 212–213, 269–270 nn.14,17; purchased foods, 90, 96, 171, 232, 271 n.14. *See also* Cultural/religious traditions; Fit livelihood ethic
Cook, O. F., 10, 13, 148
Cooperative landholding, 74–75
Corvée labor. *See Mit'a* labor
"Crazy Yabar." *See* Yabar, Luís Angel
Crop-cultivator ties. *See* Agroecology
Crop rotation, 61, 130, 135, 168, 169. *See also* Sectoral fallow commons; *specific topics*
Crop storage, 38, 39, 112, 256 n.16
Crop theft, 184–185, 209
Culinary norms. *See* Consumption practices
"Cultivated Plants of South and Central America" (Sauer), 14
Cultural aesthetic. *See* Naming customs
Cultural ecology, 262 n.27. *See also* Environmental geography; Regional political ecology
Cultural/religious traditions, 89–90, 104, 158, 229–230; change in, 7, 8, 22, 23–24, 213–217; and diversity rationales, 182–184, 187, 209–210, 215, 270 nn.3,20; and farm spaces, 136, 137, 265 nn.18,19; Inca, 37, 38, 210; and peasant innovation, 224, 225; Protestantism, 215–216; and short-term migration, 154. *See also* Consumption practices; Diversity rationales; Fit livelihood ethic
Cuzco City: as administrative center, 35; as marketplace, 56–59, 78–81, 83, 100, 115, 170, 214–215

Deforestation. *See* Logging; Vegetation, non-crop
Demography. *See* Population
Denevan, William M., 227
Departmental Federation of Cuzco Peasants (FDCC), 73
Diamond, Jared M., 227
Diet. *See* Consumption practices
Differentiation-hybridization, 33
Diverse crops. *See* Biodiversity; *specific crops*
Diversity rationales, 186–217, 268 n.36; broad nature of use-categories, 197–198, 269 n.7; and consumption practices, 187, 200–201; and cultural change, 213–217; and cultural/religious traditions, 182–184, 187, 209–210, 215, 270 n.20; and ecological adaptation theory, 186–187, 188, 199–201; ecological rationales, 186–187, 188; and farm spaces, 196–197; and fit livelihood ethic, 187, 213; and gender roles, 189; geographical rationales, 187–188; and high-yielding potato varieties, 7–8; and language, 189–196, 268–269 nn.3,12; maize, 206–212, 269 nn.15,16, 270 n.18; and naming customs, 189–193, 195–196, 203–204, 268 n.30, 268–269 nn.3,9,10,12; potatoes, 190–191, 195–204, 268–269 nn.3,7,9–12; quinoa, 206, 212–213; and reciprocity, 183–185, 189, 191, 193–194, 215, 269 nn.5,6; and seed management, 198–199, 202, 208–209, 264 n.3; ulluco, 205–206, 269 n.13; and villager-peasant relations, 204, 269 nn.11,12
Dual production, 34, 36, 58, 84, 106, 120, 270 n.2

Early Planting farm spaces, 4, 120, 140–145; Colquepata area, 162, 164, 165–166, 167, 170, 265 n.22, 267 n.14; mosaic pattern, 124; Northern Paucartambo Valley, 153, 155, 157, 159, 160–161; Southern Paucartambo Valley, 173; and Spanish colonialism, 124
Earth Mother, 182–183, 184
Earth Summit, 226, 229
Eastern Andes, geography, 26–27, 254 n.1
Ecological adaptation theory, 8–9; and agroecology, 17–18; critique of, 18–21, 223, 253 n.17, 264 n.13, 265 n.26; and diversity rationales, 186–187, 188, 199–201; and Inca agriculture, 37; and mosaic farm space pattern, 124–125; and peasant innovation, 223; verticality, 19–20, 253 n.17
Ecological complementarity. *See* Verticality
Ecologically specialized communities, 118, 264 nn.7,16. *See also* Territories
Ecology. *See* Agroecology; Ecological adaptation theory
Economic development, 3, 254 n.24, 263 n.35; and food imports, 79, 96, 220, 232, 271 n.14; functional dualism model, 73, 260 n.3; geographically uneven, 22–23, 73, 77–78, 105–106, 219; and Land Reform, 73, 164–166, 260 n.3; Modernization Theory, 22, 24, 254 n.23, 270 n.4. *See also* Postreform agriculture; Sustainable development; Terms of trade
Economic Diverse Crops. *See* Biodiversity; *specific crops*
Education, 216–217

Elevation. *See* Landscapes; Verticality
Encomienda system, 49, 257 n.22
Environmental geography, 15–25; agroecology, 16–22; social theory, 22–25, 270 n.5
Environments. *See* Climate; Landscapes; Soil fertility; Vegetation, non-crop
Ericsson, Sven, 13, 57
Erwinia, 175
Estates. *See* Haciendas
Ethnic identity, 5, 44–45, 55, 102–104, 263 n.32. *See also Comida del indio*; Socioeconomic differentiation
Ethnobotany, 15
Eucalyptus, 132
Ex situ conservation, 9–10
Extinction crisis, 3, 5, 6, 218
Extrafamily labor, 43, 87–88, 94, 96–97, 222–223

Faena, 88
Fano, Hugo, 14
Farmers. *See* Peasants
Farm knowledge, 26, 89, 262 n.21, 265 n.25. *See also* Diversity rationales; Naming customs; Peasant innovation; Seed management
Farm spaces, 4, 21–22, 108, 117–147, 264 nn.5,14; and conservation, 145–147; and cultural/religious traditions, 136, 137, 265 nn.18,19; distinct nature of, 118–119, 264 n.8, 265 n.23; and dual production, 106; and elevation, 108, 118, 119, 123, 124, 125, 128, 132, 138, 142, 146, 246 n.7, 263–264 n.2; and field perceptions, 146, 265 n.25; and landrace grouping, 110–111, 196–197, 264 n.3; and landrace loss, 219; "manufacture" of, 265 n.24; mosaic pattern, 122, 123–126, 146, 173, 178, 264 nn.13,14; and peasant innovation, 4, 126, 131, 135, 141–142, 145–147, 222; and resources, 85–86; and seed management, 108, 111–112, 142–143; system cohesion, 119–123, 146–147, 264 nn.9,10. *See also* Early Planting farm spaces; Hill farm spaces; Land tenure; Oxen Area farm spaces; Valley farm spaces
Fasciola hepatica. See Liver fluke
Fava bean. *See* Broad beans
FDCC (Departmental Federation of Cuzco Peasants), 73
Federación Departmental de Campesinos de Cusco, 73
Field balks, 180–181, 268 n.34
Field fallow, 130, 135, 140, 265 n.17. *See also* Sectoral fallow commons
Field inputs, 82–83, 86, 99, 261 n.8
Field terracing, 35, 36–37, 170–171, 255 n.14, 268 n.24
Financial credit, 80, 82, 99–100, 102, 153, 164
Fit livelihood ethic, 20–21; continued belief in, 65, 66, 89, 96; defined, 60, 252–253 nn.13,18; and diversity rationales, 187, 213; and ecological adaptation theory, 18; and Inca agriculture, 41–42; and Land Reform, 70, 73, 75; late colonial/republican period, 60, 62–64, 259–260 nn.42,43; Northern Paucartambo Valley, 157; and peasant innovation, 224; and Spanish colonialism, 53–54; and status, 104–105, 263 n.34. *See also* Consumption practices
Floury potatoes, 69–70, 104, 157; ancient cultivation, 32; and diversity rationales, 187–188; Hill farm spaces, 128–129; postreform cultivation, 82, 85, 95; seed management, 109, 110, 111, 115, 117; Southern Paucartambo Valley, 172, 175–177, 268 nn.27,31
Folk classification. *See* Naming customs
Fonseca, César, 148
Forasteros, 52–53, 257–258 nn.29,30
Forced resettlement. *See Reducción*
Frankel, Sir Otto, 5
Frijol. See Beans, common

Gade, Daniel W., 14–15, 49
García Pérez, Alan, 78
Garcilaso de la Vega, El Inca, 40–41, 45, 47, 255 n.9, 256 n.16, 259–260 n.43
Gardens, 21, 85–86, 153, 230, 261 n.15
Gender roles: in farm work, 85, 87, 97, 166, 267 n.19; and diverse crops, 63, 108–110, 117, 181, 188–189, 198–199, 208–209, 230–231, 271 n.10
Genetic erosion, 3, 5–6, 15, 16, 22, 56, 219, 233, 252 n.11. *See also* Landrace loss; *specific topics*
Genetics, 31, 33, 172, 180, 252 n.11, 268 n.26. *See also* Hybridization
Geographical rationales, 4–5, 187–188. *See also* Fit livelihood ethic; Place-based analysis
Geomorphology: disturbances, 30; landforms, 35, 48, 107, 125, 126, 152
González Holguín, Diego, 54
Green Revolution, 83, 85, 153. *See also* High-yielding potato varieties
Griñon barley, 80, 84, 116, 140, 169; Southern Paucartambo Valley, 178, 180
Grobman, Alexander, 37–38
Guamán Pomo de Ayala, Felipe, 40, 42–43
Guano-giving. *See* Soils/fertilizers
Guinea pigs, 90
Guitarrero Cave, 27, 32

Habitat diversity, 28–30, 255 n.5. *See also*

Vegetation, non-crop
Haciendas, 57–60, 61, 258 nn.31,37; ex-owners, 97–98, 262 n.26; and Land Reform, 76–77; peasant resistance, 53–54, 62–63, 74, 165. *See also* Territories
Hak'u papa. See Floury potatoes
Hanansaya Ccollana Chocopía. *See* Colquepata area
Harlan, Jack R., 5, 17–18
Hawkes, Jack G., 5, 14, 148
Haynachu. See Intercropping
Herds. *See* Livestock-raising
Herrera, Fortunato, 14
High-yielding potato varieties (HYVs), 153, 157, 162, 176; defined, 7; and farm spaces, 120, 135, 137, 138, 139, 140, 141; and landrace loss, 7–8, 220; postreform cultivation, 77, 82, 83, 84, 85, 86, 94–95; and seed management, 115–116, 270 n.20; and socioeconomic differentiation, 94–95
Hill farm spaces, 4, 121, 126–132, 264 n.15; cohesion, 123, 264 n.16; Colquepata area, 172; and crop theft, 185; Southern Paucartambo Valley, 173
Holguín, González, 60
Horkheimer, Hans, 10
Housing, 68, 107, 266 n.3
Huaqanqa. *See* Northern Paucartambo Valley
Huaranca. *See* Colquepata area
Huari Empire, 33–34
Huaynapata. *See* Southern Paucartambo Valley
Huerta. See Gardens
Humana. *See* Southern Paucartambo Valley
Hurca, Daniel, 259 n.42
Hybridization, 33, 190–191, 201, 206, 211. *See also* Genetics
HYVs. *See* High-yielding potato varieties

Improved potato varieties. *See* High-yielding potato varieties
Inca agricultural renown, 2, 8, 9, 36, 40, 44, 251–252 n.3, 255 nn.13,14
Inca period, 34–44; commoner agriculture, 36, 40–44, 256 nn.17,18; imperial agriculture, 35–40, 255–256 nn.13,14; *maway* potato cultivation, 141; Paucartambo Andes control, 34–35, 255 nn.10, 11, 256 n.18; religion, 37, 38, 210
Indian demographic collapse, 48, 49, 52–53, 151, 224, 257 n.28, 258 n.35
"Indian food." *See Comida del indio*
Indians. *See* Indian demographic collapse; Peasants; *specific topics*
Indigenismo movement, 58, 62–63, 259 n.42
Indigenous Communities, 50, 53, 57, 61, 62–65, 71, 75, 257 n.24, 259 n.42
Insecticides, 82, 112, 115, 143
In situ conservation, 9–10, 228–231, 271 nn.8–11,13
Intellectual property rights. *See* Biodiversity
Intercropping, 108, 178, 223–224, 262 n.18, 269 nn.5,6; and conservation, 172, 230–231, 271 nn.8,9; maize, 207, 208, 211–212, 268 n.21; and reciprocity, 185, 193–194; Valley farm spaces, 135–136, 137
International Potato Center (CIP), 14, 69, 148, 231–232
Interregional crop exchange, 33–34
Irrigation, 37, 76, 82–83, 122, 255 n.14, 264 n.8

Jalca, 264 n.15
Jornal, 87. *See also* Labor-time
Juan, Jorge, 257 n.23

Kanchón. See Gardens
Kañiwa, 131–132
Kawsay. See Fit livelihood ethic
Kayotaka. See Liver fluke
Kellkaykunka Pass, 57
Kidney beans. *See* Beans, common

Labor conflicts, 74, 266 n.5
Labor-time: Colquepata area, 166–167, 170–171; and farm spaces, 140, 143, 144; and landrace loss, 87–88, 106, 156, 220; and Land Reform, 71, 74; late colonial/republican period, 61, 62, 236, 259 n.41; Northern Paucartambo Valley, 154, 155, 156, 158–159; postreform, 87–88, 94–95, 262 nn.17, 18, 266 n.10; and pre-European agriculture, 31, 41, 42–44; Southern Paucartambo Valley, 175–176, 268 n.28; terms for, 88, 97, 153, 262 n.20
Landcover. *See* Farm spaces; Vegetation, non-crop
Landrace loss, 16, 22, 56, 61–62, 69–70, 84–88, 224, 225, 263 n.33, 267 n.12; Colquepata area, 166–168, 171–172; and commerce, 221; extent of, 218–219; geographical unevenness, 4, 149, 150, 220–221, 229; and high-yielding potato varieties, 7–8, 220; Northern Paucartambo Valley, 155–156, 158–159, 160; peasant perceptions, 181–185, 187–188; and scissors effect, 219, 232, 270 n.1; Southern Paucartambo Valley, 180. *See also* Biodiversity; Genetic erosion; *specific topics*
Land Reform (1969), 3, 68–77, 260 n.2; administration, 70–71; Colquepata area, 161–162, 164, 168; de facto evasion of, 74–77, 260–261 nn.5–7; and farm spaces, 145; and gender roles, 87; and intercropping, 135;

and landrace loss, 69–70; negative effects of, 24, 69, 72–73, 260 n.3; and peasant perceptions, 183; and peasant resistance, 73–74; and scissors effect, 219; and seed management, 109; and socioeconomic differentiation, 91; uncertainty of, 101, 263 n.30; and verticality, 71–72. *See also* Postreform agriculture
Landscapes: agricultural-elevational above 3550 meters, 61, 63, 113, 172, 173, 178–-180, 196; agricultural-elevational below 3550 meters, 37, 38, 83, 93, 158, 160, 178, 243; agricultural-elevational 2700–4100 meters, 11, 42, 111, 118–119, 123–126, 169, 234, 242; biophysical-elevational, 27–30, 235, 252 n.7, 255 nn.4,5, 263–264 n.2, 264 n.15, 266 n.2; cultural meanings, 4, 35, 120, 138–139, 182, 187–188, 203–204, 255 n.11, 264 nn.5,12, 265 n.19, 269 n.9. *See also* Farm spaces; Naming customs; *specific farm spaces*
Landslides, 30
Land tenure: Colquepata area, 167, 265 n.22, 267–268 n.20; Inca period, 42, 43–44, 256 n.17; late colonial/republican period, 61, 62–65, 259–260 nn.42,43; peasant perceptions, 182–183; postreform, 86–87, 261–262 n.16; *reducción*, 49–51, 151, 257 nn.25–27; and socioeconomic differentiation, 93–94. *See also* Land Reform (1969)
Land use. *See* Farm spaces; Land tenure
Language: and diversity rationales, 189–196, 268–269 nn.3,12; and socioeconomic differentiation, 54, 92. *See also* Naming customs
Larson, Brooke, 51
Late colonial/republican period (1776–1969), 55–65; economic decline, 56, 258 nn.35,36; haciendas, 56–60, 258 n.37; socioeconomic differentiation, 62, 66, 91, 259 n.41, 262 n.22; subsistence farming, 60–65, 259 nn.40–43, 263 n.31
Leguía, Augusto B., 57
Lewis, Martin W., 227
Livelihood ethic. *See* Fit livelihood ethic
Liver fluke, 162, 267 n.15
Livestock-raising, 19, 238, 258 n.36, 262 n.23; colonial, 49, 124; Colquepata area, 162, 164, 168, 267 n.15; and farm spaces, 108, 121–122, 134, 138, 139, 142, 178, 223; and fertilizer, 130, 135, 140, 168;
Inca, 36; late colonial/republican period, 56, 57, 151; Northern Paucartambo Valley, 153, 157, 266 n.9; and peasant innovation, 176, 223; postreform, 71, 83–84, 86, 238; and sectoral fallow, 50–51, 123, 179
Logging, 77, 153, 220, 228
Loma, 126, 203. *See also* Landscapes; Vegetation, non-crop
Lost Crops of the Inca (National Research Council), 251 n.3
Lupine. *See Tarwi*

Maize: ancient cultivation, 32; Colquepata area, 166, 167–168, 171; diversity rationales, 206–212, 265 n.18, 269 nn.15,16, 270 n.18; Inca cultivation, 37–38, 39, 43, 255–256 nn.13,14; late colonial/republican cultivation, 60; naming, 190; Northern Paucartambo Valley, 83, 155, 157–161, 268 n.18; postreform cultivation, 83, 84, 89, 93–94, 104; preparation, 269–270 n.17; reciprocity beliefs, 185; religious significance of, 37, 137, 209–210, 265 nn.18,19; seed management, 110, 112, 115, 208–209, 269 n.15; Southern Paucartambo Valley, 172, 178, 265 n.18, 268 n.33; Spanish colonial cultivation, 49; Valley farm spaces, 132, 133, 135, 265 n.18. *See also specific topics*
Majopata. *See* Northern Paucartambo Valley
Manú National Park, 1–2, 153, 266 n.3
Manure. *See* Soils/fertilizers
Maravier, Diego de, 50, 61
Markets. *See* Commerce
Marx, Karl, 253 n.13
Masa unit, 262 n.20. *See also* Labor-time; *Topo* unit
Mashua, 54–55, 128, 185
Massua. *See* Mashua
Maway papa potatoes. *See* Precocious potatoes
Mayer, Enrique, 20, 264 n.8
Mejía Xesspe, Mi Toribío, 60
Men. *See* Gender roles
Micca. *See* Colquepata area
Microenvironmental specialization. *See* Ecological adaptation theory
Migration: permanent, 33, 52, 53, 75–76, 224, 257–258 nn.29,30, 260 n.7; short-term, 49, 80–81, 95, 105, 153–155, 158, 159, 220, 230–231, 261 n.12, 266 n.4
Mining, 36, 45, 95, 153, 261 n.11; Potosí, 47, 56
Misconceptions, 5–9, 15; biodiversity crisis, 7, 254 n.25; ecologically noble savage assumptions, 226, 227–228, 270–271 nn.6,7; "foods left behind," 7, 8, 22, 251–252 n.3. *See also* Ecological adaptation theory; Inca agricultural renown
Misquihuara. *See* Colquepata area
Mit'a labor: Inca period, 35, 36, 43; under Spanish colonialism, 45, 47, 48, 49, 52,

256 n.19
Modernization Theory, 22, 24, 254 n.23
Mollomarca. *See* Southern Paucartambo Valley
Money potatoes. *See* High-yielding potato varieties
Morales Burmúdez, Francisco, 78
Morocho, 38
Muhu. See Seed management
Murra, John V., 36, 37, 256 n.16

Naming customs, 18–19, 26, 255 n.11; and diversity rationales, 189–193, 195–196, 203–204, 268–269 nn.3,4,9,10,12; and farm spaces, 138–139; "folk" classification, 195–197, 206–207, 230, 269 n.7, 270 n.18; and seed exchange, 115; ulluco, 205, 269 n.13; and villager-peasant relations, 19, 102, 204, 263 n.32, 269 n.12
National Research Council, 12, 251 n.3
Native crop loss. *See* Landrace loss
Ninamarca. *See* Colquepata area
Nongovernmental organizations (NGOs), 231–232, 271 n.11
Northern Paucartambo Valley, 151–161, 267 n.12; consumption practices, 157; environments, 152–153, 154–155, 266 nn.2,3,6; fencing, 157, 266 n.10; Inca period, 35; and Land Reform, 73, 74, 76–77; late colonial/republican period, 59–60; livestock-raising, 157, 266 n.9; maize cultivation, 83, 155, 157–161, 268 n.18; postreform period, 83; potato landraces, 155–157; Scottish immigrants, 151, 266 n.1; seed management, 112; short-term migration, 153–154, 159, 266 nn.4,5; socioeconomic differentiation, 156–157; under Spanish colonialism, 53; villager-peasant relations, 99, 100, 101, 103

Oats, 267 n.14
Oca, 128
Ochoa, Carlos, 14, 15, 69, 148–149, 177
Orconpuquio. *See* Colquepata area
Oxalis tuberosum. See Oca
Oxen Area farm spaces, 135, 137–140, 265 nn.20,23; Colquepata area, 169, 172, 267 n.19; mosaic pattern, 123–124; Northern Paucartambo Valley, 153; Southern Paucartambo Valley, 173; and Spanish colonialism, 124, 125, 126

Pachamama. See Earth Mother
Palacio Pimental, Gustavo, 58, 62
Papa chawcha potatoes. *See* Precocious potatoes
Papa lisas. See Ulluco
Papa ruk'i potatoes. *See* Bitter potatoes
Pata pata. See Field terracing
Patria, 57
Patron-client relations. *See* Villager-peasant relations
Paucartambo Andes: geography, 10–11, 26–30, 254 n.1, 255 n.11; extent of biodiversity, 4, 10–14; "Indianness," 12–13. *See also specific topics*
Pavón, Joseph, 54–55
Peas, 60, 66
Peasant Communities, 71, 75, 76, 85, 92, 260 nn.5,6; government recognition, 101, 263 n.30. *See also* Territories
Peasant innovation, 5, 7, 65, 85, 109–111, 112, 131, 135, 141–142, 145–147, 162, 166, 175–176, 178, 218, 221–225, 254 n.21, 265 n.24, 268 n.28, 270 n.2; and cultural/religious traditions, 224, 225, 270 nn.3–5
Peasants: defined, 16; demographic collapse, 48, 49, 52–53, 151, 257 n.28, 258 n.35; hacienda work conditions, 58; immigration, 52–53, 75–76, 154, 257–258 nn.29,30, 260–261 n.7; lack of civil/human rights, 251 n.1; resistance, 53–54, 62–63, 73–75, 80, 165, 261 n.11, 266 n.5; short-term migration, 80–81, 153–154, 159, 220, 261 n.12, 266 n.4. *See also* Dual production; Socioeconomic differentiation; *specific topics*
Peasants' Union, 75
Pesticides, 82, 99, 115, 143, 261 n.8
Phytophtera infestans. See Potato Late Blight
Pineapple, 153
Pisum sativa. See Peas
Pizarro, Hernando, 47
Place-based analysis, 4, 21, 106, 108, 147, 187–188, 220–221, 231, 265 n.26. *See also* Farm spaces; *specific places*
Place, sense of. *See* Geographical rationales
Plant disease: and language, 191–192, 269 n.4; potatoes, 32, 175, 266 n.8; and seed management, 264 n.4; wheat, 59, 259 n.38. *See also* Potato Late Blight
Political economy. *See* Colonialism; Commerce; Economic development; Land Reform (1969)
Political impacts, 33–34, 56, 77–78, 101, 261 n.9. *See also* Inca period; Land Reform (1969); Colonialism
Polo de Ondegardo, Juan, 43
Population: colonial period, 47–48, 50, 52, 53, 151, 165, 257 n.28; late colonial and republican period, 56–57, 61; post-1969 rural, 76, 151, 152, 154, 172, 237; post-1969 villages, 13, 48, 97, 237. *See also* Indian demographic collapse

Poroto. See Beans, common
Postreform agriculture (1969–1990), 77–106, 261 n.9, 263 n.35; barley cultivation, 79–80, 261 n.10; consumption practices, 89–90, 95–96; farm knowledge, 89, 261 n.21; field inputs, 82–83, 86, 99, 261 n.8; and government policies, 77–78; labor-time, 87–88, 94–95, 262 n.17; land tenure, 86–87, 261–262 n.16; resource availability, 84–90, 94; and short-term migration, 80–81, 261 n.12; terms of trade, 78–79, 261 n.8; tools, 88–89. *See also* Socioeconomic differentiation; *specific topics*
Potatoes: ancient cultivation, 31–32; Colquepata area, 162, 268 n.26; diversity rationales, 190–191, 195–204, 268–269 nn.3, 7,9–12; Early Planting farm spaces, 140, 141–142; ecological adaptation theory, 8; Hill farm spaces, 127–129, 131–132; hybridization, 190–191, 201; Inca cultivation, 38–39, 43, 256 n.16; late colonial/republican cultivation, 56, 57, 58–59, 60, 61–62; Northern Paucartambo Valley, 153, 155; Oxen Area farm spaces, 137–138; postreform cultivation, 79, 81, 82–83, 90, 94–95, 261 nn.9,13; reciprocity beliefs, 185; seed management, 109–110, 112, 115; Southern Paucartambo Valley, 172, 173–177, 180, 268 nn.27,28,30–32; Spanish colonial cultivation, 49; Spanish descriptions of, 45; and ulluco, 205, 269 n.13. *See also specific topics*
Potato Late Blight, 141, 143, 153, 155, 156, 264 n.4, 266 n.8
Potato Leaf Bug, 143
Potosí, 47, 56
Precocious potatoes: ancient cultivation, 32; disease resistance, 156, 266 n.8; Inca cultivation, 43, 141; late colonial/republican cultivation, 56, 61–62, 259 n.40; Northern Paucartambo Valley, 155–156, 157, 161, 267 n.12; seed management, 110, 111
Prestatarios, 82. *See also* Financial credit
Production zones, 20, 264 nn.8,14. *See also* Farm spaces
Protestantism, 215–216
Pseudomonas solanaceum, 266 n.8
Puna, 126, 203
Purchased foods, 90, 96, 171, 232, 271 n.14

Qolqe papa potatoes. *See* High-yielding potato varieties
Quechua. *See* Peasants; *specific topics*
Quencomayo Valley. *See* Colquepata area
Quinine bark, 57, 266 n.1
Quinoa: ancient cultivation, 33; Colquepata area, 61, 166, 167, 168–169, 171, 268 n.21; diversity rationales, 206, 212–213; Hill farm spaces, 131–132; Inca cultivation, 39; late colonial/republican cultivation, 56, 60–61; Oxen Area farm spaces, 138, 139, 140; postreform cultivation, 89, 90, 93–94; seed management, 110–111, 112; Southern Paucartambo Valley, 172, 178, 181; Spanish descriptions of, 45; Valley farm spaces, 133, 135, 136. *See also specific topics*
Quispe, Miguel, 63, 259 n.42

Raimondi, Antonio, 13, 56
Rain forest. *See* Vegetation, non-crop
Reciprocal labor exchange, 43, 87, 97, 119–120, 122
Reciprocity, 183–185, 189, 191, 215, 269 n.6; and intercropping, 193–194, 269 n.5
Redford, Kent H., 227–228
Red qompis ulluco, 180
Reducción, 49–51, 151, 257 nn.25–27
Regional political ecology, 16, 252 n.9, 254 n.25, 268 n.36. *See also* Environmental geography
Religion. *See* Cultural/religious traditions
Republican period. *See* Late colonial/republican period
Rhoades, Robert, 14
Rice, 80, 89, 90, 153
Ripening period, 38
Ritual. *See* Cultural/religious traditions
Roquechiri. *See* Colquepata area
Roquepata. *See* Colquepata area
Ruiz, Hipólito, 54–55
Ruk'i potatoes. *See* Bitter potatoes
Russian Potato Collecting Expedition, 13

Saint's Days. *See* Cultural/religious traditions
Salaman, Redcliffe, 28, 188, 228
Salomon, Frank, 41, 252 n.13, 253 n.18
Sauer, Carl, 10, 11, 14, 15, 59, 227
Sayllapata. *See* Colquepata area
Schmidt, Ella, 7
Scottish immigrants, 151, 266 n.1
Sectoral fallow commons, 86, 169; Colquepata area, 50–51, 64–65, 121, 169; Hill farm spaces, 123, 129, 134, 147; Oxen Area farm spaces, 139, 140, 265 nn.20,21; and *reducción,* 50–51; Southern Paucartambo Valley, 179–180
Seed management, 107–117, 176; and commerce, 116–117; and diversity rationales, 198–199, 202, 208–209; and farm spaces, 108, 111–112, 142–143; landrace grouping, 110–111, 264 n.3; maize, 110, 112, 115, 208–209, 269 n.15; seed exchange, 113–115, 177, 268 n.31; seed purchase, 115–

116. *See also* Diversity rationales
Seeds of Tomorrow, 6–8, 15, 22, 177, 251 n.3; ecological adaptation theory in, 8, 18, 199; and socioeconomic differentiation, 97
Sendero Luminoso. See Shining Path
Sharecropping, 86–87, 95, 99, 263 n.31
Sheep. *See* Livestock-raising
Shining Path, 14, 100, 188, 232 n.12, 263 n.29
Short-term migration, 80–81, 153–154, 159, 220, 261 n.12, 266 n.4. *See also* Migration
SINAMOS (System for Social Mobilization), 164, 165
Sipascancha. *See* Colquepata area
Slash-and-burn. *See* Swidden techniques
Social theory, 16, 22–25
Social units, 85
Socioeconomic differentiation, 90–106; and *comida del indio,* 258 n.34; and consumption practices, 95–97; and Early Planting farm spaces, 142; and land tenure, 93–94, 239; late colonial/republican period, 62, 66, 91, 259 n.41, 262 n.22; Northern Paucartambo Valley, 156–157; and production practices, 54, 62, 91–95, 139, 156, 225, 229; and Spanish colonialism, 54–55, 258 nn.32,34; status indicators, 23, 24, 92–93, 96–97, 103–105, 214, 225, 254 n.24, 262 n.23. *See also* Villager-peasant relations
Soil erosion, 51, 162, 165, 170–171
Soil fertility, 51, 60, 76, 105, 119, 140, 154, 264 n.17; Colquepata area, 168, 169; Hill farm spaces, 130–131; and landrace loss, 86, 168, 220; and Land Reform, 76–77; Oxen Area farm spaces, 140; Valley farm spaces, 135, 207–208. *See also* Soils/fertilizers
Soils/fertilizers, 79, 82, 86, 99, 130, 169, 261 n.8; Early Planting farm spaces, 141, 143; manure, 83, 130, 135, 140, 168; and seed management, 115; and villager-peasant relations, 99
Solanum ajanhuiri, 131–132
Solanum andigenum. See Floury potatoes
Solanum canasense, 31
Solanum curtilobum. See Bitter potatoes
Solanum goniocalyx. See Floury potatoes
Solanum juzepczukii. See Bitter potatoes
Solanum leptophyes, 31
Solanum phureja. See Precocious potatoes
Solanum stenotomum. See Floury potatoes
Solanum tuberosum subsp. *andigena. See* Floury potatoes
Solanum tuberosum subsp. *tuberosum. See* High-yielding potato varieties
Solanum x *chaucha. See* Floury potatoes
Solanum x *curtilobum. See* Bitter potatoes
Solanum x *juzepczukii. See* Bitter potatoes
Soncco, Isidoro, 63
Sonqo. *See* Colquepata area
Soqra. See Potato Late Blight
Southern Paucartambo Valley, 172–181; barley cultivation, 173, 178–181; field balks, 180–181, 268 n.34; Inca period, 35; and Land Reform, 68–69, 73, 75; maize cultivation, 172, 178, 265 n.18, 268 n.33; post-reform period, 81; potato cultivation, 172, 173–177, 180, 268 nn.27,28,30–32; under Spanish colonialism, 53; villager-peasant relations, 104, 178
Spatial organization. *See* Farm spaces; Verticality
Special Statute of Peasant Communities, 71, 75, 261–262 n.16. *See also* Land Reform (1969)
Step-step terracing. *See* Field terracing
Subsistence. *See* Fit livelihood ethic
Sustainable development, 5, 9, 226–227, 228–229, 270 n.6. *See also Seeds of Tomorrow*
Swidden techniques, 154, 158, 159, 266 n.6
System for Social Mobilization (SINAMOS), 164, 165

Tarwi, 39, 86, 136; and farm spaces, 135, 138, 139; Southern Paucartambo Valley, 178, 181
Tayasqa. See Oxen Area farm spaces
Temperature. *See* Climate
Temporary migration, 80–81, 153–154, 159, 220, 261 n.12, 266 n.4. *See also* Migration, short-term
Teqte beer, 96
Terms of trade, 78–79, 261 n.8
Territories: colonial period, 44–49, 256 n.20, 257 n.22; haciendas, 53, 56–60, 70, 262 n.22; hacienda-community conflicts, 53–54, 73, 164–165, 269 n.11; Inca period, 34–36, 41–43, 256 nn.17,18; indigenous communities (pre-1969), 20, 53–54, 57, 61–65; late colonial/republican period, 56; Peasant Communities, 2, 4, 20, 70–72, 74–75, 260 n.5, 262 n.26, 264 nn.7,16, 267–268 n.20; Peasant Communities and conflicts, 101, 120–121, 183; Peasant Communities and family land, 85, 86, 91, 106, 118, 124, 167, 204, 261 n.16. *See also* Land Reform (1969); Land tenure; *Reducción*; Verticality
Tiwanaku Empire, 34
Toledo, Francisco de, 45, 48, 49–50
Tools, 42, 88–89, 138, 239, 262 n.20, 264 n.10
Topo unit, 50, 88, 256 n.17, 257 n.25
Totara. *See* Northern Paucartambo Valley
Transplanting, 213
Transportation, 1–2, 35, 47, 57, 98, 100–101,

148–149, 151, 154, 161, 166, 266 n.4, 267 n.14
Tres Ventanas, 31–32
Tribute system, 49, 61, 124, 257 n.28, 258 nn.30,32
Troll, Carl, 30
Tropaeolum tuberosum. See Mashua
Tupac Amaru II, 56

Ulloa, Antonio de, 257 n.23
Ulluco: ancient cultivation, 32–33; diversity rationales, 205–206, 269 n.13; Early Planting farm spaces, 140, 141, 142; Hill farm spaces, 127–128; Northern Paucartambo Valley, 153, 161; Oxen Area farm spaces, 138, 139; postreform cultivation, 81, 82, 89, 104; preparation, 269 n.14; seed management, 109–111, 112; Southern Paucartambo Valley, 172, 180. *See also specific topics*
Umamarca. *See* Southern Paucartambo Valley
Unión Campesinos. *See* Peasants' Union
United States National Academy of Sciences, 5–6, 7
Urinsaya Ccollana Kalla. *See* Colquepata area
Urubamba Valley, 32, 34, 57, 60, 79, 136, 197

Valley farm spaces, 4, 132–137, 153, 173, 265 n.18; Colquepata area, 167–168, 268 n.21
Vargas, César, 13, 148, 188
Vavilov, Nikolai, 10, 13
Vegetation, non-crop, 30, 86; cloud forest, 29, 152, 154, 155, 235; degradation of, 77, 80, 153, 154, 181, 220; Early Planting farm spaces, 143–144; grassland-moor, 29, 126, 128, 235; Hill farm spaces, 126, 128; Oxen Area farm spaces, 123–124, 138; Pilcopata-Qosñipata rain forest, 2, 80, 220; shrubland/shrub savanna, 29, 132, 138, 181, 235; thorn scrub/savanna, 29, 134, 235; use of, 30, 36, 57, 86, 89, 107, 112, 153, 159, 170, 240, 261 nn.15,17, 264 n.17, 268 n.34; Valley farm spaces, 132, 134, 136; weeds, 144, 154, 158, 168; wild crop relatives, 27, 31–33, 191, 255 n.2. *See also* Gardens; Wetlands
Velasco Alvarado, Juan, 70, 78, 260 nn.2,3
Verticality, 19–20, 21–22, 59–60, 71–72, 124, 146, 222, 253 n.17, 258 n.31, 259 n.39, 264 n.13, 265 n.23, 266–267 n.11, 269 n.10
Vicia faba. See Broad beans
Villager-peasant relations, 97–105, 262 nn.26,27; Colquepata area, 99, 103, 164–165, 267 n.16; and credit, 99–100; and cultural change, 214; and diverse crops as ethnic status indicators, 104–105, 263 n.34; and ethnic identity, 102–103, 263 n.32; and field inputs, 99; and naming customs, 19, 204, 269 nn.11,12; and regional government, 101; sharecropping, 263 n.31; and Shining Path, 100, 263 n.29; and short-term migration, 154; Southern Paucartambo Valley, 104, 178; and transportation, 98, 100–101
Villages. *See* Cities and villages; *specific areas*
Virgen del Rosario. *See* Southern Paucartambo Valley
Viscachone. *See* Colquepata area
Vizcarra Rojas, Abraham, 74
Vizcarra Rojas, Edgar, 74

Wachu beds, 130–131
Wanuchiy. See Guano-giving
Waqa yapuy. See Oxen Area farm spaces
Weberbauer, Augusto, 30
Weeds. *See* Vegetation, non-crop
Wetlands: agriculture, 162–166, 267 n.18; environments, 142, 162–166, 267 n.13; non-crop use, 162, 164; non-crop vegetation, 142, 143, 144, 162
Wheat, 66, 79, 90, 168, 180, 271 n.14; black stem rust, 59, 259 n.38; late colonial/republican cultivation, 57, 59, 61, 258 n.37, 259 n.38; Spanish colonial cultivation, 49, 52, 257 n.23. *See also* Purchased foods
Women's roles. *See* Gender roles
Work routines, 119–120, 175, 205, 222, 264 n.9, 271 n.10. *See also* Labor-time; *specific types of work*

Yabar: family, 58; Luís Angel, 13, 252 n.8
Yacovleff, E., 14
Yapo, Carlos, 53

Designer: U.C. Press Staff
Compositor: Prestige Typography
Text: 10/12 Times Roman
Display: Helvetica
Printer & Binder: Bookcrafters, Inc.